TELECOMMUNICATIONS TRANSFORMATION: TECHNOLOGY, STRATEGY AND POLICY

Telecommunications Transformation: Technology, Strategy and Policy

Edited by

E. Bohlin

School of Technology Management and Economics, Research Group on Industrial Dynamics, Chalmers University of Technology, Göteborg, Sweden

and

S.L. Levin

School of Business, Department of Economics, Southern Illinois University at Edwardsville, Edwardsville, Illinois, USA

IOS Press

Ohmsha

Amsterdam • Berlin • Oxford • Tokyo • Washington, DC

ISBN 90 5199 366 8 (IOS Press)
ISBN 4 274 90205 6 C3055 (Ohmsha)
Library of Congress Catalog Card Number 97-77020

Publisher
IOS Press
Van Diemenstraat 94
1013 CN Amsterdam
Netherlands

Distributor in the UK and Ireland
IOS Press/Lavis Marketing
73 Lime Walk
Headington
Oxford OX3 7AD
England

Distributor in Germany
IOS Press
Spandauer Strasse 2
D-10178 Berlin
Germany

Distributor in the USA and Canada
IOS Press, Inc.
5795-G Burke Center Parkway
Burke, VA 22015
USA

Distributor in Japan
Ohmsha, Ltd.
3-1 Kanda Nishiki-cho
Chiyoda-ku
Tokyo 101
Japan

PRINTED IN THE NETHERLANDS

List of Contents

Part III: Network Evolution, Strategy, and Structure

Part IV: Current Issues in Pricing

Part V: The Internet: Demand and Policy

Part VI: Networks, Competition, and United States Policy

Preface

From the time of its privatization in 1990, TELUS Communications (formerly Alberta Government Telephones), the incumbent telephone company in Alberta, Canada, has been one of the most innovative and ardent advocates of regulatory reform in Canadian telecommunications. Many of the concepts and proposals advanced by TELUS in the regulatory forum have been instrumental in shaping Canadian regulatory policy over the same period. The genesis of TELUS' views on regulatory reform, in turn, can be traced in good part to papers presented at various conferences, seminars, and meetings of the International Telecommunications Society (ITS).

And so it was, in 1994, when I was Vice President for Regulatory Affairs at TELUS, that we decided to host a North American meeting in Calgary of the ITS. Our proposal was accepted by the ITS Board at a meeting in Sydney, Australia, but with one modification: the conference would be held in 1997 rather than 1996. In retrospect, the extra year was no small blessing. It gave us more time to prepare and allowed us to assemble a conference program of absolutely the highest calibre and one much broader in scope that we had conceived at first. As it turned out, during the course of 1995, we had the outstanding opportunity to combine this meeting of the ITS with the annual meeting of the International Council for Computer Communication (ICCC), with which TELUS had also worked in the past. Thus was born Global Networking '97, or GN '97 as we came to call it.

The conference, held in Calgary in June 1997, was an unqualified success, drawing approximately 500 attendees from around the world and a great many favorable comments and reviews. The conference included the presentation of over 100 papers, panel discussions, debates, technical demonstrations, and two live, interactive video conferences between Calgary, Brussels, Geneva, Madeira, and Moscow. As with all such conferences, of course, one of the major benefits was the continuing informal discussion among the conference attendees.

The distinguishing feature of this conference, however, was bringing together the members of these two groups: the economists, lawyers, and policy analysts of the ITS and those involved with technology and social applications from the ICCC. This provided opportunities for interaction that are not normally available and which took place both within the sessions and outside them. In an area such as telecommunications, where technology, economics, policy, and applications are so closely intertwined, this was truly an opportunity not to be missed.

This book is a fitting conclusion to the GN '97 conference. Although the conference proceedings were published and distributed at the time of the Calgary meeting, Professor Bohlin and Professor Levin have invited the authors of twenty-four of the most distinguished papers from the conference to expand and revise their papers for inclusion in this volume. As at the conference, the papers in this volume explore the consequences of the rapid changes taking place in telecommunications from the perspectives of technology, strategy, economics and policy. The volume also includes a group of interesting and timely papers on the internet and one of the first comprehensive evaluations of the United States Telecommunications Act of 1996.

I am sure that the readers of this book will find it immensely worthwhile, and I fervently hope that this multidisciplinary exploration of telecommunications not only continues but flourishes with creative abandon.

Bohdan S. Romaniuk
President
TELUS Advertising Services

Calgary
September 1997

Acknowledgements

The editors would like to thank the International Telecommunications Society (ITS) and the International Council for Computer Communication (ICCC) for organizing the Global Networking '97 (GN '97) conference in Calgary, Alberta, Canada, June 15-19, 1997, from which these papers were drawn. This volume would not have been possible without the support and encouragement of the Program Committee and the Publications Committee of the GN '97 conference, and thanks are extended to the Chairs of the respective committees, Dr. Dorothy Phillips and Professor Philip H. Enslow Jr. In addition, the conference would not have been possible without the generous financial support of TELUS Corporation, the GN '97 host. The numerous conference sponsors also were crucial to the success of the conference. The editors would, in particular, like to recognize the primary role played by Bohdan Romaniuk, at that time Vice President of Regulatory Affairs at TELUS, without whose initiative and leadership the GN '97 conference would not have taken place.

The editors received financial support for this volume from the Swedish Transport and Communications Research Board (Kommunikationsforskningsberedningen, KFB) in Stockholm and from the Swedish Information Service in the United States. The editors also received support from Chalmers University of Technology (Chalmers tekniska högskola) and from Southern Illinois University at Edwardsville, and they are particularly grateful for the excellent facilities and cooperation available to them at Chalmers University.

For unending logistical support and arrangements for the publication of this volume, the editors are deeply appreciative of the hard work of Irene Bonifacio and Pete Desrochers at TELUS. For preparation of the final manuscript, the authors are indebted to the efforts of Eva Burford and Jon van Leuven.

List of Tables

List of Figures

Biographical Information

James H. Alleman is an Associate Professor in the Interdisciplinary Telecommunications Program at the University of Colorado. His research is in industrial organization and telecommunications policy, with emphasis on pricing, costing, regulation, international issues, and infrastructure. He was Chair of the International Telecommunications Society from its inception in 1987 until 1996. Dr. Alleman founded Paragon Services International, a telecommunications call back firm, and has a patent pending on the call back technique. He is also a Senior Advisor with Hagler Bailly Consulting, an economics and litigation practice that focuses on telecommunications and energy issues.

Moonsik Bae is a senior member of the research staff of the Technology Strategy Planning Group at ETRI (Electronics and Telecommunications Research Institute) in Taejon, Korea. He graduated from Seoul National University with B.E. and M.E. degrees in industrial engineering in 1981 and 1983, respectively. His research interests include strategic R&D planning and technology assessment.

Joseph P. Bailey is a Ph.D. Candidate in Technology, Management and Policy at MIT and a Research Assistant in the Research Program for Communications Policy at the Center for Technology, Policy and Industrial Development. His research focuses on Internet economics, electronic commerce, and the economics of interoperability. He received an M.S. in Engineering-Economic Systems from Stanford University in 1993 and a B.S. in Electrical Engineering from Carnegie Mellon University in 1992.

Erik Bohlin is Assistant Professor in the School of Technology Management and Economics, Research Group on Industrial Dynamics, at Chalmers University of Technology. He currently publishes in the area of telecommunications strategy and policy, and he is a Board Member of the International Telecommunications Society. He obtained his graduate degree in Business Administration and Economics at the Stockholm School of Economics and his Ph. D. at Chalmers University of Technology.

Laurence Caby is a researcher at CNET, the national telecommunications research center in France. She received her MBA and Ph. D. from the University of Paris Dauphine. Dr. Caby conducts research and publishes in the area of marketing and telecommunications services. She also teaches at the Sorbonne University and formerly was on the faculty of the National Institute of Telecommunications in France.

Chuck C. W. Chan, P. Eng., is with the Network Architecture Planning team of TELUS Communications Inc. Mr. Chan's primary responsibility is ATM-related broadband network architecture planning and xDSL-related deployment planning. He is currently leading a team finalizing commercial deployment plans for ADSL in Alberta. He has worked on a number of broadband evolution and design projects with Stentor Resource Centre Inc, a research consortium sponsored by the incumbent telecom service providers in Canada. Mr. Chan graduated from the University of Kansas in Lawrence, Kansas, with a M. Sc. in Electrical Engineering.

Barbara A. Cherry is Director of Public Policy Studies at Ameritech and Adjunct Professor in the Department of Communications Studies at Northwestern University. Dr. Cherry has worked in the telecommunications industry for over fourteen years, including a decade with AT&T, engaged in legal and public policy matters. She is regularly invited to address industry and academic groups on telecommunications policy and universal service. She received her Ph.D. in communications studies from Northwestern University, J.D. from Harvard Law School, M.A. in economics and law from Harvard University, and her B.S. in economics from the University of Michigan.

Massimo G. Colombo is Professor of Applied Economics at Università di Pavia. He also teaches the Economics of Technical Change at Politecnico di Milano. Professor Colombo is a member of the scientific Advisory Board of DRUID (Danish Research Unit in Industrial Dynamics) in Copenhagen and CIRET (Centro Interdipartimentale di Ricerca nell'Economia delle Telecomunicazioni) in Milan. He is the author of numerous publications in international scientific journals in the areas of industrial economics, economics of innovation, and economics of telecommunications.

Livio Cricelli is a Ph.D student in Industrial Engeneering at the University of Rome "Tor Vergata" in the Department of Production, Systems and Computer Science. He graduated from the Faculty of Engineering at the University of Naples. His main interest is on systems analysis and its application to economic modeling, especially focusing on telecommunications systems.

James E. Freeman, P.Eng., is Manager of Network Technology & Business Strategy with TELUS Communications Inc. Mr. Freeman's technical areas of specialty include ADSL, Broadband Access, MMDS, and Local Number Portability. Mr. Freeman graduated from the University of Alberta in Edmonton, Canada, with a M.Sc. in Electrical Engineering in 1991. In December, 1994, he completed his MBA, specializing in Finance and Business Strategy. Mr. Freeman has spoken at international conferences in the past on xDSL technology and has co-authored six technical papers.

Robert M. Frieden serves as Professor of Telecommunications at Penn State University where he teaches courses in law, regulation and new technologies. He also provides legal, management and market forecasting consultancy services in such diverse fields as personal and mobile communications, satellites, and international telecommunications business development. He authored the *International Telecommunications Handbook* (Artech, 1996), a comprehensive desktop reference, and has published dozens of articles in academic and trade periodicals. Mr. Frieden holds a B.A., with distinction, from the University of Pennsylvania (1977) and a J.D. from the University of Virginia (1980).

Paola Garrone is a Researcher at the Politecnico di Milano in the Department of Economics. Her main research interests are in industrial economics and the economics of technical change. She has published numerous studies of the telecommunications industry.

Massimo Gastaldi is Assistant Professor at the University of Rome "Tor Vergata" in the Department of Production, Systems and Computer Science. He graduated from the Faculty of Statistical Science at University of Rome "La Sapienza". His main interest is in transportation planning and control and systems analysis and its application to economic modelling, especially focusing on telecommunications systems.

Steven Globerman is currently the Ross Distinguished Professor of Canada-U.S. Business and Economic Relations at Western Washington University in Bellingham, Washington. He is also Professor of Economics at Simon Fraser University in Burnaby, British Columbia. Dr. Globerman has published over 100 articles and monographs and has consulted widely in the public and private sectors. His current research and consulting are focused in the area of telecommunications. Among the organizations he has advised are the Canadian Radio-television and Telecommunications Commission, Bell Canada, and B.C. Telephone. Dr. Globerman earned his B.A. degree from Brooklyn College, an M.A. from the University of California at Los Angeles and his Ph. D. from New York University.

Christine Jaeger is Professor of Economics at Reims University in France and a researcher at Prisme, a group of researchers from France Telecom (CNET) and the French National Center for Scientific Research (CNRS). Dr. Jaeger received her Ph. D. in economics from the University of Reims and DEA in economics from the University of Paris. Her current research focuses on the sociology and economics of telecommunications.

Petros Kavassalis is a researcher with the Centre de Recherche en Gestion at the Ecole Polytechnique in Paris and a consultant for the Internet Telephony Consortium led by MIT. He is a civil engineer by training (National Technical University of Athens, Greece) and he received his Ph. D. from Paris-Dauphine University in Sciences de Gestion. He publishes in the area of telecommunications economics and policy, in particular Internet economics and efficient organization in decentralized communications networks.

Robert Kraut is Professor of Social Psychology and Human Computer Interaction at Carnegie Mellon University, with joint appointments in the Department of Social and Decision Sciences, the Human Computer Interaction Institute, and the Graduate School of Industrial Administration. Dr. Kraut formerly was the director of the Interpersonal Communications Research Group at Bellcore. His research concentrates on the design and social impacts of information technologies in small groups, in the home, and between organizations.

Donald J. Kridel is Assistant Professor of Economics at the University of Missouri-St. Louis, where he has been teaching since 1993. Prior to joining the faculty at UM-St. Louis, Professor Kridel was Director-Strategic Marketing at Southwestern Bell where he had held various positions since 1982. He publishes in the area of applied econometrics and regularly addresses industry and academic groups on telecommunications demand analysis. Professor Kridel earned his Ph.D. in economics from the University of Arizona.

Thomas Y. Lee is a Ph. D. candidate in the Technology, Management, and Policy program at the Massachusetts Institute of Technology. His research focuses on knowledge-based systems and knowledge representation for the integration of disparate (heterogeneous and distributed) data sources. After receiving his Masters degree from MIT on role-based access controls for computer-based patient records, he contracted with the Defense Advanced Research Projects Agency, supporting the data and network security activities of the Federal Networking Council. He received a B. A. in Political Science and a B. S. in Symbolic Systems/Artifical Intelligence from Stanford University.

Dale E. Lehman is Senior Economist at Southwestern Bell Telephone Company and Associate Professor of Economics at Fort Lewis College. He has held academic positions at 10 universities and consulted extensively in the telecommunications and information industries. He publishes regularly in these areas and frequently makes presentations in the areas of regulatory policy and information services. He has a Ph. D. in economics from the University of Rochester and a B. A. in economics from SUNY at Stony Brook.

Nathan Levialdi is a Researcher at the Institute for Systems Analysis and Computer Science of the National Research Council. He graduated from the Faculty of Economics at the University of Rome "La Sapienza". His main interest is in industrial economics, especially focused on telecommunications systems.

Stanford L. Levin is Professor of Economics at Southern Illinois University at Edwardsville, where he has been teaching since 1972 and was chairman of the Department of Economics from 1986-94. Professor Levin served as a Commissioner on the Illinois Commerce Commission, the public utility regulatory agency in Illinois, from 1984-86. He publishes in the area of industrial organization and is regularly invited to address industry and academic groups on telecommunications policy. His Ph. D. in economics is from the University of Michigan, and his B. A. in economics is from Grinnell College.

Ding Lu is a Lecturer in the Department of Economics & Statistics and Director of the Center for the Study of Transitional Economies at the National University of Singapore. He teaches courses in industrial organization, comparative economic systems, and public finance. He has published books and journal articles on issues related to economics of regulation, government-business relations, and telecommunications studies.

Jing Lu is Assistant Professor of Economics at York University in Toronto, Canada, where she has been teaching since 1993. Professor Lu obtained her Ph. D. in economics from the University of Western Ontario, Canada, and her B. A. in economics from Fudan University in China. Professor Lu publishes in the area of empirical macroeconomics and institutional economics.

Jeffrey L. Magenau is an associate attorney with Halprin, Temple, Goodman & Sugrue in Washington, D. C. Mr. Magenau practices telecommunications law with an emphasis on issues related to the introduction of competition into the telecommunications industry. Previously, Mr. Magenau was an intern with NTIA and the FCC. Mr. Magenau has published articles on satellite radio services, telemedicine, and the U. S. Communications Decency Act. Mr. Magenau is a graduate of American University's Washington College of Law (J. D. 1996) and Boston University (B. A. 1985)

William F. Maher, Jr., is a partner in the law firm of Halprin, Temple, Goodman, & Sugrue in Washington, D. C., where he practices U. S. and international communications law. Mr. Maher was formerly Associate Administrator of the National Telecommunications and Information Administration (NTIA) of the U.S. Commerce Department and Special Counsel in the Common Carrier Bureau of the U. S. Federal Communications Commission (FCC). Before his legal career, Mr. Maher was an engineer at Bell Labs. Mr. Maher is a graduate of Harvard Law School (J. D. 1984), the Massachusetts Institute of Technology (S. M. 1980), and the University of Notre Dame (B. S. E. E. 1978).

Gordon Meeberg is currently Business Development Director at TELUS Multimedia, a wholly owned subsidiary of TELUS Corporation. Prior to joining TELUS Multimedia, Mr. Meeberg held several positions within TELUS Corporation, including Strategy Development Advisor within the corporate strategic planning group and various project management and staff functions within the Information Technology organization. He received a Bachelor of Science degree in computer engineering from the University of Alberta in 1987 and a Master of Business Administration from the same institution in 1993. Mr. Meeberg is a professional member of the Association of Professional Engineer Geologists and Geophysicists of Alberta (APEGGA).

John B. Meisel received a B. A. from St. Lawrence University in 1973 and a Ph. D. in economics from Boston College in 1977. He is Professor of Economics at Southern Illinois University at Edwardsville, where he has been employed 20 years. While on sabbatical from Fall, 1989, to Summer 1990, he was a staff member of the Office of Plans and Policy at the Federal Communications Commission. He has also worked as a consultant to the Illinois Commerce Commission. His research interests are in the fields of industrial organization and regulation.

Euisun Paik is head of the R&D Planning Section at ETRI (Electronics and Telecommunications Research Institute) in Taejon, Korea. He graduated from Yonsei University with a B. A. in library & information science in 1979 and subsequently received an M. A. in management from Yokohama National University in 1985. Since 1994, he has been attending a Ph.D. program in Technology Management at KAIST. His research interests include strategic R&D planning and the techno-economics of telecommunications.

Brad Poulos is Director of Telesat's Business and Government Services division, responsible for the sales and marketing of its telecommunications services portfolio, and General Manager of the company's DirecPC (TM) business unit. He has held numerous marketing, sales and operations positions since he began his career with Telesat in 1983. Mr. Poulos is on the board of the Cybertech Council and is a member of several industry associations. He holds a Masters of Business Administration degree from the Richard Ivey School of Business at the University of Western Ontario and an Electronics Technologist Diploma from the DeVry Institute of Technology.

Paul Rappoport received his Ph. D. in economics from the Ohio State University in 1974. He is currently an Associate Professor at Temple University. Over the last 20 years he has performed numerous economic and statistical analyses for private and public agencies. He has developed forecasting models for electricity demand and for telecommunications services. For the last 8 years, Dr. Rappoport has been associated with the National Telecommunications Demand Study. With PNR, he has led the development of unique and comprehensive databases to support and define research in the telecommunications and energy industries. Dr. Rappoport's most recent published works focus on carrier choice by consumers, switching behavior and internet demand modeling.

Knut Erik Solem is Professor of Environment, Technology and Social Change at the Norwegian University of Science and Technology (NTNU), which is located in Trondheim, Norway. He is a Member of the Board of Directors of NTNU's Centre for Environment and Development (SMU) and also a Member of the Board of Directors of the Alternativ Framtid/ProSus (Oslo). He is a Governor of the International Council of Computer Communication (ICCC) and a Member of the Environmental Information Task Force of the European Environment Agency (EEA/EU) in Copenhagen. Dr. Solem holds degrees from the Universities of Oslo, Norway, and Manitoba, Canada, as well as Leicester and Oxford Universities in England, where he carried out research on political and economic integration. He has written extensively on these subjects as well as on energy, science and technology, forecasting, long-term planning, and strategic questions, and he has taught and lectured widely.

Barbara P. Sorce recently received her masters degree from the Interdisciplinary Telecommunications Program at the University of Colorado, Boulder, Colorado. While completing the masters program, Ms. Sorce was appointed Lead Graduate Teacher from 1994-1996, worked as a research and teaching assistant, and received several grants for special projects. Ms. Sorce held the International Telecommunications Society's Secretariat position from February 1995 till late 1996, when the Society headquarters was relocated to Europe. Ms. Sorce served on two Global Networking '97 conference committees.

Pablo T. Spiller is the Joe Shoong Professor of International Business, Professor of Business and Public Policy, and Chair of the Business and Public Policy Group at the Haas School of Business of the University of California, Berkeley. Professor Spiller is also a Principal at the Law & Economics Consulting Group, Inc. He has studied economics at the Hebrew University (B. A., M. A.) and obtained his Ph. D. at the University of Chicago in 1980. Prior to joining Berkeley, he was the William B. McKinley Professor of Economics and Public Utilities at the University of Illinois. Before that he was Senior Research Fellow at the Hoover Institution at Stanford University and Assistant Professor of Economics at the University of Pennsylvania. Professor Spiller has received numerous awards and has written more than 60 scholarly articles and three books in the areas of Industrial Organization, Antitrust and Regulation. He has served as economic consultant on regulatory issues and antitrust to the Federal Trade Commission, the Department of Transportation,the World Bank, the United Nations, the InterAmerican Development Bank, foreign governments and private companies. He has contributed to the design and implementation of regulatory reforms in Argentina, Bolivia, Colombia, Dominica, El Salvador, Guatemala, Jamaica, Malaysia, Mexico, New Zealand, Norway, the U. S, and Uruguay.

Charles Steinfield is Professor of Telecommunication at Michigan State University. He holds a master's degree and doctorate in Communications Theory and Research from the Annenberg School for Communication at the University of Southern California. Dr. Steinfield was a member of technical staff at Bellcore as well as a visiting researcher at CNET, the national telecommunications research center in France. He has also been a visiting faculty member at the Institut National des Telecommunications in France and at the Helsinki School of Economics and Business Administration. He conducts research in the area of organizational uses of communication technologies.

Neal Stolleman joined Bellcore in 1990, where his responsibilities included demand/cost analysis and simulation modeling of Local Number Portability, PCS, BISDN, ADSL-based video service and network services. He currently is Director-Software Pricing. From 1982 to 1990 Dr. Stolleman was a regulatory policy economist at GTE Service Corporation. He obtained a Ph. D. in economics from the CUNY Graduate Center and has presented original research at U. S. and European conferences on alternative regulatory frameworks, bypass of the public switched network, price theory and econometric methods.

Zixiang (Alex) Tan is currently Assistant Professor of Information Studies at Syracuse University in New York. He has been working in the telecommunications filed for more than ten years. Dr. Tan has published many journal articles and a book on telecommunications policy, regulations, and new technology applications. Dr. Tan received his Ph. D. in telecommunications policy and management from Rutgers University.

Tim Tardiff is a Vice President with National Economic Research Associates (NERA). He received a B. S. with honors in Mathematics from the California Institute of Technology and a Ph. D. in Social Science from the University of California, Irvine. At NERA since 1984, he has conducted research and participated in regulatory proceedings on pricing policies for competitive telecommunications markets, including pricing access services to competitors; actual and potential competition for telecommunications services, including new rate structures; the market potential for mobile telephony; and the terms and conditions for interconnection and unbundled elements, including evaluations of cost models.

Lester D. Taylor is currently Professor of Economics and Professor of Agricultural & Natural Resource Economics at the University of Arizona. He has a Ph. D. in economics from Harvard University and taught at Harvard and the University of Michigan before taking up residence in Arizona in 1972. During the spring semester of 1996, he taught at Charles University in Prague in the Czech Republic. His research in telecommunications covers more than 20 years, and he has published extensively on telecommunications demand, pricing, and costing. His most recent book, *Telecommunications Demand in Theory and Practice*, was published in 1994.

Svein Ulset is a researcher at the Foundation for Research in Economics and Business Administration at Bergen, Norway, and holds a doctoral degree in Business Administration from the Norwegian School of Economics and Business Administration. His research is mainly within the field of economic organization and strategy, where he has published several papers, articles and reports, recently with a special focus on the telecommunications industry.

Dennis Weisman is currently Associate Professor of Economics at Kansas State University. He has published numerous journal articles on the economics of regulation and is the co-author of a 1996 book for the MIT Press (co-published with the AEI Press), *Designing Incentive Regulation For The Telecommunications Industry*. He also serves on the editorial boards of the *Journal of Regulatory Economics* and *Information Economics and Policy*. Dr. Weisman received his Ph. D.in economics from the University of Florida and both his M.A. in economics and his B.A. in mathematics and economics from the University of Colorado.

Jisuk Woo is a Research Fellow at the Korea Information Society Development Institute where she conducts research and consultation in the area of international cooperation and negotiations on telecommunications policy. Dr. Woo has published several articles on political communication, journalism, and information policy. She has also presented numerous papers in the area of visual communication, political communication, and legal and policy issues on information technology. Dr. Woo received her B. A. in Mass Communication from Yonsei University in Korea, her Ph. D. in Communication from the Annenberg School for Communication at the University of Pennsylvania, and her J. D. from the New York University School of Law.

Edward E. Zajac's appointment from Bell Laboratories to an AT&T task force on pricing policy in 1965 made him switch careers from engineering mechanics to economics. In 1968, he started Bell Laboratories' first economics research group, and, in 1978, was made its Director of Economics Research. He is the author of the books *Fairness or Efficiency: An Introduction to Public Utility Pricing* and *Political Economy of Fairness*, as well as numerous academic articles on regulatory economics. From December 1983 until July 1991, he was Head of the Economics Department at the University of Arizona, where he is currently Professor of Economics.

Part I

Overview

Telecommunications Transformation: Technology, Strategy and Policy
E. Bohlin and S.L. Levin (Eds.)
IOS Press, 1998

CHAPTER 1

Telecommunications Transformation: Technology, Strategy and Policy

Erik Bohlin
School of Technology Management and Economics,
Research Group on Industrial Dynamics,
Chalmers University of Technology, S-412 96 Göteborg, Sweden

Stanford L. Levin
Department of Economics, Southern Illinois University at Edwardsville
Edwardsville, Illinois 62026-1102, USA

1. Introduction

One cannot help but be impressed by the pace of change in telecommunications around the world. Hardly a day goes by during which one cannot read about new technologies or new initiatives in one country or another to introduce competition into telecommunications and to reduce government control over the industry. The transformation of telecommunications is also resulting in the transformation of economies and in the way of doing business around the world.

Most books, let alone individual articles, address only one aspect of this transformation. The papers in this volume, however, have been selected to provide a broader perspective on the changes in the telecommunications industry. The papers approach this transformation from the perspectives of technology, network evolution, and business structure and strategy. They address some resulting issues in pricing, and they explore the changes being brought about by the rapid rise of the Internet. Finally, a group of papers discusses the United States Telecommunications Act of 1996 and the way in which this new legislation will affect the development of the telecommunications industry in the U.S.

The implications of this transformation extend beyond just the United States and the other countries that are specifically incorporated into the analysis in these papers. Not only do the authors, from around the world, bring an international perspective themselves to the analysis of these issues, but a comparison of countries can be particularly informative. For example, the lessons from the United States' experience with its new legislation and with local telecommunications competition can be compared to developments in Europe and Japan. This experience and comparison will provide valuable lessons for other countries around the world as they liberalize their telecommunications markets and reduce regulatory oversight. Furthermore, the lessons from telecommunications will find a new audience as countries begin the process of deregulating and introducing competition into other utility sectors, including electricity and natural gas.

The book is divided into five main sections, each of which we will address in the following. The order of the sections is from generic issues to increasingly applied and defined subject matter. Also, the order of the sections suggests a chain effect: technology change impacting network structure, market structure and pricing. New technology is also a fundamental force behind the Internet developments and changes in legislation, such as the United States Telecommunications Act of 1996.

2. The Impacts of New Technology

In the first main section, there are four papers that address the impacts of new technology. The order of the papers reflects the logic of going from user applications toward technologies that realize the potential of future usage patterns, ending with an increasingly general vision of the information society.

G. Meeberg addresses the future of consumer devices in an age of multimedia. Meeberg identifies two generic future devices – the smart television and the video computer – and compares them in terms of operating characteristics, from both a technological and a market point of view. Meeberg also offers some advice to network operators for managing the future multimedia network. In sum, the paper by Meeberg offers an analysis of the development of the multimedia future from the consumer's perspective.

J. Freeman and C. Chan offer a different perspective, namely that of networking technologies. In particular, they address one important bottleneck for the provision of multimedia solutions, namely the local loop and some new solutions for increasing the capacity of the local loop. They provide an overview of digital subscriber loop (DSL) technologies in terms of history, the range of DSL solutions and the reasons for the interest in DSL. The paper also suggests some long-term social implications of the utilization of DSL, and more specifically Internet usage.

B. Poulos analyzes another critical network technology for future broadband networks, that of satellites. Thus, while the previous paper dealt with wired network solutions to broadband constraints, Poulos looks at wireless solutions to the same problem. This paper explores the role that satellites have in the provision of broadband services and looks at future scenarios for satellites. In particular, it suggests various technical trends in satellites that contribute to their growing importance, such as increased performance in power and frequencies, more concentrated spot beams, signal compression, and non-geostationary orbits.

M. Bae and E. Paik, finally, provide a visionary statement as to the future development of information and communications technologies and services in the 21st century. It suggests trends in services, terminals and networks, all pointing towards an increasingly integrated, multimedia-oriented, intelligent and personalized information society. The paper is a fitting conclusion to the section on new technology, pointing towards the fundamental changes in the society that new information and communications technologies will bring.

3. Network Evolution, Strategy and Structure

This section deals with the evolution of networks from a structural and strategic perspective. The papers discuss structure and strategy from analytical as well as empirical perspectives, beginning with two papers that attempt to capture the dynamics of change with rigorous models, followed by papers that describe the changing industry structure empirically and qualitatively.

L. Cricelli, M. Gastaldi and N. Levialdi address how an international telecommunications market is affected by the level of vertical integration of the carriers and by competitive forces. A simulation model shows how final consumer prices and intermediate tariffs are reduced as the number of vertically integrated firms increase (i.e. competition). Furthermore, the model shows that the firms have an incentive to integrate depending on the final market structure.

N. Stolleman has a completely different structural model: instead of traditional economic optimization as in the previous paper, Stolleman brings chaos theory to bear. The model shows how the efficiency of the telecommunications system will evolve and how the evolution will depend, in part, on interconnection arrangements and market entry policies. Generally, he shows that regulatory constraints have potentially destabilizing effects on market structure and negative consequences on long-run network operating efficiency.

P. Spiller and S. Ulset return to the question of international networks, but from a very different perspective from that of Levialdi et al., namely that of transaction cost analysis. This paper presents a typology of organizational arrangements (contract, hierarchy and hybrid) and hypotheses based on theory. Empirically, it addresses the nature of various international alliances and collaborations. Some hypotheses are confirmed. Specifically, Spiller and Ulset show that hierarchical arrangements include a high level of non-redeployable and non-tradable assets, while contractual arrangements do not.

M. Colombo and P. Garrone focus on the strategic positioning and competence base of telecommunications operators in a multimedia era. The authors argue that it is crucial, for telecommunications operators that want to participate in the key activity of service provision in multimedia, to develop a network set-up of cross-industry alliances. The paper also includes an analysis of an extensive database, showing that telecommunications operators have targeted the service provision function and that they have developed networks of cross-industry alliances.

C. Steinfield, L. Caby, C. Jaeger, and R. Kraut also bring out findings on corporate collaborations and relationships, although with a different focus – on users of telecommunications. A number of issues are explored in a comparative study of the United States and France, although the central question addressed is whether electronic networks lead to greater outsourcing, i.e. induce vertical disintegration. However, this hypothesis could not be confirmed with the data available.

P. Kavassalis, T.Y. Lee and J. Bailey offer some analytical tools to address the emerging industry structure and network status of digital communications networks. In particular, the paper addresses the separation of infrastructure facilities and service provision, and argues that there will be a sustainable market for service provision which is independent of the underlying physical network technologies. This argument derives its force from the emerging Internet model which demonstrates the technical and functional robustness of a technology-independent bearer service abstraction. This paper concludes the section on network evolution, strategy and structure by capturing in an integrated fashion the long-term technological trends and their potential market structure implications, offering a thought-provoking perspective on the future path of network and market evolution.

4. Current Issues in Pricing

This section includes two papers, each of which addresses current 'hot' topics in telecommunications pricing: interconnection schemes and international pricing.

D. Lehman and D. Weisman offer a refined solution to optimal interconnection pricing which provides a trade-off between several conflicting objectives. Generally, interconnection pricing is fraught with difficulties due to the complex situation at hand: static, dynamic and technical efficiency concerns; imperfectly competitive markets; subsidy-laden retail structures; information asymmetries; discrimination and other public objectives. The solution provided differs from many other schemes as it simultaneously addresses a number of incentive problems inherent in interconnection, although with the consequence that none of the incentive problems is solved optimally.

J. Alleman and B. Sorce take up the increasingly problematic international pricing rules which were designed in a non-competitive era. Specifically, the problem is how to settle payments between operators in different countries when, in effect, the operator that terminates the incoming call 'exports' the final connection. Also, countries and operators that work as transit stations are affected. The traditional solution has been based on negotiated rates (so-called accounting rates) that are separate from the charges paid by users. The problem is that the accounting rates are not cost-based, which has created significant deficits for some countries and players at the same time that the recipients of the surpluses want to keep the status quo. The paper discusses the reasons for this problem and then examines different settlement policies, demonstrating the superiority of cost-based pricing rules.

5. The Internet: Demand and Policy

The section on the Internet provides a number of perspectives on this growing field. The first three papers in the section are predominantly oriented toward economic analysis, while the latter three bring up different foci: intellectual property, security and censorship issues.

P. Rappoport et al. deal with demand for Internet and on-line access. The paper employs two large and representative surveys of Internet usage in the United States. The paper is mostly descriptive but also includes estimates of price elasticities and a cluster analysis of users. Among some general findings are that Internet usage is strongly associated with age, education and household income.

S. Globerman asks whether there are any particular reasons why there should be government intervention in the Internet. Globerman then reviews the externality literature and asks what properties the Internet has that would contribute to such externalities so as to justify government intervention. The main conclusion of the paper is that the case for government intervention is weak and that further development of the Internet is best left to market forces.

J. Lu brings a forceful argument that the Internet will not cause disintermediation, i.e. reduce the need for intermediaries, but rather that the structure of the Internet will contribute towards a strong demand for intermediaries to perform market-coordinating activities. In cyberspace there are a number of sources of uncertainty: quality uncertainty, identity uncertainty and contract uncertainty. The demand for intermediaries arises from the so-called 'lemons' problem, i.e. that a market for a particular good breaks down due to fundamental information asymmetries over quality. Lu develops a model to illustrate this problem.

J. Woo takes stock of the copyright problems faced in Internet. Internet as a medium is built upon digitized information and interactivity, but copyright laws were originally developed for the printed word. The Internet poses a number of challenges, such as: the exclusive right to copying (what is a copy in a digitized world?), fixation of the copy on tangible expression, original authorship (what is original and who is the author?), fair use of the copyrighted material. The challenges are not easily resolved and Woo suggests that it may be necessary to consider more deeply what the purpose of the copyright law has been and what interests it is supposed to protect.

K. Solem ventures to take a broad view on the fundamental impacts of Internet on security, censorship, freedom of speech and equality. Internet presents new security problems such as Logic Bombs and Information Crime, but Internet can promote the spread of international understanding as well. Critical issues are, though, freedom of speech and equal access to information.

Z. Tan and D. Lu take a different outlook and examines the possibilities for governments to impose national boundaries in cyberspace. Four categories of strategies for government content regulation of Internet are presented, and these models are addressed in the context of how Asian countries have sought to control information on the Internet. The authors develop case studies of content control in China and Singapore and find that active state regulation of Internet content is indeed feasible. Cyberspace is 'borderless' but only partially so. The final note, then, of the section on Internet perhaps does not sound a minor key but, at the same time, is not exuberant in terms of traditional Western values of free speech. Technological opportunities are conditioned by the complex web that we call society.

6. Networks, Competition and United States Policy

The last section in the book provides a wide-ranging review of the United States Telecommunications Act of 1996 (TA96) from diverse standpoints. It represents one of the first major academic-based reviews on this landmark legislation – a legislation that will continue to draw attention and contention for years to come. The section starts with a broad analysis of the merits of TA96, moves into discussions of specific pricing and regulatory issues, and culminates with a general assessment of the winners and losers.

E. Zajac gives a critical review of TA96, suggesting that the legislation contains conflicting objectives which over time will create a need for further reform and law revisions. This conclusion is based upon an analysis of TA96 which focuses upon how the legislation deviates from policy principles that put economic efficiency at center stage. There will be unfavorable treatment of various interest groups, providing opportunities for political entrepreneurs to leverage discontent over fairness issues into revised legislation. In short, Zajac argues that TA96 by its very design contains time-bombs of fairness concerns that will reshape the legislation in the future.

J. Meisel addresses economic efficiency aspects of the decisions reached by the Federal Communications Commission (FCC) on local telephone competition. Various issues are covered, such as: the pricing methodologies for interconnection, unbundled network elements, collocation, and resale. Moreover, institutional concerns are treated, such as the division of power between federal and state regulators and the positions of primary stakeholders. This paper has several implications for FCC's implementation of TA96.

T. Tardiff discusses the costing and pricing issues which arise from the requirements in TA96 that incumbent local exchange carriers provide unbundled network elements and make retail services available for resale at a discount. The paper gives stylized examples of cost estimates made by contending parties, and the reasons for major differences in cost estimates by entrants and incumbents are analyzed. The costing and pricing issues that are examined in the paper have emerged in court proceedings, and Tardiff raises some general concerns about the increased dependence of the regulation and competitive process on arbitration in judicial settings.

B. Cherry analyzes the recent FCC Universal Service Order and the prospects for achieving universal service in a competitive environment. The analysis is based upon a set of principles for sustainability coupled with the administrative process of rule-setting (cf. bilateral and unilateral rules), developed by Cherry and others in previous works. Cherry finds that the application of the FCC order will be problematic in several respects when confronted with these principles.

W. Maher and J. Magenau address the increasing concerns about access charges in the United States as raised by the emergence of Internet, competition in the local loop and FCC's recent reviews on access charges. Specifically for Internet, the access charges differ from that of local telephony, and Maher and Magenau raise concerns about the long-term impact of rate differentials in the access charge system. They argue instead that a fair and efficient access charge system should allow greater flexibility for access providers to recover the costs incurred by access customers, since there are considerable difficulties of distinguishing among the various users of telecommunications networks.

R. Frieden lastly provides a broad overview of the potential winners and losers in the wake of TA96. It addresses the opportunities and the barriers created by TA96 for incumbents and new entrants in various fields: local telephony, interexchange telephony, cable television, Internet access and other market segments. The short-term and long-term analysis of winners and losers is based on a qualitative appraisal of the synergies, cost savings, and market integration opportunities that the legislation facilitates. In the end, though, Frieden emphasizes that exploiting opportunities created by changed legislation and regulation will not be sustainable, but that players should focus on the fundamental changes going on in the telecommunications industry. The final paper, then, provides a fitting conclusion to this book – an integrated analysis of the transformation process induced by changes in technology, strategy and policy.

Part II

The Impacts of New Technology

Telecommunications Transformation: Technology, Strategy and Policy
E. Bohlin and S.L. Levin (Eds.)
IOS Press, 1998

CHAPTER 2

The Future of Consumer Devices in a Multimedia Universe

Gordon Meeberg

TELUS Multimedia, Floor 8, 10025 – Jasper Ave, Edmonton, Alberta, T5J 1S6, Canada

Abstract. In a future converged digital world, it is important for multimedia network and service providers to understand consumer devices and their relative impacts on the choice of delivery platforms and technologies. In this light, several models for understanding device convergence (both from a technology and consumer adoption perspective) will be explored in order to shed understanding on key drivers in determining future winning consumer devices and the services to be delivered to those devices. Based on these models, a forecast will be made on what the characteristics of winning devices will be and how multimedia network and service providers need to tailor their response.

1. Introduction

Surfing the Web on your TV? Watching movies on your PC? E-mail on your phone? Which service on which device? From the perspective of a provider of network or services to consumers, much effort can be expended understanding which devices consumers will have, what the attributes of those devices will be, and hence what services can be economically delivered. To assist in this endeavor, it is meaningful to establish a framework for device convergence and consumer choice, and within this framework examine the characteristics which will determine winners and losers.

Rather than an exhaustive evaluation of consumer electronic devices, this paper will focus on the two at the heart of convergence, the PC and the TV. Many will suggest that these two devices will ultimately converge. However, I would suggest that while a pure technological approach would cause one to arrive at this conclusion, a thorough examination of consumer adoption models will lead to a differing conclusion – one in which two distinct yet somewhat converged devices will emerge as winners.

2. Technology-driven Model

Many attempts to characterize future devices based on their inherent technological attributes have been attempted. One such approach is depicted in Figure 1, which examines current leading network-connected devices based on two key attributes, bandwidth and interactivity.

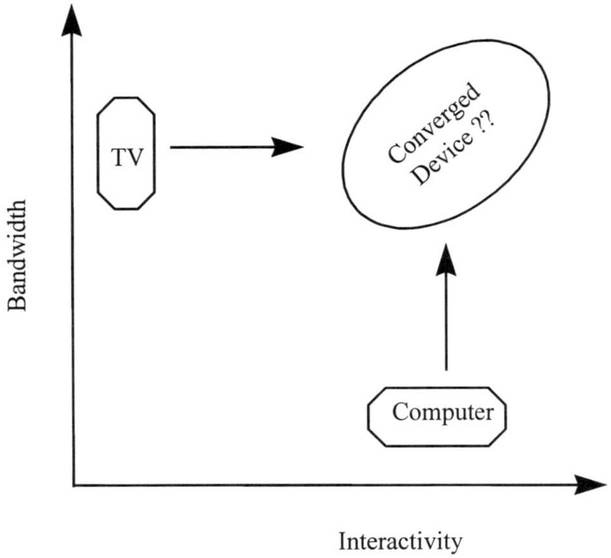

Figure 1. Device Convergence Framework

Within this model, the TV as it exists today is inherently a high bandwidth device (a typical television today can tune to over 100 6-Mhz channels), but with little inherent interactive capability. Much has been written and small fortunes have been spent over the past several years in the quest to deliver interactive TV. Yet despite numerous setbacks, the quest is still continuing – this time through Web-based products as manufacturers are incorporating World Wide Web browsing capability into their TV sets or into TV-attached peripheral devices.

For its part, the personal computer is inherently highly interactive. However, the real push today in networked computers is to increase the effective bandwidth to the computer to support more engaging graphics and video-rich applications. The once unthinkable speed of 28.8 kbps is now passé as manufacturers strive to deliver 56-kbps products and cable modems for the truly speed-hungry.

In addition, a new generation of "TV-enabled PCs" are now hitting store shelves from the likes of Compaq, Gateway 2000, ELLi, and NetTV. These devices typically feature TV tuners, remote control interfaces, extra-large VGA or SVGA monitors, and fully loaded system boxes – with a price to match, starting at around $4000 (U.S.).

The real debate centers around the nature of the "converged device", and which of the two combatants – the PC or TV – will win in pursuit of this future device. In addition to the forecast convergence of two significant consumer devices, there is also the convergence of two huge industries – the consumer electronics and computing industries. Recent announcements from industry powers such as Microsoft and Intel have heralded the death of television, pronouncing the personal computer as the ultimate convergence device in a digital world.[1]

While models such as that depicted in Figure 1 readily lead to the conclusion of a single converged device, it is worthwhile to move our analysis away from pure technological attributes, and focus on consumer behavior and characteristics of winning consumer devices.

3. Consumer-driven Model

One of the more fundamental models for consumer buying behavior is based on an historical examination of successful consumer products – and on what the key drivers behind these products really are. One factor that seems consistently to appear is price, and the axiom that a product will only achieve mass market penetration if it is priced below some mythical price barrier ($250 or $500 are frequently cited as magic numbers by consumer electronics companies).

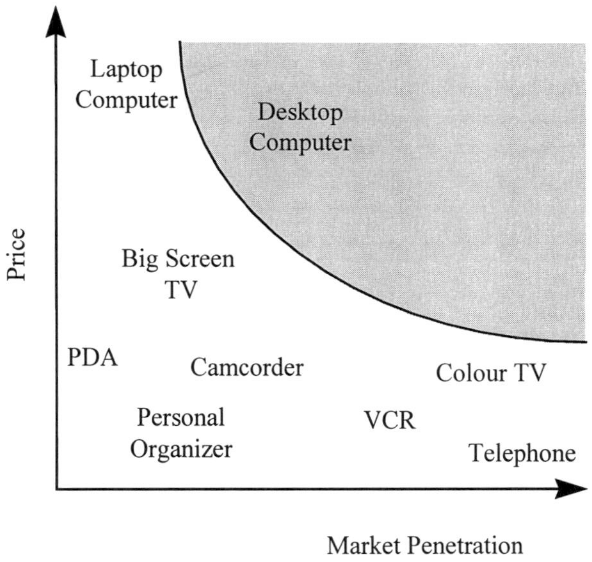

Figure 2. Price vs. Penetration of Consumer Devices

Figure 2 plots the price vs. market penetration for a number of consumer electronics items. In general, the notion of price as a predictor of device acceptance holds true, as all devices with high penetration have relatively low prices, and high-priced items have generally low acceptance. In fact, a curve can be drawn as shown in Figure 2 under which all devices should fall. This is true with one noted exception: the desktop computer.

[1] Crothers, Brooke, Gates: PC a natural for living room, *Computing*, April 8,1997.

That a consumer device with a price tag starting around the $2000 figure could achieve penetration in excess of 40% into homes is unprecedented (more on this later). In addition, price is clearly not the only predictor, as there are some low-priced devices (e.g. personal organizers) which have not seen mass market acceptance despite low prices. In order to explain some of these deviations, it is worthwhile to explore these concepts using consumer adoption theory. That is to say that consumers attach a certain utility to an object and are willing to pay a price equivalent to the perceived utility (as depicted in Figure 3).

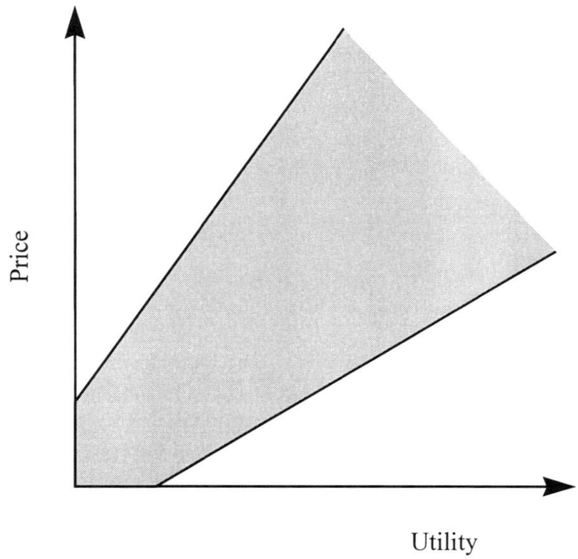

Figure 3. Perceived Utility of Consumer Purchase Behaviour

While the examination of utility[2] does provide a plausible explanation for low adoption of devices such as personal organizers, it does little to provide clarity into adoption behavior with regard to the personal computer. In order for the utility function as depicted in Figure 3 to hold true, it would have to be argued that an extremely high utility value is being placed on the PC. Yet intuitively, this is not the case. If one were to give consumers a choice of giving up either their TV or their PC, I would suggest that most would be unwilling to relinquish their TV, if current viewing and usage patterns are indicative of utility. Thus, for most consumers, the absolute utility of a PC is lower than that of a TV, and one would expect the PC to remain at a penetration rate of no more than 10-15% of homes.

Clearly there must be some other rational explanation for this behavior. I would hypothesize that the solution might be contained within some combination of the following social and demographic factors:

[2] In this context, utility theory is not used in the classical economic sense, but rather as a more general term to refer to consumer preferences.

1. *Incremental Utility* – For a consumer that already has a full complement of existing electronic products (TV, VCR, stereo...) it may be possible to attach an "incremental utility" to a PC purchase. Even though the PC may have lower absolute utility than another consumer electronics device, the consumer is not evaluating the PC purchase against pre-existent purchases, he/she is evaluating it against other not currently owned goods. In this scenario, the consumer is attaching an incremental utility to owning a PC beyond the basket of existing goods that he/she already possesses.

2. *Guilt / Fear* – Much like purchases of encyclopedias in the 1960s and 1970s, many parents have been motivated to purchase computers out of a fear that without one their child would be disadvantaged in his/her educational pursuits. In a related vein, a number of computers have been purchased by individuals who see computers encroaching into their workplace, and are gripped by a fear of career obsolescence if they do not acquire computer skills.

3. *Home Office* – Another significant sociographic trend has been the advent of the home office – whether for individuals bringing work home from the office, or for those choosing to conduct a business from their home. It is estimated that 2 million Canadians today operate a home-based or small business, with a projected growth of 25% per year to the year 2000[3]. If even one quarter of the 2 million figure are truly home-based, the resultant demand for home office equipment including personal computers is significant.

While not exhaustive, I believe the above list does offer some insight into plausible explanations with regard to consumer purchase decisions for personal computers.

A final model of consumer adoption is also important to this analysis. Product lifecycle or growth theory states that successful products go through four phases: a slow startup phase where product awareness is created, a rapid growth phase where sales flourish and prices fall, a maturity phase where growth tapers, and a decline phase. Many formulas have been developed to depict this growth, two of the more common of which are by Gompertz and Fisher Pry. Figure 4 shows the adoption of consumer products in Alberta over the past 30 years fitted to a Gompertz curve.

The goodness of fit (percentage figure in brackets under each of the labels in Figure 4) of each of the various data points to a Gompertz curve is actually quite remarkable. In every case, the data can be plotted to a Gompertz curve with a greater than 90% goodness of fit. In addition to the observation that consumer adoption of home electronics follows a very predictable adoption curve, it is also meaningful to note that in every case of a successful consumer electronics device, at least six years were required from initial product introduction until the product had achieved a 10% penetration into the marketplace.

Another point worthy of note is the slope of the Home PC adoption curve, which is noticeably flatter in the growth phase than any of the adoption curves for consumer electronics devices. This again is likely to be indicative of the impact of price point on adoption for personal computers.

---

[3] Corporate Literature, SOHO Business Group, 1997.

4. Future Device Characteristics

Rather than a single converged device emerging as has been suggested by some, I would suggest that an examination of the underlying social and consumer behavior factors behind PC adoption into the home would cause one to arrive at a differing conclusion.

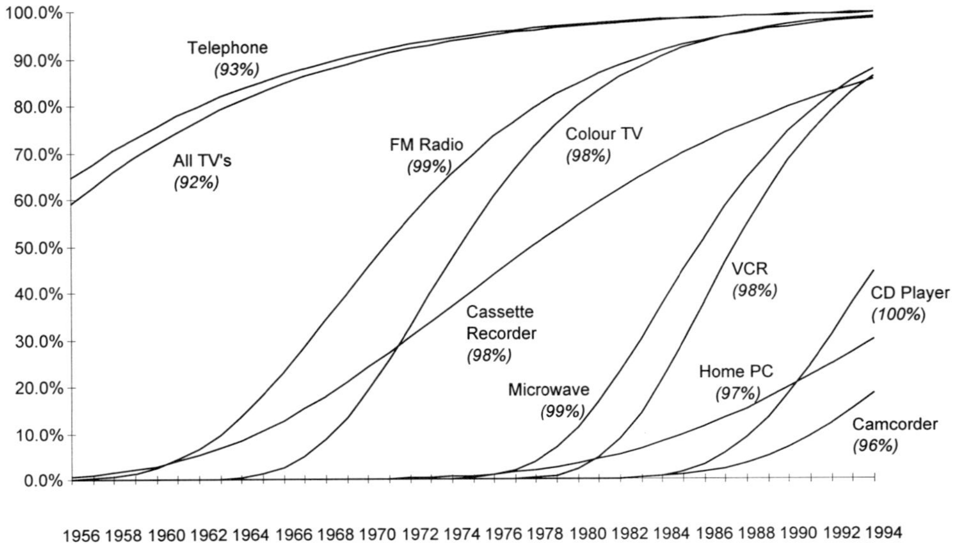

Figure 4. Fitted Alberta Consumer Adoption Curves

First off, price cannot be discounted as a significant factor. It is intuitively apparent that a device containing all the functionality of both a PC and a TV would have a price tag in excess of either of these devices taken individually (this has certainly been the case for any of the hybrid devices on the market today). As was seen in analyzing consumer buying behavior, the PC is only consumer device to date that has achieved significant penetration at a high price point. This can be partially explained by the social and demographic factors examined earlier; however, these same factors will not contribute to adoption of hybrid PC/TV devices. There is no underlying rationale for a buyer purchasing a device for a home office environment to consider a PC/TV hybrid. Nor is there rationale for those motivated by guilt or fear to ante up extra dollars for a hybrid device. In fact, in this context it is interesting to note that the fastest-growing segment of the consumer PC market is for devices priced at under $1000 (U.S.), with the average price at retail in March of 1997 dropping to $1400 (U.S.)[4]

Secondly, there has not yet been a successful example of a hybrid device in the marketplace. A recent attempt at combining TVs and VCR into a single product has met with very

[4] Lanctot, Roger C., MMX Computer Sales Dip, *Computer Retail Week*, May 2,1997.

limited market success. None of the devices depicted under the consumer adoption curves for successful devices in Figure 4 are hybrid devices. Hybrid devices generally tend to deliver the worst of both worlds rather than the best of either.

Thus, it is my assertion that in the future there will continue to exist two somewhat converged, although still distinct, devices in the home. Table 1 identifies the key defining characteristics of these two devices – dubbed the Smart Television and the Video Computer (note that these names are designed more to identify the devices' origins, rather than to attribute distinct attributes to the two devices).

Table 1. Characteristics of Future Devices

	Smart Television	**Video Computer**
Location	• Family Room	• Home Office / Den
Viewing Distance	• 8 – 12 feet	• 12 – 18 inches
Hardware Characteristics	• 25"+ display screen • 16 MB RAM • no hard drive • MPEG-2 video • non-expandable • 6 Mb/s+ network connection	• 15 – 17" display screen • 32 MB+ RAM • 6 GB+ hard drive • MPEG-2 video • expandable • 1.5 Mb/s+ network connection
Input / Output Devices	• I/R remote control • I/R keyboard (optional) • DVD • smart card	• mouse • tethered keyboard • CD-ROM (DVD?)
Target Applications	• Broadcast TV • Movies • Interactive TV Listings • Interactive Games • Shopping	• Word Processing • Home Banking • Finance Management • Information Retrieval • Education • Interactive Games
Focus	• adding interactivity to broadcast video • "Entertainment – centric" applications	• adding texture to information through video • "Information -centric" applications
Price	• $500 – $1000	• $1000 – $1500

The Smart Television is clearly a derivative of today's television, with enhancements to enable a more engrossing interactive experience. It should be noted that the functionality identified will not necessarily be integrated into the television – at least not at the outset. Rather it may very likely take the form of a peripheral or set-top box which provides the indicated functionality to existing analog and future digital TVs. In this latter case, the price stated is assumed to be inclusive of the cost of both the television and the peripheral – implying a very low price for a successful peripheral device.[5]

For its part, the Video Computer is a near cousin of today's personal computer but with enhanced video processing capability. One may argue that it is not vastly different from

[5] In this context, it is interesting to note that both Sony and Philips/Magnavox have recently reduced the priced of a WebTV box to $250 (U.S.) – in a box that begins to deliver some of the noted functionality.

many of today's high-end machines, and yet the biggest difference is probably in price, which is significantly reduced from current levels.

Although the two devices have a number of similarities, there are still a number of key distinctions including viewing distance, screen size, modularity, core focus and price. The Smart Television is an entertainment-centric device, with enhancements simply to provide a more engrossing interactive experience, tied around a broadcast video delivery. The Video Computer is still information-centric, with enhancements designed to add more texture to information through short video and other multimedia clips.

The one exception to this general device classification is in applications – where interactive games are identified as key on both machines. In this case, the PC is more an incidental games machine – one would have great difficulty justifying the expenditure on a PC merely as a gaming machine relative to the $200 – $300 gaming consoles available from companies such as Sony, Nintendo and Sega. In aggregate, games for such consoles still outsell PC-based games, although the gap is narrowing.[6]

5. Summary and Conclusion

As an owner of network, or someone with service delivery aspirations, it is easy to get focused on device wars. In reality, many early trials and initial service delivery approaches will require a network or service provider to seed the marketplace with consumer devices, typically through some form of lease or rental arrangement, in order to overcome the "Chicken and Egg" scenario.[7] However most, if not all, companies in the service delivery business will maintain that in the long term consumers must purchase devices, not companies.

In light of this, and based on the examination of technical and social factors in consumer device adoption, there are four guidelines that network and service providers should follow:

1. ***Don't get caught in the latest fad.*** Any successful new consumer device takes at least 6 years to achieve a market penetration of 10%. This is supported by the analysis on growth curves examined earlier. Thus, even if a successful device were to emerge today, it would not have market significance for at least 6 years.

2. ***Hybrid devices will not win***. As discussed earlier, there has not yet been a successful example of a hybrid device in the marketplace, nor is there likely to be one any time in the near future. Beware the Swiss army knife of consumer devices – it cannot be delivered cost-effectively and will never be more than a niche product.

3. ***Labels are not relevant, features are***. It doesn't really matter what label is placed on a future converged device – whether it's called a TV or a PC or some other name. The focus of service providers should be on features of devices and the applications they en-

[6] According to The NPD Group, games for consoles accounted for 55% of all entertainment software sold in 1996.

[7] The "Chicken & Egg" scenario refers to the conundrum that occurs when a service can't be delivered because consumers have not yet purchased devices to receive the service, and consumers are not buying the devices because services are not yet available.

able, not labels. Winning devices will be those that have services which deliver perceived value – not those with the right label.

4. ***Let the consumer decide – become a device agnostic.*** Betting on a winning device is a losing game for a network provider or service provider. Bet on services, not devices. Let the consumer vote on winning devices with his/her pocketbook.

Telecommunications Transformation: Technology, Strategy and Policy
E. Bohlin and S.L. Levin (Eds.)
IOS Press, 1998

CHAPTER 3

An Overview of xDSL Technology and its Implications for High-Speed Networking

James Freeman and Chuck Chan*

TELUS Communications Inc.
Rm. 866 10044 108 Street, Edmonton, Alberta, Canada, T5J-3S7

Abstract. This paper describes xDSL technology, how it enables high-speed internetworking, and how this in turn will affect other technologies and society. The history and variety of xDSL technology are discussed, and the reasons for interest in xDSL are examined. A network topology using xDSL is presented, and associated technology choices are illustrated. In conclusion, the impacts of xDSL on internetworking and society are postulated.

1. Introduction

xDSL technology will be an important part of network evolution plans for telephone companies into the 21st Century. There are many factors that make such technology attractive, and its deployment will have many technological and social implications.

In this paper we will first define what Digital Subscriber Loop (xDSL) technology is, and discuss some of its attributes. We will then focus on what is creating interest in xDSL by examining the internet and how it fulfills consumer needs. Next, we will focus on how xDSL will allow service providers to meet these needs. We will draw on our experience trialing xDSL technology at TELUS in order to offer some insight into the issues that concern service providers. Finally, we will focus on the longer-term implications of xDSL for other technologies and ultimately, for society.

2. Technology Overview

2.1 xDSL in the 1990s

Telephone companies around the world have long been looking at increasing the bandwidth to residential customers in order to support a variety of multi-media services. A number of potential access network technologies have been considered since the late 1980s, but rela-

* The authors wish to acknowledge the kind review of this paper by Brian Kraus, also of TELUS Communications.

tively few telephone companies have made large-scale investments in residential broadband infrastructure. In the early 1990s, most believed that video-on-demand (VOD) would be the "killer application" of the multimedia era, and xDSL technology first emerged during this time as a possible solution to providing such services. However, because of the high cost, limited bandwidth, immaturity of the technology, and soft consumer response, telephone companies began to focus on higher-bandwidth, fibre-rich solutions such as Fibre-to-the-Curb (FTTC) and Hybrid Fibre/Coax (HFC).

With the mass appeal of the internet becoming clear in 1995, uncertain demand for a broad suite of multimedia services and the high first-cost investment required for fibre-rich access solutions, telephone companies once again became interested in xDSL as a potential access solution. The technology had matured somewhat over the intervening period and was improving in both cost and performance. The primary attraction of xDSL to the telephone companies is that it leverages investment in the existing copper infrastructure and defers the majority of investment until an interested customer is identified. Deployment of xDSL allows telephone companies to enter the market rapidly and flexibly by leveraging past investment in over 750 million copper access lines globally.

2.2 The Many Flavours

Our discussion of xDSL as an access technology encompasses many high-speed transmission technologies operating over twisted-pair copper. These are summarized in Table 1.

Table 1. Attributes of the Main xDSL Technology Choices

Technology	**Description**
ADSL	"Asymmetric" – rates vary from 1.5 to 8 Mb/s downstream, and 64 to 640 kb/s upstream; originally conceived for VOD delivery
HDSL	"High bit-rate" – typically a 4-wire symmetric 1.5 Mb/s technology; in commercial use since the late 1980s for business customers
RADSL	"Rate-Adaptive" – an asymmetrical, rate-adaptive technology capable of 2.2 Mb/s downstream and 1.1 Mb/s upstream today
SDSL	"Symmetric" – similar to HDSL, except that only one pair is required, and rates tend to be lower (768 kb/s)
VDSL	"Very-High Bit-rate" – rates vary from 13 to 52 Mb/s downstream, and up to 6 Mb/s upstream; a key enabler for FTTC. Fully symmetric uses for VDSL are also being explored, particularly for business applications

A bandwidth-distance trade-off exists on all transmission media, and copper is particularly restrictive. All xDSL technologies employ either single-carrier or multicarrier modulation schemes such as Carrierless Amplitude/Phase (CAP) or Discrete Multitone (DMT). These are variations of Quadrature Amplitude Modulation (QAM), and effectively achieve spectral efficiencies of up to 7 bits/s/Hz. Given this, we can plot the bit-rate-distance trade-off for copper, assuming a 6 dB noise margin and a bit-error rate of 10^{-7}, and map the various xDSL technologies against this, as is done in Figure 1.

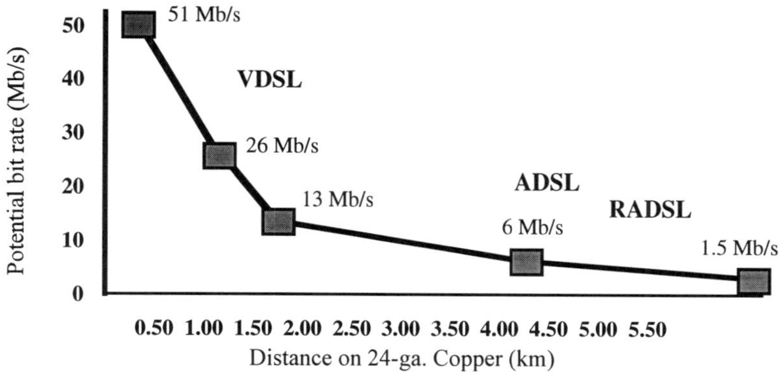

Figure 1. Reach on 24-gauge Copper of Various xDSL Technologies

3. The Consumer Perspective

What is driving the interest in xDSL technology? The growing need for internetworking. Customers want access to the internet, and they want it at speeds much faster than they typically experience today. Phenomenal growth is predicted in the number of devices and users accessing the internet between now and the year 2000. According to some estimates, the number of devices on the Web is predicted to increase from 12.6 million worldwide in 1996 to 233.3 million by year-end 2000, for a monthly increase of 5%. The number of users on the Web could conservatively increase from 16.1 million at year-end 1995 to 163 million by year-end 2000, representing a 3.9% monthly increase [1].

The business and consumer markets use the internet to meet three needs: *communication*, *information*, and *transactions*. For communications needs, the traditional choice on the internet has been e-mail, although the emergence of new voice compression technology combined with standardization of multimedia transactions with ITU H.323 is making the internet a possibility for live voice traffic. The use of the internet for telephony is being supported by an increasing number of hardware and software products, and xDSL technology, by increasing the access bandwidth, will support duplex audio at qualities much higher than those experienced today with 28.8 kb/s modems.

For information needs, the World Wide Web is by far the network of choice; in fact, the success of the WWW has led some to suggest that it is the long-sought "killer application" of the multimedia era. This success is due largely to the emergence of GUI browsers such as the Netscape Navigator and MS Explorer, but the TCP/IP protocol suite combined with HTTP for transaction management and HTML for creation of Web pages are also critical enablers to the Web's growth. With sites increasingly sporting JAVA enhancements, the Web is beginning to "come alive", and high-speed access enables full-motion video, not simply graphics that move.

The last need met by the internet, the need to transact, is in its ascendancy. Internet commerce has long been discussed, but with the development in 1996 of the Secure Electronic Transactions standard, the major impediments to internet commerce, fears of security, will slowly diminish. Studies projecting the use of the internet for commerce provide mixed messages. A recent poll of 1,500 companies worldwide found that only 1 in 20 companies currently use the internet for business transactions, and only 1 in 8 plan to do so in the next few years.[2] Another study suggests that use of the internet for business transactions in the US will increase from 11% today to over 44% by 1999.[3] This growth is supported by a third study which suggests that 53% of all Web users in Canada and the U.S. spent some portion of their time looking for specific product information on the Net, and 15% have actually purchased products on the Web. This is up significantly from the fall of 1995, when the numbers were 35% and <5% respectively.

This same survey also points to a narrowing of the gender-gap of on-line users, and suggests that women now account for 42% of on-line users, up from 34% in the fall of 1995.[4]

The need for internetworking is not limited to access to the internet. Increasingly, companies are using internet-like technology for corporate intranets in order to reduce costs and increase availability of information in their organizations. This trend combined with interest in tele-working, particularly in the U.S., has created a second opportunity for xDSL technology. It may be used to extend secure access to the corporate LAN to the employee's residence. Used thus, employees can then perform all tasks they normally would at the office – e-mail, access to file servers, access to intranet information – from home.

4. The Service Provider Perspective

We have described the growth of internetworking and the needs it meets for consumers. We will now focus on the technical issues faced by a service provider using xDSL technology to provide solutions for customers. We will do this by first discussing our experience trialing xDSL technology at TELUS.

4.1 Trials of xDSL Technology

Use of xDSL technology for high-speed access to corporate LANS or the internet is now under trial by all major telephone companies in the world, and is in commercial service in some cities in Canada and the U.S. TELUS' experience trialing ADSL is similar to those of most companies and provides a good introduction to deployment issues.

TELUS began trial of xDSL technology by initiating Project RACE (Residential ATM/ADSL for Competitive (market) Entry) in November 1995. The first phase of the trial involved 25 TELUS employees using ADSL for high-speed access to the internet from their home. The equipment used was from Performance Telecom Inc. and was capable of 1.5 Mb/s downstream and 64 kb/s upstream to customers served over 26-ga. copper loops up to 5.5 km long. This phase of the trial was completed in June, 1996, and TELUS continued trial activity with ADSL equipment from Paradyne Inc. In this second trial phase, the feasibility of LAN Extension from the TELUS corporate LAN to employees at home was examined. The Paradyne equipment was also capable of 1.5/64 performance, but was tested under different loop conditions. Both trials illustrated the feasibility of creating customer services using ADSL. The trials did, however, illustrate the influence that variables such as computer processor type, RAM availability, and download software had on overall throughput.

4.2 A Network Using xDSL Technology

Shown in Figure 2 is a network that uses xDSL as an access technology. While xDSL over copper is used as the physical layer for network access, choice of an appropriate layer 2/3 protocol must be made. This will depend primarily on customer traffic requirements but is also influenced by investments made in other areas of the network. A thorough analysis of the options available to the service provider will consider not only initial costs to deploy the overall network, but also service activation and assurance issues that influence customer quality of service and network operating costs.

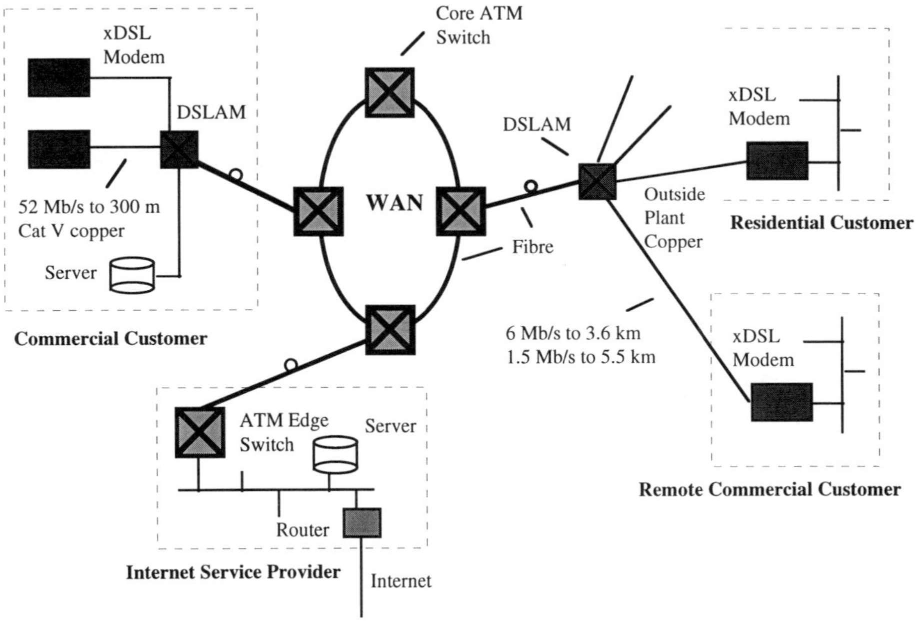

Figure 2. A Network Access Topology Showing Possible Application of xDSL and DSLAMs

Many telephone companies favor use of ATM for delivery of services not only to business customers but also to residential consumers as well. The choice of ATM is premised on the belief that ATM will best meet a variety of traffic requirements from customers in the most bandwidth-efficient way. This view is typical of the telecom industry but is challenged by an alternative view expressed primarily by the computer networking industry. This view advocates continued use of IP packets and technologies that currently comprise the internet, such as layer 3 routers, in order to meet internetworking needs. Development of improved routing protocols that support cut-through switching in order to speed transmission of certain types of data flows combined with bandwidth management protocols such as the ReSerVation Protocol, or RSVP, are attempts to improve quality of service and obviate the need for investment in ATM. Regardless of the choice of layer 2 or 3 approach, xDSL technology, as a physical layer technology, is ready to support either paradigm. We will focus on the choice of ATM as the method used to meet packet transport requirements.

4.3 ATM and xDSL Technology

When we consider the use of ATM and xDSL for internet access, we must decide how far into the network we choose to deploy ATM. There are two possibilities. In the first scenario, we use ATM only for wide-area transport, and some other networking protocol such as ethernet for network access. In the second scenario, we extend ATM to the customer.

In the first scenario, some sort of aggregation and segmentation of incoming access traffic is required. One simple way to achieve this is to obtain xDSL technology with standard interfaces such as 10BaseT, and use them essentially as modems between the customer and wire centre. Next, an ethernet hub may be used to concentrate traffic onto one segment, and a router with ATM output card may then be used to interface to the wide-area network. For added security (and cost), an ethernet switch may be used in place of an ethernet hub to create a number of virtual LANS. Use of a switch permits preferential segmentation of traffic with different destinations and therefore improves security.

xDSL manufacturers are now developing plug-in shelves for xDSL cards that aggregate traffic over the back-plane and produce a variety of outputs, including ATM at OC-3c (155 Mb/s) rates. The Digital Subscriber Line Access Multiplexer (DSLAM) is one of the major building blocks for ATM over ADSL network. Basically, the DSLAM acts as the aggregator for collecting IP/ PVC traffic from the local loops and transporting the traffic to the service provider via the core ATM backbone. The telephone company uses the DSLAM to create an open architecture using PVCs, and allows connection of a customer using xDSL technology to the service provider of his/her choice. With the addition of ATM SVCs, a customer could select between a number of potential service providers, or between services such as internet access and access to the corporate LAN.

In the second scenario, ATM is extended directly to the customer, and the customer premise equipment then terminates not only the xDSL line but may also perform protocol conversion on the ATM cells. A number of interface options are typically available, including ATM25 and 10BaseT. It should be noted that built-in bridge/router capability is often available from xDSL vendors such that a number of devices may share the same xDSL modem connection. The extension of ATM to the customer premise using xDSL technology allows for better management of quality of service on the network side of the connection.

In all cases, xDSL over copper lines has absolutely no effect on the line's use for regular telephony. Unlike voice-band modems, which require dial tone and depend on the public switched network for connectivity, a copper line with xDSL modems may be used simultaneously for regular voice and high-speed data traffic. Signal splitters commonly known as POTS splitters are installed at both the customer and central-office ends of the loop in order to allow the data traffic to "hitch a ride" on the copper loop. Given the low quality of customer inside wire (usually Cat I/II equivalent), rewiring with Cat V wiring is often advantageous. Figure 3 illustrates the configuration of wiring at the customer premises.

4.4 IP, Ethernet and ATM

If a service provider chooses to use ATM as part of the layer 2 solution, then the issue of adaptation of IP packets into ATM cells must be considered. By 1998, ATM solutions using Multiprotocol over ATM (MPOA) will be available to solve this problem. MPOA will provide layer 3 routing capability for all routable protocols such as IP or IPX. Today, the issue of adaptation is dealt with through RFC 1577 (classical IP over ATM), and ethernet LAN emulation is dealt with using RFC 1483. For LAN extension services, the LAN Emulation 2.0 (LANE) specification for ATM will allow easy bridging between an xDSL modem at the remote site and the corporate LAN.

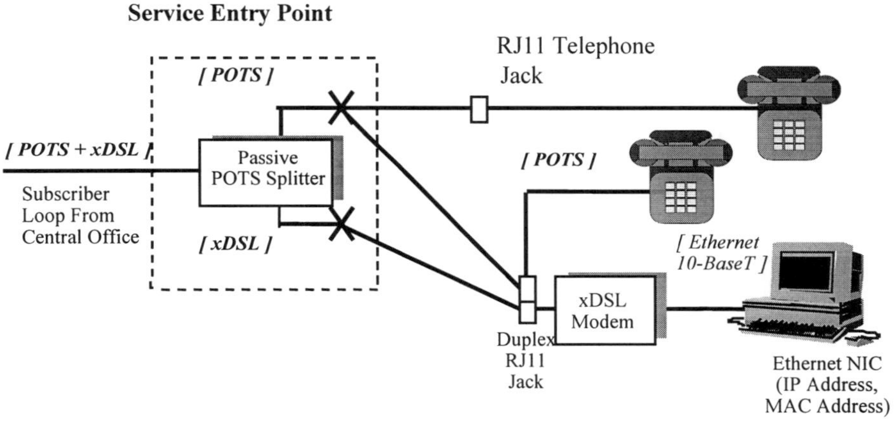

Figure 3. xDSL Technology in the Customer Premises (Assumes Ethernet Termination)

5. Implications of xDSL Technology

The fates of xDSL and the internet are inextricably linked. With the popular emergence in 1995 of the World Wide Web, demand for internet access has increased exponentially. Growing interest in the graphical side of the internet has prompted demand for higher-speed access from both business and residential customers alike.

5.1 Technological Implications

Increasing access bandwidths by using xDSL technology will affect many aspects of internetworking.

First of all, higher-speed access will enable the creation of "richer" sites on the internet. Today, videostreaming over the internet is possible, with products from Progressive Networks, VDOnet Inc., or Vxtreme Inc. However, this software is designed to work over today's dial-up 28.8 kb/s modems, and manages only 2-15 frames/s in a small window on your computer screen (TV is 30 frames/s in North America). With xDSL technology, higher frame rates are possible, enabling inclusion of visual and audio content over and above simply graphics and text. With the higher rates of VDSL, a compressed video clip could be downloaded and played locally instead of depending on streaming through the network. This minimizes dependency on network connection over long periods of time.

Second, demand for intelligent agents will increase. Given the richness of sites enabled by xDSL combined with the continued growth in the number of such sites, a need for helper applications that scan the internet for material of interest will grow. These "personal assistants" scan the internet and return with information of personal interest to the user. Today, such agents simply respond to a list of customized preferences, and scan not the whole internet but rather news wires and aggregation agencies for current news. Screen-saver software such as PointCast is a current example of such an agent. In a way, this is like tuning to the TV channel of interest, and having content and advertising of interest delivered. These agents are part of the so-called "push" model for delivery of services.

Third, with richer sites, more storage media will be required. This is true not only for service providers but is also true for information consumers. In spite of the advances in compression technology, audio and visual files still consume large amounts of storage. For example, a 10-minute video clip compressed using MPEG-2 and designed for playback at 4 Mb/s still requires 300 Mbytes of storage with no redundancy. The development of DVD-ROM, as a next generation of optical read/writable storage peripheral for PCs, will meet these requirements.

Further, while xDSL does address the problem of the access bottleneck, it does not address issues of transport or server congestion. A recent study suggests that 85% of the perceived congestion problems on the internet are attributable to poor configuration of server and inadequate client caching.[6] Better server configuration and increased "mirroring" of popular content will bring material closer to customers and also drive the need for more storage.

Finally, new browser technology will evolve. Higher-speed access will create new demands on the capability of browser technology to support audio and visual elements to interactions. Further, higher-speed access to premises will enable a business or place of residence to connect more devices to the network. Conceivably, a TV, PC, and display phone could simultaneously connect to the network using xDSL technology. New devices such as Network Computers or Set-top Boxes that depend on network connection will also improve in performance. Different devices have different interface, display, and storage requirements, creating challenges for browser designers.

These technology advances will have implications for how people use the internet.

5.2 Social Implications

High-speed access to the internet or corporate LAN will affect how we live, work, and play in many ways. First of all, high-speed access enables distance learning. Now educational institutes can assemble visually rich material, and even create virtual classrooms where students remain at home but interact through video-conferencing. This concept builds on the internet chat rooms that exist today. The power of this idea is that students from anywhere in the world may be brought together to learn and share experiences; no longer would geographical locale decide who participates.

Next, high-speed access will affect the way we build or renovate our homes. People today have as many as seven networks in their homes: one for powering, one for heating and air-conditioning, a third for home security, another for home computer usage, a fifth for audio systems including stereos and intercoms, one for cable TV, and one for the telephone.[5] The ability to provide high-speed access to the home combined with acumen for integration of services offers a huge opportunity for service providers, and for simplification of the home wiring problem.

The idea of teleworking has already been discussed. Teleworking supported by high-speed xDSL technology offers the prospect of lower operating costs and higher productivity. No longer do employees need to commute to work, and people juggling careers and family will be given more flexibility. One motivation for telework, particularly in the U.S., is reduction of air-borne pollutants that result from commuter traffic. High-speed access is good for the environment.

We have also touched on the idea of internet commerce. Doing business over the internet means that every business, no matter how large, has a global audience. The internet reduces transaction costs significantly; it is the ultimate global market-maker. Lower transaction costs mean lower prices for goods and services to consumers.

Finally, networked gaming is a growing business. The ability to play multimedia-rich games with partners anywhere in the world opens new vistas for game manufacturers such as Sega, Nintendo, and Sony.

5.3 Is there a Dark Side?

There is growing fear that a gulf exists between those prepared for the information age and those who are not. If it is true that information will be the single most important measure of wealth creation in the next century, then we should pause to consider the implications of high-speed access in such a time. As we have suggested, high-speed access will enable true multimedia experiences on a scale previously possible only with travel. The *accessibility* to high-speed access for all citizens is one that policy makers will have to grapple with over the next decade. Some fear that people will simply cocoon themselves in their homes or place of work, and interact with the world through a PC. The irony is that the internet is the ultimate way to interact with many societies quickly. Distance no longer separates people. High-speed access will make this an even richer interaction.

One might also speculate on the impact such technology might ultimately have on employment; is there a net increase or decrease in overall employment created because of such technology? While the effects on net employment may be uncertain, it is reasonable to suggest that shift in employment from some industries towards others is inevitable. This shift is central to the societal revolution the Information Age is creating.

6. Summary and Conclusions

xDSL technology enables high-speed access, and high-speed access in turn has the power to revolutionize many aspects of our lives. The power to conduct business transactions, learn, or communicate with people in full motion and sound anywhere in the world has powerful implications for society. It is by overcoming the bottleneck of network access that we will enable the true vision of the information age.

References

[1] IDC Forecasts Unprecedented Growth in Global Usage of the Internet and the World Wide Web, *PR Newswire via Individual Inc.*, January 6, 1997.

[2] G. Rowen, The Internet not Exactly Main Street for Doing Business, *The Globe & Mail Daily*, February 18, 1997, B1.

[3] E.R. Greenberg, AMA Survey Predicts Sharp Rise in Business Internet Use, *Newsbytes via Individual Inc.,* April 11, 1997.

[4] S. Bressler, 'Startling Increase' in Internet Shopping Reported in New Commercenet/Nielsen Media Research Survey, *Business Wire via Individual Inc.*, March 12, 1997.

[5] D. Ticoll, This New House, *tele.com*, March 1997, 43-44.

[6] A. Doan, Web Forces Performance Issues, *InfoWorld*, (infoworld.com), April 14, 1997.

Telecommunications Transformation: Technology, Strategy and Policy
E. Bohlin and S.L. Levin (Eds.)
IOS Press, 1998

CHAPTER 4

The Evolving Role of Satellites in Providing Current and Future Broadband Services

Bradley J. Poulos

Telesat Canada
800 – 100 Sheppard Avenue East, Toronto, Ontario, M2N 6N5 CANADA

Abstract. This paper will explore the role that satellites have and continue to play in the provision of broadband solutions, and provide insight into plans for the next generation of satellites. It will also examine how hybrid terrestrial and satellite-based solutions might solve the problems presented by universal service requirements and limited reach of terrestrial alternatives, as we move into a world in which high-speed communications becomes as critical as plain old telephone service (POTS) is today.

The power of wireless communications is just now being unleashed. The success of satellite-based broadband applications being used today is an important precursor to the next generation of satellites that will for the first time include on-board switching in addition to their traditional role as a "bent pipe". They will enable broadband applications to extend globally – any time, anywhere – whether the target is mobile or stationary.

Some of the key determinants of success will be the availability of terminal equipment in sufficient quantities and at a reasonable price, a global partnership to allow sharing of R&D and other valuable resources, and the use of open systems to create an environment which fosters innovation.

1. Celebrating 25 Years of Satellite Communications The Role of Satellites to Date

The world was introduced to domestic satellite service by Telesat Canada 25 years ago this year, with the launch of its inaugural spacecraft, Anik A1. Canada was a natural place for this to begin as it had and continues to have the correct combination of economic and geographic qualities to take advantage of the unique capabilities of satellites. Its sparsely populated and huge expanses of land, need to serve the people and military installations in the Arctic, and strong domestic telecommunications industry justified and aided the development of the first generation of satellites and their associated ground-based equipment and systems.

Until now satellites have acted as simple repeaters in the sky, performing little differently than those ubiquitous microwave repeaters dotting the Trans Canada Highway. The architecture is often called "bent pipe" to illustrate what it is that they *do not* do as much as what they *do*.

As with any space-based endeavour, operating a satellite is risky business. The cost of a catastrophic failure is measured in the hundreds of millions of dollars, and it simply makes sense to ensure that all systems are the most reliable possible. To keep the weight low and to allow the loading of maximum fuel, satellite designers were challenged to adopt new technologies to incorporate weight and space-saving components. However, these advances have been applied mostly to better telemetry, power and control technology, allowing satellite life to more than double and raw communications capacity to increase. The capabilities of satellite communications *payloads* have not changed much in 25 years. While Telesat's Anik E series satellites have much higher power and bandwidth capabilities than the A series, and they will last twice as long, the basic functions remain the same.

The "bent pipe" allows the simple reception and retransmission of signals. This architecture does not include switching; therefore the most compelling argument for the use of satellite has been when a need existed to broadcast information to multiple locations within the footprint of the satellite. Point-to-point services have been primarily for communications with remote locations where no other viable communications system is available, or when the distance-sensitive costs of wireline services were higher than the distance-insensitive costs of satellite-based service. Competition in the telecom industry has pushed costs down to the point that within Canada there are very few cases where point-to-point traffic is carried on satellite, outside of those to remote and underserved areas.

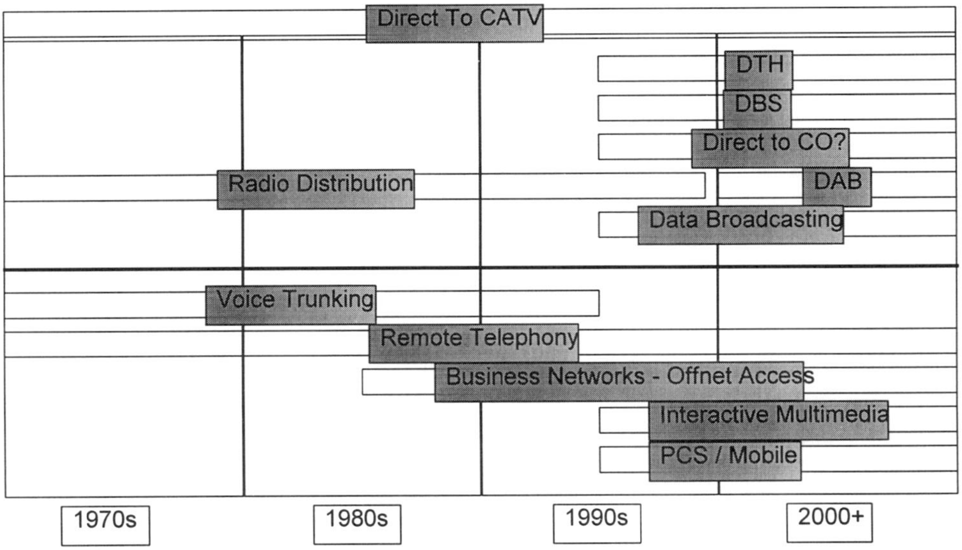

Figure 1. The Evolving Role of Satellites

The introduction of higher power direct broadcast service (DBS) satellites has allowed this traditional role to expand to include Direct-to-Home (DTH) entertainment and information services, and nascent satellite-based multimedia services. The success of the industry in the US and in Europe is well documented. Telesat's success with its DirecPC multimedia product will be complemented by a made-in-Canada DBS solution some time in 1998.

All of this development occurred in the midst of ever-expanding – in terms of both reach and bandwidth – terrestrial or wireline networks. Without them, even greater reliance on satellite and other wireless communications technologies would occur, as evidenced around the world. Developing nations are leapfrogging straight to systems that are all-wireless in architecture. Satellites are feeding central offices which then provide level 1 access using various wireless technologies. This is allowing rapid deployment of millions of lines of additional service annually, something that would be cost-prohibitive were fibre to be used.

2. Satellite's Traditional Competitive Advantage

The most obvious competitive advantage of satellite is its ability to provide communications, at virtually any speed, from anywhere, to anywhere. The inherent broadcast architecture makes it ideal for delivery of the same information to multiple locations (distribution to cable, DTH/DBS, large file delivery applications), and this is the mainstay of virtually every satellite company in the industry today. While less compelling than the broadcast traffic argument, satellite's reach has allowed the provision of non-broadcast services to locations that could not be served otherwise.

There are other less obvious capabilities of satellite that have enabled it to gain a significant market share in certain applications outside of broadcast applications.

There are numerous satellite access methods that involve the use of specialized ground segment equipment that bursts or pulses its transmission to the satellite, effectively allowing the sharing of the satellite's bandwidth between locations.

Demand Assigned Multiple Access (DAMA) systems allow the number of trunks in a mesh network to be less than the number of potential routes. Unlike a terrestrial network where the physical connection must be in place at all times, satellite connections may be made and broken as required simply by establishing and dropping the radio link. Similar to the way a PBX assigns a long-distance or local trunk line to an extension when demanded, DAMA systems set up radio links between locations only when traffic demands them. The net result is that space segment costs can be reduced by employing intelligent ground equipment that can assign radio frequency channels as required, provided that the traffic between locations is not continuous.

Time Division Multiple Access (TDMA) systems have been in use for about 15 years in satellite networks, most notably in Very Small Aperture Terminal (VSAT) data communications. Similar to DAMA in that they use satellite bandwidth only when needed, TDMA systems typically burst into a high-speed channel multiple times during each session or call. These systems were originally designed to take advantage of the bursty nature of interactive data traffic (such as SNA/SDLC connections). The communications channel usually runs at speeds of at least 64 Kb/s, thus providing instantaneous throughputs at least equivalent to terrestrial networks. They achieve economic advantage by allowing multiple locations to share a single RF link due to the intermittent nature of such transmissions. By providing instantaneous throughput equivalent or in many cases superior to terrestrial networks, TDMA systems provide similar response times, while using resources only as and when required. Further averaging across time zones increases the effectiveness of this access method, as traffic peaks typically move from East to West throughout any given day.

Finally, in applications where data transfer is asymmetrical, networks can be provisioned asymmetrically over satellite as most ground segment equipment, and the satellite itself, have the required flexibility built in. The World Wide Web (WWW) is a very current example of this. Typically an end user issues a few bytes of information and receives hundreds or thousands of times as much information in response to their request. When provisioned over satellite this application shares a single low-speed TDMA link for inbound (request) transmissions, and a higher-speed Time Division Multiplexed or Statistically Multiplexed outbound link for responses. The advantage over traditional land-based services is obvious, and carriers have taken notice. They are addressing this through the various Digital Subscriber Line (xDSL) technologies that are also asymmetrical.

3. Satellite's New Competitive Advantage: "Smarter" Payloads

The competitive advantage of satellite, and thus the role that it is able to carve out for itself in the telecom industry, will change as the satellite payloads become "smarter". The ability to switch between satellite spot beams, and between satellites themselves, will allow re-use of frequencies many times over what can be accomplished today, and lead to orders-of-magnitude increases in the available bandwidth of satellites. As costs continue to fall the economic break-even distance for point-to-point satellite links will shorten. At the same time the need for ubiquitous or universal access to broadband services will expand the number of locations that fall beyond the physical or economically viable reach of wireline carriers.

One of the difficulties faced by governments in developed nations, where telecom competition has been or is being introduced, is the potential destruction of the model that allowed the provision of universal POTS. By bundling lucrative urban markets with less well-paying rural ones, we have been able to enjoy almost universal access to basic telecom services. As we move into a world in which high-speed communications becomes as critical as POTS is today, this issue becomes much more difficult to manage. In many areas the cost of upgrading the basic infrastructure prohibits the introduction of broadband services. Hybrid terrestrial and satellite-based solutions can solve the problems presented by universal service requirements, and by the limited reach of terrestrial carriers.

While competing firms are reducing their costs and could ultimately threaten this advantage, there will always be between 10 and 30% of the market that can't be served using terrestrial-based services. Satellite will remain a viable and robust alternative in broadband applications for the foreseeable future.

4. Satellite Broadband Applications for Today and Tomorrow

Satellites are being used today to provide a multitude of broadband applications, in both rural and underserved areas. Certainly the phenomenal success of Direct Broadcast Satellites such as Hughes' DirecTV/USSB and Echostar's DISH network represent the most widely recognized and understood of these, but there are many others that do not fit the mass-market, broadcast model.

For three years Telesat has been a leader in the development and testing of ATM-based applications over satellite. This work was done to verify both that the protocol is developed in a way that will allow satellite as a transmission medium, and that the *applications* can be developed that will ride on the network.

Tele-robotics applications have been demonstrated that allow the control of equipment in potentially dangerous situations from a distance. This is relevant in mining and other resource-based industries, and search and rescue applications.

In some cases the geographic separation of locations drives not only the need for satellite as a transmission medium, but the need for the application itself. Tele-health applications have been used by specialists in urban areas to perform diagnosis or consultation with less experienced medical staff in remote areas. In the future it may be possible to extend this to the performance of medical procedures, eliminating the sometimes dangerous transportation of patients to urban medical centres, and its resultant disruption of their families.

Of more widespread interest is Telesat's DirecPC™ service. Based on technology developed by Hughes Network Systems, DirecPC provides multimedia applications direct to the PC at speeds up to 3 Megabits per second. There are three different means of implementing DirecPC into a network today:

i) as a real-time transmission medium to the PC for video, text-based or any multimedia content;
ii) for non-real-time delivery of electronic files or "packages"; and,
iii) as a high-speed overlay to a dial-up Internet Access service.

By marrying the latter with the TDMA capabilities of VSAT services, graphically intense Intranet applications are extended anywhere cost-effectively. When the speed of this system is increased to 24 Mb/s later this year, even faster performance will be possible.

Other firms are using VSAT and other satellite-based services to provide Internet backbone extension where no facilities exist for an Internet Access provider to provide local dialup service. Where the cost of extending the Internet backbone into a community is not cost-justified, and the costs of dialing long distance into an IAP render that option unfeasible, Canadian Satellite Communications Inc. (Cancom) will provide a link to a local Point-of-Presence (POP). A trial is currently underway in Quebec which uses a similar satellite service to connect to a Cable Head End, and cable modems are the local access method. This allows Cable companies in more remote areas to offer high-speed information services like their urban brethren.

5. Trends in Satellite Communications. A Look at Advanced Satellite Systems

There are a number of trends that are contributing to satellite's growing importance in the communications which businesses are conducting globally. All of these lead ultimately to either lower transmission costs, or lower end-user terminal costs, and thus will aid in the further proliferation of satellite services in a broadband environment.

This includes direct-to-the-end-user services such as DTH television or information services, as well as those delivered to remote central offices for local telephone companies or other service providers, who then use local broadband services for 'the last mile'.

5.1 Higher Power / Higher Frequency

In the early 1980s, the typical North American backyard dish owner required a 3.7 m (10 feet) diameter or higher antenna to receive acceptable television signals. Those dish antennas were used to receive signals in the 4 GHz band, from satellites with transponder amplifiers in the range of 5 watts. Today, DBS antennas are as small as 40 cm (15 inches) in diameter because they're receiving signals from DBS satellites with 250-watt transponders (when phase-combined), at frequencies 3 times that of the early DTH systems.

The ability to use small, easy-to-install antennas with mass-produced low-cost receivers has expanded the market for DTH service by a factor of 10 in just a few short years.

5.2 Spot Beams

One of the keys to success in the effort to make satellites pervasive is reduction in the cost and size of user terminals. By concentrating power ever more intensely on smaller geographic areas, future satellites will be able to transfer more information for a given size of antenna than today's spacecraft (Figure 2). In addition, the ability to reuse precious frequency spectrum many times over will help to reduce overall system costs (on a per-bit basis).

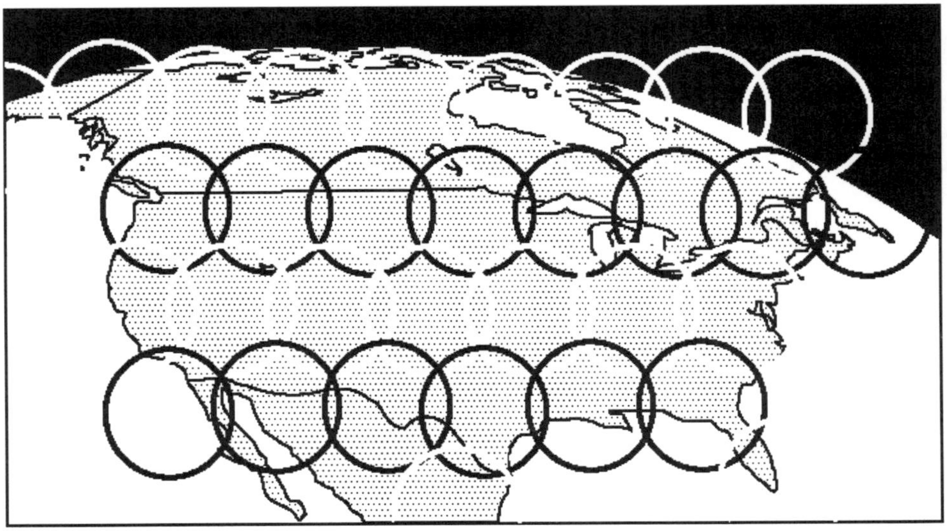

Figure 2. Spot Beams

5.3 On-board Processing and Inter-satellite Links

Distributed computing is increasing the need for Peer-to-Peer or full mesh networks. It used to be that a network simply had to connect a remote location to a central processing site, and this star configuration is perfectly suited to satellites. The growth of the Internet and other distributed networks is rendering star networks ineffective – especially over satellite due to the double-hop required and its resultant latency – so satellites must change if they are to continue to compete. Two developments will enable satellites to continue to be a significant contributor to the broadband communications industry.

On-board Processing will allow signals to be switched between spot beams in the satellite. Without on-board switching, all signals would have to pass through a single gateway station, or expensive switching stations would have to be built on the ground within the footprint of each spot beam. In either case, without On-board Processing a second satellite hop would be required for any remote-to-remote site transmission.

Similarly, in the case of global systems, the ability to communicate between satellites will eliminate the need for signals to pass through ground-based gateways, yielding the same benefits in terms of increased efficiency of satellite capacity usage and resultant lower costs.

5.4 Signal Compression

Fundamentally, in the digital world we now find ourselves, once proven technically adequate, competing communication technologies are evaluated on their cost per bit. Advanced compression techniques often yield bandwidth savings greater than 90%, increasing the amount of information that can be transmitted for a given slice of spectrum. While not unique to satellite technology, these advances have allowed cost reductions that enabled the development of DBS television services, and more cost-effective voice service to remote areas.

5.5 Non-geostationary Orbits

Geostationary satellites – those in orbits with the same 24-hour period as the earth – have the advantage of using fixed-position antennas, and covering potentially vast areas of the globe. The distance required to accomplish this is approximately 23,000 miles from earth which results in high (approximately ¼ second) latency, and in large amounts of signal loss during transmission.

There are dozens of exciting initiatives under way that will lead to satellites being launched in lower orbit. Some of these are backed by some of the largest companies in the telecom industry, and they promise to be a force just as the new millennium arrives. The advantage is that lower orbits result in much lower propagation delays, an important consideration in interactive applications. In addition, depending on the type of system and bandwidth required, more convenient antenna systems may be used.

The field is roughly divided into two categories – Low Earth Orbiting LEO's, and Medium Earth Orbit MEO's. With LEO's very low latency is provided, and potential bandwidth is maximized, but the number of satellites required is increased. MEO's are a compromise between LEO's and conventional geostationary satellites in terms of their benefits and tradeoffs.

5.6 Convergence

Society, especially in the developed economies, is changing how it spends its leisure time. Over time the line between computer and television will continue to blur. While the ultimate looks and capabilities of the information appliance that will eventually supplant these two special-purpose devices are unknown, it is safe to say that any service provider that can deliver both entertainment and information services on a single platform will have an advantage over its competition. As satellite technology is in the forefront of DBS services, the addition of multimedia services like DirecPC makes the satellite package an even more compelling alternative.

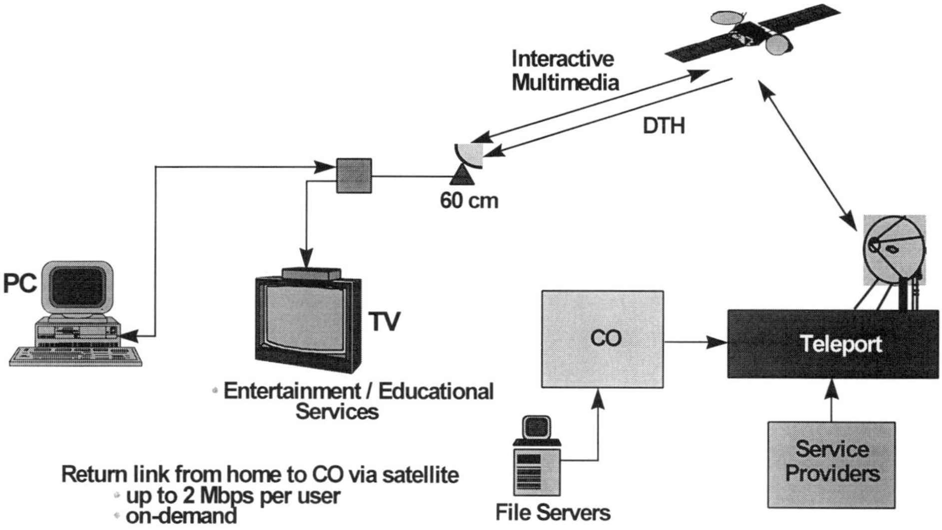

Figure 3. Interactive Multimedia with DTH

6. What Remains?

A number of key issues remain to be addressed before the role of satellite is assured in the delivery of broadband services on a widespread basis. Paramount is the availability of terminal equipment in sufficient quantities, small enough to be easily installed, and at a reasonable price. The target of Telesat and its industry partners is a terminal capable of duplex 384 kb/s using a 75 cm (30 inch) antenna, at a price under $1,000.

Even if this goal is not met at the outset, the use of open systems to create an environment which fosters innovation and competition will help to enable it. Witness the experience with pricing for DBS set-top boxes over the past year for evidence of what competition can do for hardware prices.

This is a high-stakes game with high barriers to entry. The capital muscle required to play in the global satellite multimedia arena eliminates all but the largest of telecom firms. To allow smaller, innovative and fleet-footed firms to participate, a global partnership to allow sharing of R&D and other valuable resources is a prerequisite.

This is why virtually all of the major broadband satellite initiatives today are consortiums led by one or more of the giants.

Finally, a greater understanding of the user traffic characteristics for wideband/broadband applications is required. Many of the applications don't exist today, so this is an interesting exercise.

While there are challenges to be faced before this becomes reality, will the rewards justify the investment of capital and brain trust? There is no doubt that satellites will continue to play an expanding and ever more important role as the Information Skyway becomes the predominant passing lane in the Information Highway.

References

[1] C. Blythe, The Next Space Race. Canadian Telecom, Toronto, ON, February 1997.

[2] A. Grami, K. Gordon and A. Shoamanesh, The Role of Satellites in the Information Superhighway. In: ICC '95.

Telecommunications Transformation: Technology, Strategy and Policy
E. Bohlin and S.L. Levin (Eds.)
IOS Press, 1998

CHAPTER 5

A Future of Information Technologies (Vision 2015)

Moonsik Bae and Euisun Paik

ETRI (Electronics and Telecommunications Research Institute), 161 Kajong-Dong, Yusong-Gu, Taejon 305-350, Korea

Abstract. Information technology (IT) is being applied to all aspects of life. Information, as a result, has become a capital good. Furthermore, the IT sector itself is undergoing explosive growth. Indeed, we will most likely find the future information technology quite multimedia-oriented, quite intelligent, quite personal, quite human but interrelated. The purpose of this article is twofold. First, some evolutionary processes will be discussed regarding the future of information services. Second, the related development scenario for the future terminal system and network will be examined. This article presents the technology development scenario to help us provide the new information services and construct a network for leading the transition to the new services.

1. Introduction

This new age of the Information Society brings with it new complexities and choices. The start of a new century seems to convey dramatic change. The biggest technological change and resulting social change will be the convergence of computers and communications. We are already becoming aware of this tendency as more and more consumers use multimedia PCs and go on line to use the Internet.

The next 20 years will mark the full growth of the multimedia age, freeing people from constraints of time, place and information type. People will communicate in the combinational media that best express their needs. In the long run, our society will become a virtual society which is made possible by a comprehensive information infrastructure based on broadband transmission and switching of voice, data and video with broadband network. The broadband network will be achieved by integrating separate networks.

Information technology (IT) is an all-inclusive term that encompasses computers and telecommunications in all their forms, whatever their use. It includes mainframe computers, supercomputers, and microcomputers. IT includes the public telephone network, TV and radio broadcasting, credit cards, paging systems, facsimile machines, and cellular telephones, as well as the software that controls them all [1].

IT is being applied to all aspects of life and is undergoing explosive growth. IT is needed to serve our needs which will be more diversified as time passes. New services with high reliability will be provided to effectively meet the diversified needs of customers. At the same time, new networks and terminal systems will be developed to work with these new services.

Technology is the helpmate of human beings. From prehistoric days, people have been making and using technology to do more things, better things, newer things, and things faster than before. This is because technology enables competition, which is the cornerstone of human development and economic progress [6]. Technology competition is arising not only from traditional technologies, but also from new entrants to a specific technology. Many technologies have begun a transition to the new aspect of rapid change. However, increased customer needs with human creativeness are reinforcing the process of accelerated technological changes.

There are some forecasting techniques to be categorized into families: direct, correlative, and structural. Direct methods are those that directly forecast parameters that measure functional capacity or some other relevant characteristic of the technology. Correlative methods relate development of the subject technology to the growth or change of one or more elements in contexts thought to be analogous. Structural methods forecast by formally considering the interaction between technology and context [5].

This article outlines the future vision of IT including information services, terminal systems and networks. The forecasting framework, service evolution, and evolutionary process of terminal systems and networks are described. To clarify the future development trends and evolutionary process for IT, we use a brainstorming method which directly utilizes the empirical knowledge of many professionals (40 persons) who specialize in the information technologies.

2. Forecasting Framework

There are many variations of information technology forecasting, concerned with how technologies will progress in the future.

Successful management of any project, especially information technology forecasting, requires planning which is totally systematic. Technology forecasting deals with causal elements of any sort: social, economic, or technological. However, the effects of interest are new technologies and incremental changes in existing technologies [5]. Thus, the focus of the analysis is on societal demand and service-providing which can be realized by the technology itself.

Information technologies are often more complex, so they usually require more planning and more analysis of their core technology structure. We can establish a systematic R&D planning for IT as the features and technological possibilities of information services are effectively defined and predicted. This forecasting is not implemented solely by a special engineer; it requires many kinds of engineers, including system analysts in various fields.

The impetus for this forecasting effort comes from the engineers and experts in the IT field who have called for a future vision in information technologies. In fact, because there may be many variations of the forecasting model for the future information services and technologies, an expert-based brainstorming method is used in the phased forecasting for IT.

The forecasting method includes several phases, each of which is intended to give careful consideration to defining the features and realization time for each service and technology. The first phase is intended to establish a conceptual foundation for the types, providing times and features of information services. Other phases are focusing on the development of a general framework for the terminals and networks.

Figure 1 presents the basic phased process of IT forecasting. There are three forecasting phases: service evolution, development process of terminal system, and network evolution. They are all based on a systematic transition from one forecasting phase to the next, until the full forecasting is complete. We present the evolutionary process by dividing it into 3 stages per phase; stage 1 is for the time to 1997, stage 2 from 1998 to 2002, stage 3 from 2003 to 2015.

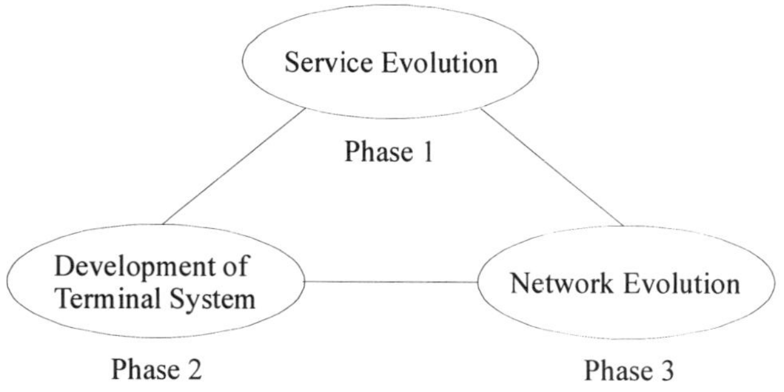

Figure 1. Phased Process of IT Forecasting

3. Vision of Future Services

3.1 The Need for an Evolution

Experience has generally shown demand-led investment to be much more efficient than supply-driven investment. An example of investment being demand-led has been the rapid development of mobile telecommunication. Most forecasts of the take-up of mobile services over the past decade underestimated the demand for service. The obvious reason for the success of mobile telecommunication has been the added service dimension brought about by this technology. At the same time the development of mobile services has been built upon the existing capabilities of the fixed networks [4].

Similarly, the rapid growth in the utilization of Internet services has been driven by consumer demand built on the foundation of the computer networks. Over the past three years, access to the Internet has become vastly easier with Internet service providers popping up [3]. The number of Internet users in the world is growing very rapidly. The number of Internet hosts is estimated to be nearly 16 million in January 1997, and has been increased by four times over the two previous years. The reasons for a substantial increase in Internet users are the very low price, the convenient usage and the universal access.

In light of this, many information services have been created to meet the increasingly diversified customer needs. Recently, customer needs are more and more regarded as an important factor for adoption of new services. IT will largely alter the way we live or the way we go about living in the coming decades.

3.2 Service Evolution

The requirements for IT including telecommunications will become more and more diversified and complicated in the years leading up to the 21st century. In compliance with this trend, new services will be provided with technology development. Service-providing depends heavily on customer needs as the technologies develop.

In this study, we classify the information services into seven categories from a viewpoint of service application. Table 1 illustrates the types and features of application services by each developmental stage.

Table 1. Types and Features of Application Services

Year		Stage I (~ '97)	Stage II ('98~2002)	Stage III (2003~2015)
Fields	Home	- Intelligent Home	- Advanced Intelligent Home	- Omnipotent Home (Virtual Office)
	Industry	- Video Conference - Tele-Groupware - Electronic Bank/Cash	- Distributed Production - Facilities Network - VEOS(Virtual Environment Operating System)	- Realistic Groupware - Space Communications
	Adminis-tration	- Environment Monitoring - Multimedia Civil Services	- Portable Civil Services - Intelligent Consulting (e.g., Tax & Law) - Intelligent Environment Monitoring	- Realistic Integrated Administration Services
	Public	- Digital Library/Museum - Multimedia Traffic Information	- Intelligent Vehicle - Multi-Card	- Voice & Letter Recognition/ Interpretation - Personal Digital Library
	Education	- Tele-Education	- Intelligent Tele-Education	- Realistic Tele-Education
	Welfare	- Tele-Medicine	- 3-D Virtual Tele-Medicine - Portable Terminal for the Handicapped	- Conquest for the Handicapped
	Entertain-ment	- Network Game - VOD	- Realistic Tourist Agency - 3-D Network Game	- Interactive Realistic Movie - Self-devotion Cyber Journey
Features		- Prosumer Services (Producer+Consumer) - Multimedia Processing - Universalization - Integration & diversification	- Advanced Video & Voice Processing - Wireless & Portable - Personal & Intelligent - Virtual Society - Advanced Prosumer Services - Realtime Groupware	- Humanized Services - Realistic Services - Cosmic Services

4. Evolutionary Process of Information Technology

4.1 Terminal System

The most significant factors in the provision of IT services are the state of the terminal system and network development. Large-capacity memory devices and digital technology have been put to practical use, making it possible for terminal systems to be portable and intelligent. Along with technological changes, we are seeing customer needs that are more and more diversified for terminal systems in recent times. Some of these are: (1) importance of intelligent user interface; (2) demand for mobile computing; (3) demand for a composite terminal; and (4) demand for portable and wearable terminals.

Figure 2 presents the evolutionary process of the terminal system which is resulting from technological change and user needs for information services.

Terminal systems are classified into two types, portable and fixed. In portable terminal systems, Internet MM PDA with a 64-kbps wireless data connection will be developed into a 3-D portable Internet terminal with 2-Mbps broadband wireless public network connection. In about 2005, a wearable computer will be realized to combine the functions of portable picture, 3-D PDA and VSAT. Finally, humanized realistic PDA will be developed with the features of 10-Tbps wireless network connection equipment, dialogue style voice I/O, and the sense of touch and vibration codec.

In the case of fixed terminal systems, network computers and multimedia PCs will be changed into 3-dimensional terminals in about 2002. Digital TV and Internet TV will be integrated into composite TV with MPEG-2 codec and gigabit level broadband interface, and will finally be developed into hologram-type composite terminals with high-resolution projection.

4.2 Network

In order to cope with the enormous growth of the traffic needed, the network itself needs to be developed. The evolution of applications and the increase of users are much faster than the rate of increase in available bandwidth.

Network evolution is being studied not only from the technical but from the service standpoint as well, to determine the kind of services that should be provided and how they can be provided to be of greatest use to the greatest number of customers. The network system will form the basic infrastructure for improved communications efficiency in industrial and social activities.

The new telecommunications network must be economical enough so that customers can afford to pay for new services, including video service, with no significant increase in payment for them [2]. It must therefore take full advantage of the rapidly advancing new technologies described in this article.

A diversity of information types including voice, data and images will characterize the coming information society. The new information network must be able to meet these new and equally diverse requirements for non-voice services. Achieving this objective will require overall adoption of digital technology and enhancement of network capabilities. Thus, technology capable of providing digital information services economically should be becoming increasingly available at a time when customer needs for such services are rapidly expanding.

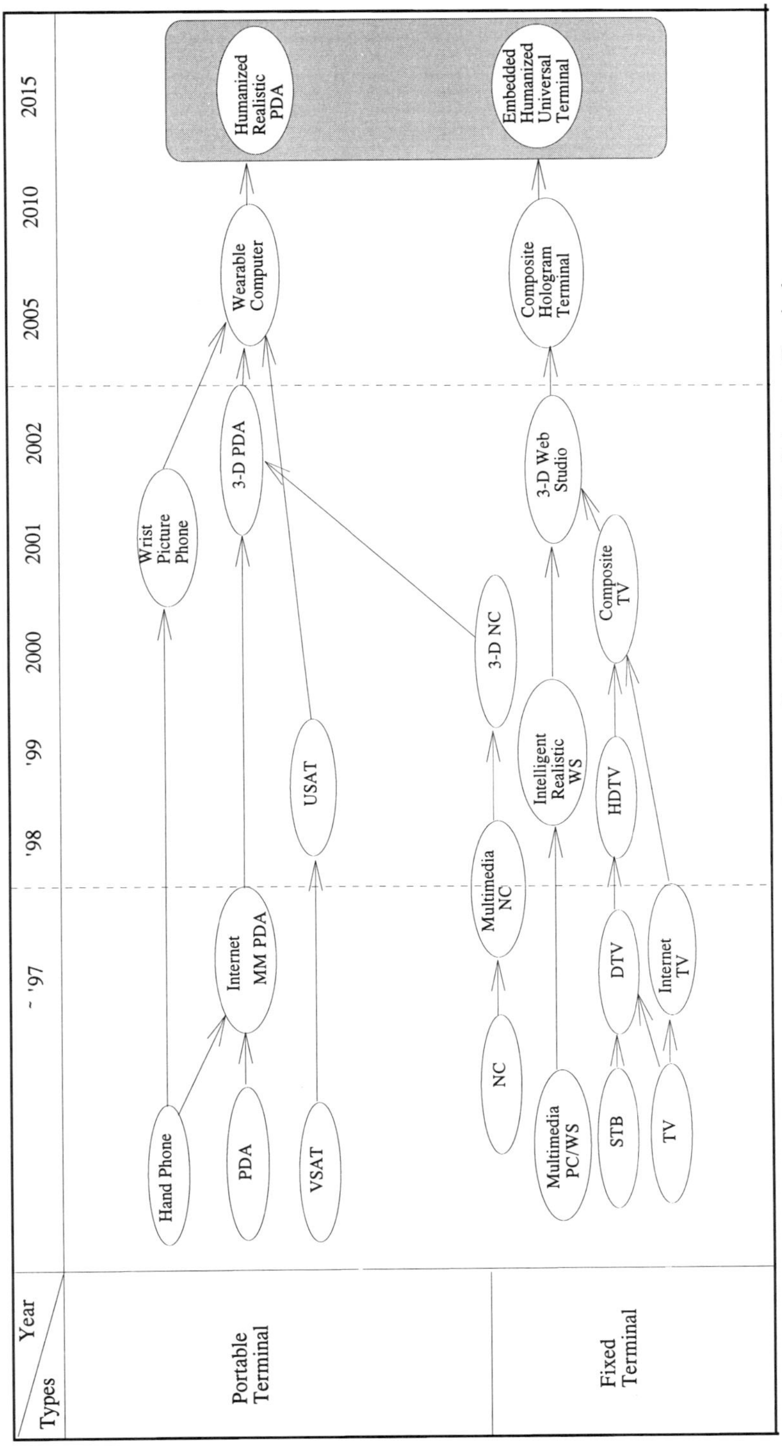

MM PDA: Multimedia Personal Digital Assistant, VSAT: Very Small Aperture Terminal, USAT: Ultra Small Aperture Terminal
NC: Network Computer, STB: Set-Top Box, DTV: Digital TV

Figure 2. Evolutionary Process of Terminal System

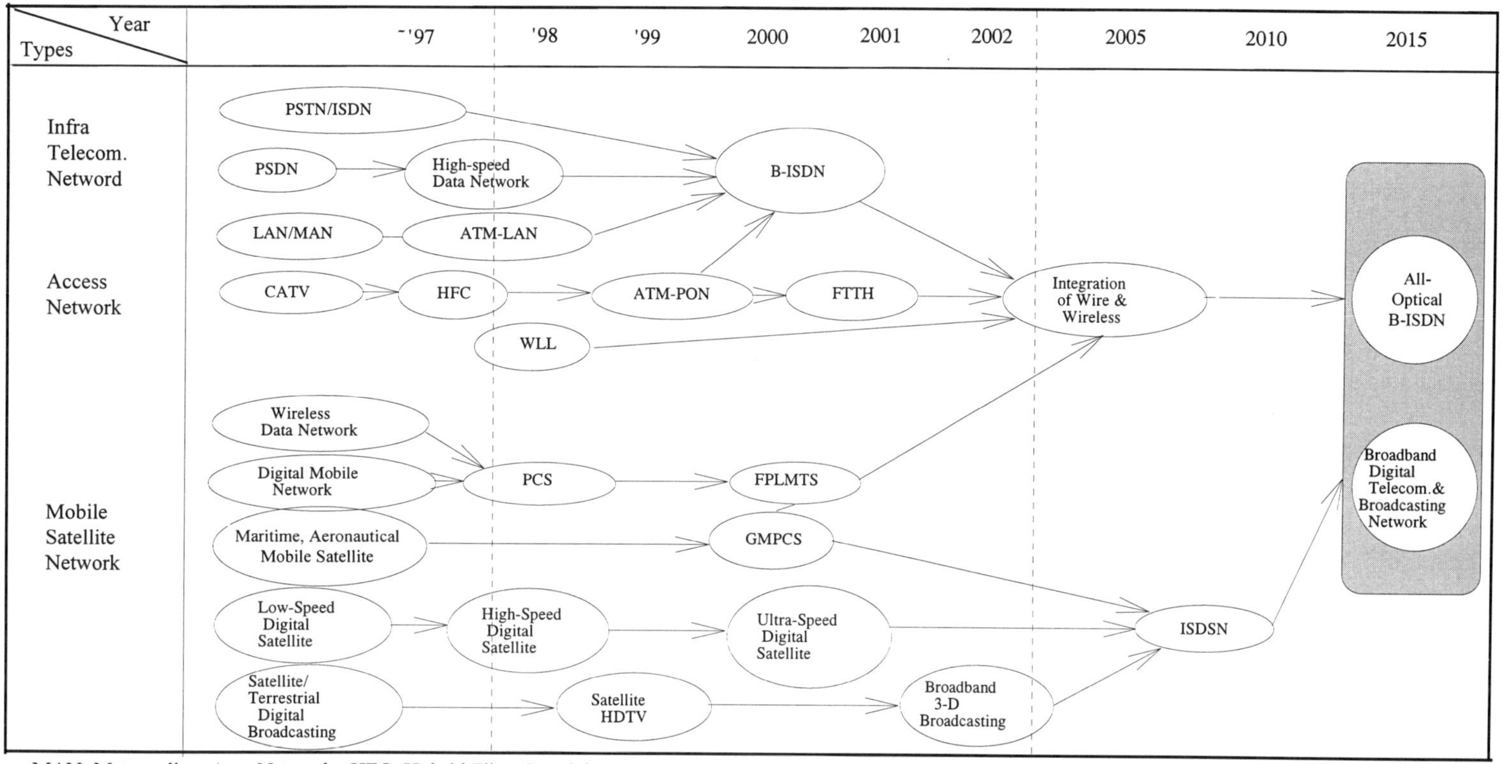

MAN: Metropolitan Area Network, HFC: Hybrid Fiber Coaxial
ATM-PON: Asynchronous Transfer Mode - Passive Optical Network, FPLMTS: Future Public Land Mobile Telecommunication System
GMPCS: Global Mobile Personal Communication System, ISDSN: Integrated Services Digital Satellite Network

Figure 3. Evolutionary Process of Network

Figure 3 shows the evolutionary process of the network. At present, separate specialized networks are provided for PSTN, PSDN, ISDN, CATV, Internet, and mobile network including satellite network. However, it would be far more efficient to supply all the varied services through an integrated network than by an individual network for each application service. As network evolution progresses, these separate networks will be gradually integrated into a broadband digital network.

Integration of wire and wireless network will be required to provide the variety of services referred to as ISDN, FTTH and FPLMTS. The introduction of an integrated wire and wireless network will establish a basis to meet the demands of the future, providing a network infrastructure for a wide variety of information. The most significant development will then occur in mobile communication with the widespread penetration of portable services down to the customers. The full potential of the integrated network to carry a new service can only be realized when communication capability is extended to each customer.

The first step in network evolution can be characterized as enhancing the capabilities of the individual network. The features of the evolutionary process in the first step can be explained as the following: appearance of individual networks such as N-ISDN, digital mobile network, mobile data network, low-speed digital satellite network, and high-speed data network.

The public telecommunication network provides primarily voice and data services using PSTN/PSDN. N-ISDN is being activated. ATM switching system and B-ISDN appear gradually with inter-operability between ATM networks and existing networks. PSDN is being developed into a high-speed data network with 2.5 Gbps. In digital mobile networks, the capacity to accommodate subscriber lines will be increased as ACNs (Analog Cellular Network) change to DCN (Digital Cellular Network) and digital satellite TV broadcasting services can be provided with MPEG-2.

The second step can be characterized as the integration of physical networks. Wireless Internet will be introduced. N-ISDN and individual networks including subscriber access networks will be integrated into B-ISDN. FTTH (Fiber to the Home) will be introduced with the completion of FTTC (Fiber to the Curb). FPLMTS will be made possible by the integration of mobile networks and non-GEO satellite telecommunication networks. Digital satellite services with high speed will be provided using Ka band to make satellite HDTV and 3-D broadcasting possible.

All-optical B-ISDN and broadband digital telecommunication and broadcasting networks are expected to be practical by 2015 with the construction of FTTH. Broadband satellite broadcasting networks based on ISDSN will also be realized.

5. Summary and Conclusions

Thus far, there has been a lot of analytical research that forecasts a trend of technology development. Most of this research neglects the features and exact times of providing service. In this article, we have tried to present a vision for IT evolution including the features and realization time from a viewpoint of services and related technologies. Establishment of an IT vision based on time and features is greatly needed to present guidelines for technology development.

However, there must be an ability to meet customer needs and to turn the vision into reality. Most of the professionals involved in this study suggest that the IT vision described in this article would be achievable and would provide the implementation strategies to realize a future technology. But, in an era of change, things will change, so there must be many feasibility studies before the vision becomes fact. We also have to be sure that technological changes will occur rapidly and efficiently.

References

[1] G.M. Hoffman, The Technology Payoff. Richard D. Irwin, Inc., 1994.

[2] N. Inoue, New Technologies for the Information Network System, *Telecommunication Technologies,* JARECT vol. 14, OHMSHA, Ltd., April 1984, pp. 19-43.

[3] L. Levitt, Commercial Use of the Internet. Proceedings of INET '95, Internet Society, 1995, pp. 233-240.

[4] OECD, Communications Outlook 1995. OECD Publications, 1995.

[5] A.L. Porter et al., Forecasting and Management of Technology. John Wiley & Sons, Inc., 1991.

[6] N. Sharif, Reengineering Technology Governance for Philippines 2000, *Technological Forecasting and Social Change*, **54**, Elsevier Science Inc., 1997, pp. 37-55.

Part III

Network Evolution, Strategy, and Structure

Telecommunications Transformation: Technology, Strategy and Policy
E. Bohlin and S.L. Levin (Eds.)
IOS Press, 1998

CHAPTER 6

Optimal Industry Structure in International Telecommunications: A Mathematical Model

Livio Cricelli, Massimo Gastaldi and Nathan Levialdi

Department of Production, Systems and Computer Science, University of Rome – "Tor Vergata"
Via di Tor Vergata, 00173 Rome, Italy

Abstract. The process of liberalization and privatization is drastically altering the links among international telecommunications carriers. The model presented in this paper analyses the international telecommunications market illustrating the effects of vertical mergers, under symbiotic production conditions, when vertically integrated and unintegrated carriers coexist. The effect of competition is analysed both between carriers in the final market and in the intermediate one in which carriers compete over tariffs to get an appropriate distribution of profits. In particular, the results show the relations among final market prices, intermediate tariffs and market dimensions and their implications for profits.

1. Introduction

Over the last decade, regulatory reforms and technological progress have had strong impacts on the telecommunications sector in industrialized countries. The globalization process of the international economy, with the corresponding increase of telecommunication intensity, induces a growing demand for international telecommunication services [1]. This is changing the functioning of the international telecommunication system, traditionally based upon bilateral trade cartels among the monopolists serving each national market; the increasing volume of international calls and the correspondingly growing carriers' interest involve shifting telecommunications from the traditional monopoly to a market with a growing competitive pressure [2]. The traditional scenario of telecommunication service formed by state monopoly carriers is changing into a single conduit in which there are different, exclusive, territorially competitive, companies interconnected internationally. In the USA, since 1982 when the regional operating companies were separated from AT&T, a new gradual deregulation process has started. In ensuing years, the US telecom marketplace became an exceedingly competitive arena and strategic alliances became a core future of competitive strategies. The recent deregulation of the US telecommunications industry through the landmark Telecommunications Reform Act of 1996 promises to accelerate even faster the pace of competitive dynamics in US markets. In particular, the Telecommunications Act has authorized the Bell Operating Companies (BOCs) and the Local Exchange

Carriers (LECs) to provide long-distance services competing with the existing Interexchange Carriers (IXCs) [3].

In the UK in 1984 British Telecom became a private company, and the Government allowed Mercury to operate in the long-distance market. Such transition from monopoly to competition causes new problems in determining competition pressure results. As shown by Baumol and Sidak [3] for the local telephony, the local exchange carriers' services constitute inputs for the activities of their rivals in other markets, inputs without which the rivals cannot hope to operate. The same competitive problem arises in the international telecommunication system, in which an international call utilizes the services of a telephone company at each end (final market) where the revenue is collected by the originating telephone company from the party who initiates the call. This company then compensates the other one through the tariffs (intermediate market) for the costs it incurs in handling the call. Therefore, the interconnection services in the intermediate market are deemed "bottlenecks" or "essential facilities" [4], since an international carrier could use such essential facilities to achieve monopolistic rents inducing final price distortions.

In this framework, telephone services can be represented by a form of production in which each producer has the following characteristics:

- each producer has a different competitive power in its own market;
- each producer operates both in the intermediate and in the final market;
- each producer must purchase the intermediate good from the other producers.

This market form has some parallels with the standard vertical integration model and can be seen as an extension, in which the production links stretch out in both directions. This is the fundamental complicating phenomenon that besets the deregulation of telephone services. This framework can be considered a symbiotic production form in which suppliers trade essential inputs with one another [5].

In this paper, we analyse an extension of the general model of symbiotic production focusing on the long-distance telecommunications service in which vertical integration is considered. The treatment of vertical integration has been almost an issue for competition policy in the telecommunications sector. Traditionally, vertical integration of an input supplier into downstream production has been suspected to reduce competition either in the downstream market or in the upstream market by foreclosing competitors' access to downstream firms. In our production model, we show that, under some conditions, a vertical integration may induce an increase in competition both between carriers in the final market and in the intermediate one. The presented simulation model, based on the coexistence of vertically and non-vertically integrated carriers [6], shows how final consumer prices and intermediate tariffs decrease with the number of vertically integrated firms. This result arises both from the elimination of the double marginalization phenomenon and from the peculiarity of the symbiotic production approach. By means of a numerical simulation, we show that the firms have an incentive to integrate depending upon the final market structure.

The paper is organized as follows. In the next section the basic assumptions and the model are set up. Then the results of vertical integration provided by the model are analysed and, finally, the analytical evidence using a simulation approach is presented.

2. The Model

To analyse the effects of vertical integration in the international telecommunication system, the following scenario is considered (Figure 1) based on two different final markets A and B in which there are, respectively, N_A and N_B IXCs, n of which are vertically integrated (n Min N_A, N_B), that sell their final products (long-distance telephone calls). Furthermore, there is an intermediate market in which carriers sell each other the intermediate interconnection service. In this framework, the interconnection service is sold, respectively, at tariffs t_A and t_B among the unintegrated firms, or otherwise at the marginal cost, if carriers are vertically integrated. We assume the participation of vertically integrated firms in the market for the intermediate good.

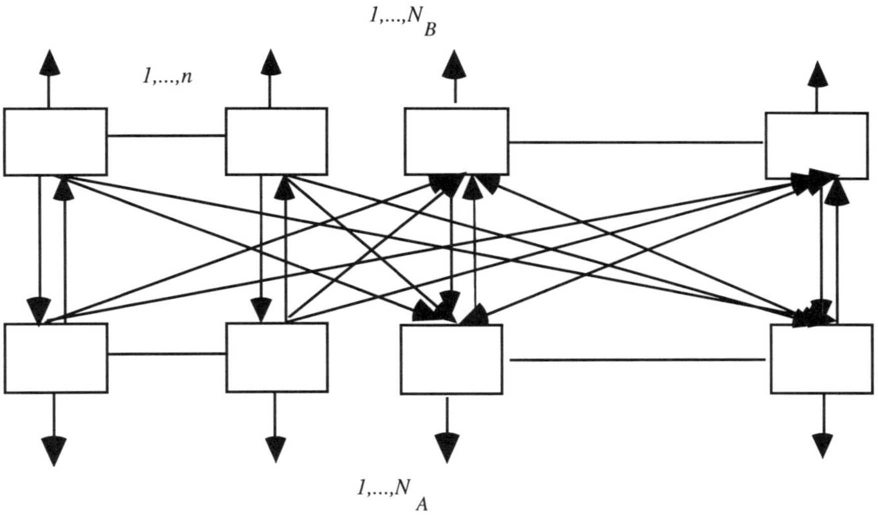

Figure 1. International Telecommunications System Network

Let:

c_A = marginal cost of final market A;
c_B = marginal cost of final market B;
c = marginal cost of the intermediate market;
Q_A = total final quantity sold in market A;
Q_B = total final quantity sold in market B;
q_i^A = final quantity sold by integrated firm i in market A;
q_i^B = final quantity sold by integrated firm i in market B;
q_j^A = final quantity sold by unintegrated firm j in market A;
q_j^B = final quantity sold by unintegrated firm j in market B.

Also assumed are the following inverse demand functions for the final good:

$$p_A = a - bQ_A \tag{1}$$

$$p_B = e - fQ_B \tag{1'}$$

The profit functions of integrated firm i and of unintegrated firm j in markets A and B are respectively given by:

$$\pi_i = (p_A - c_A)q_i^A + (p_B - c_B)q_i^B - c(q_i^A + q_i^B) \qquad i=1,2,\ldots,n \tag{2}$$

$$\pi_j^A = (p_A - c_A)q_j^A - t_B q_j^A \qquad j=n+1,n+2,\ldots, N_A \tag{3}$$

$$\pi_j^B = (p_B - c_B)q_j^B - t_A q_j^B \qquad j=n+1,n+2,\ldots, N_B. \tag{3'}$$

Notice that the above profit functions are related only to the sales of the final goods whereas, in the sequel, the profit contribution of the intermediate interconnection service will be considered.

Differentiating the profit functions and assuming

$$R_i^A = Q_A - q_i^A \qquad R_i^B = Q_B - q_i^B$$

yields the Cournot optimal equilibrium quantities:

$$q_i^A = \frac{a - c - c_A - bR_i^A}{2b} \qquad i=1,2,\ldots n \tag{4}$$

$$q_j^A = \frac{a - t_B - c_A - bR_j^A}{2b} \qquad j=n+1,\ n+2,\ldots N_A \tag{4'}$$

$$q_i^B = \frac{e - c - c_B - fR_i^B}{2f} \qquad i=1,2,\ldots n \tag{5'}$$

$$q_j^B = \frac{e - t_A - c_B - fR_j^B}{2f} \qquad j=n+1,\ n+2,\ldots N_B. \tag{5'}$$

Solving the system of N_A equations (4) and (4'), we obtain:

$$q_i^A = \frac{1}{b}\left[\frac{1}{N_A+1}(a - c - c_A) + \frac{N_A - n}{N_A+1}(t_B - c)\right] \qquad i=1,2,\ldots n \tag{6}$$

$$q_j^A = \frac{1}{b}\left[\frac{1}{N_A+1}(a - t_B - c_A) + \frac{n}{N_A+1}(c - t_B)\right] \qquad j=n+1,\ n+2,\ldots N_A. \tag{6'}$$

Similarly from N_B equations (5) and (5'), we have:

$$q_i^B = \frac{1}{f}\left[\frac{1}{N_B+1}(e - c - c_B) + \frac{N_B - n}{N_B+1}(t_A - c)\right] \qquad i=1,2,\ldots n \tag{7}$$

$$q_j^B = \frac{1}{f}\left[\frac{1}{N_B+1}(e - t_A - c_B) + \frac{n}{N_B+1}(c - t_A)\right] \qquad j=n+1,\ n+2,\ldots N_B. \tag{7'}$$

Assuming

$$Q_{NI}^{B} = \sum_{j=n+1}^{N_B} q_j^B$$

the optimal intermediate tariff is given by

$$t_A = \frac{e - c_B + nc}{n+1} - \frac{N_B + 1}{(N_B - n)(n+1)} f Q_{NI}^{B}. \qquad (8)$$

In order to obtain the optimal prices, the intermediate market is now considered. The profit functions related to this market are given by:

$$\pi_i^{tA} = (t_A - c) q_i^{tA} \qquad i=1,2,....N_A \qquad (9)$$

$$\pi_i^{tB} = (t_B - c) q_i^{tB} \qquad i=1,2,....N_B. \qquad (9')$$

Maximizing (9) and (9') and assuming

$$H = \frac{e - c_B + nc}{n+1}$$

$$K = \frac{N_B + 1}{(N_B - n)(n+1)} f$$

$$R_i^{tA} = Q_{NI}^{B} - q_i^{tA}$$

and noting that

$$Q_{NI}^{B} = \sum_{i=1}^{N_A} q_i^{tA} \qquad (10)$$

we have:

$$q_i^{tA} = \frac{H - c - KR_i^{tA}}{2K} \qquad i=1,2,....,N_A. \qquad (11)$$

Solving the system of equations (11) yields the profit-maximizing intermediate quantities:

$$q_i^{tA} = \frac{(N_B - n)(n+1)}{(N_B + 1)(N_A + 1)} \left(\frac{e - c_B + nc}{n+1} - c \right) \qquad i=1,2,....,N_A. \qquad (12)$$

Adding (12) and substituting in (8), we obtain the optimal intermediate market tariffs:

$$t_A = \frac{e - c_B - c}{(n+1)(N_A + 1)} + c \qquad (13)$$

and

$$t_B = \frac{a - c_A - c}{(n+1)(N_B + 1)} + c. \qquad (14)$$

The obtained tariffs are functions of demand function parameters, number of vertically integrated firms and, respectively, of N_A and N_B. Substituting the optimal intermediate market tariff (13) into the optimal equilibrium final quantities, (7) and (7'), we have:

$$q_i^B = \frac{e-c_B-c}{f(N_B+1)}\left[\frac{(N_B-n)}{(n+1)(N_A+1)}+1\right] \qquad i=1,2,...,n \qquad (15)$$

$$q_j^B = \frac{(e-c_B-c)N_A}{f(N_B+1)(N_A+1)} \qquad j=n+1,n+2,...,N_A. \qquad (15')$$

Such relations allow us to consider the final quantities in the market B in terms of both the demand function parameters and the dimension of the markets.

Finally, adding (15) and (15') and substituting into (1'), the optimal final price is obtained:

$$p_B = c + c_B + \frac{e-c_B-c}{(N_B+1)}\left[\frac{(N_B-n)}{(n+1)(N_A+1)}+1\right]. \qquad (16)$$

Analogously, it is possible to obtain:

$$p_A = c + c_A + \frac{a-c_A-c}{(N_A+1)}\left[\frac{(N_A-n)}{(n+1)(N_B+1)}+1\right]. \qquad (17)$$

The presented analytical results show some important relations among microeconomic variables. In particular, the following propositions hold:

Prop. 1: Intermediate tariffs and final consumer prices are negatively related to n.

Prop. 2: Intermediate tariffs and final consumer prices are negatively related to N_A and N_B.

Proof. See Appendix.

One of the main results arising from the above propositions is that vertical integration may lower final prices, despite what is stated in literature in similar conditions [7, 8, 9]. This is due to the particular features of the international telecommunication system. The existence of two final markets in "symbiotic" linkage amplifies the relevance of intermediate market tariffs, which provide a powerful contribution to profits. In this way, the growing number of vertically integrated firms involves a decreasing value of intermediate market tariffs, as shown above. Consequently, with the reduced intermediate stage costs, the firms may lower final prices. However, an increasing level of horizontal competition has a positive effect for both tariffs' and final prices' decrease. So the coexistence of more carriers in each country, even if under vertical integration, is important. In other words, a decrease in final prices may be due not only to a global market expansion, but also to a progressive hierarchic integration process. Efficiency in international telecommunications will be enhanced by promotion of competition among carriers nationally, while attempts to undermine the vertical integration trend could produce greater inefficiency. Such a result is in tune with the real evolution of the telecommunications market, in which there are both general long-distance tariff decreases and an increasing horizontal and vertical concentration process.

To evaluate the real advantages of the vertical integration process, we can consider the implications for maximum profits of the determined optimum level of microeconomic variables. More specifically, substituting such values into the profit function, we obtain the optimal profit of a vertically integrated firm arising from the sale of the final good only:

$$\pi_{VI} = \frac{\frac{(a-c-c_A)^2}{b}\left(N_B + \frac{N_A+1}{n+1}\right)^2 + \frac{(e-c-c_B)^2}{f}\left(N_A + \frac{N_B+1}{n+1}\right)^2}{(N_A+1)^2(N_B+1)^2} \tag{18}$$

Adding (3) and (3') and executing the right substitutions, we obtain the joint maximum profit for two unintegrated firms operating in each final national market:

$$\pi_{NI} = \frac{\frac{1}{b}(a-c-c_A)^2 N_B{}^2 + \frac{1}{f}(e-c-c_B)^2 N_A{}^2}{(N_A+1)^2(N_B+1)^2}. \tag{19}$$

The following proposition holds:

Prop. 3: The final market profit of unintegrated firms is independent of the number of integrated firms.

This is due to the compensation between the reduction of final prices and the reduction of intermediate tariffs. Using (18) and (19), we obtain the final market joint profit variation of two firms that, starting from the existence of n vertically integrated firms, decide to integrate:

$$\Delta\pi_{MF} = \frac{\frac{(a-c-c_A)^2}{b}\left[\left(\frac{N_A+1}{n+2}\right)^2 + 2N_B\frac{N_A+1}{n+2}\right]}{(N_A+1)^2(N_B+1)^2} + \frac{\frac{(e-c-c_B)^2}{f}\left[\left(\frac{N_B+1}{n+2}\right)^2 + 2N_A\frac{N_B+1}{n+2}\right]}{(N_A+1)^2(N_B+1)^2} \tag{20}$$

The following proposition holds:

Prop. 4: The final market profit increase arising from vertical integration is always positive, and is decreasing with an increase in the number of vertically integrated firms.

Proof. See Appendix.

At this point, to determine the real convenience of vertical integration, it is necessary to consider the profit variation attributable to the intermediate market. Adding (9) and (9'), after the appropriate substitutions, we have the intermediate market joint maximum profit of two firms operating into different final markets:

$$\pi_{MI} = \frac{(a-c-c_A)^2(N_A-n)}{b(N_B+1)^2(N_A+1)(n+1)} + \frac{(e-c-c_B)^2(N_B-n)}{f(N_A+1)^2(N_B+1)(n+1)} \tag{21}$$

From (21), it is possible to determine the intermediate market joint profit variation of two firms that, starting from the existence of *n* vertically integrated firms, decide to integrate:

$$\Delta \pi_{MI} = -\frac{(a-c-c_A)^2}{b(N_B+1)(n+2)(n+1)} - \frac{(e-c-c_B)^2}{f(N_A+1)(n+2)(n+1)} \tag{22}$$

As a consequence the following proposition holds:

Prop. 5: The intermediate market joint profit variation is always negative and is, in absolute terms, negatively related to the number of integrated firms.

Proof. See Appendix.

Adding (20) and (22) and assuming

$$A = \frac{(a-c-c_A)^2}{b} \qquad B = \frac{(e-c-c_B)^2}{f}$$

we obtain the total joint profit variation of two firms that decide to integrate:

$$\begin{aligned}\Delta \pi = & \frac{A(N_A+1)[2N_B(n+2)(n+1)-(N_A+1)]}{(N_A+1)^2(N_B+1)^2(n+2)^2(n+1)} \\ & + \frac{B(N_B+1)[2N_A(n+2)(n+1)-(N_B+1)]}{(N_A+1)^2(N_B+1)^2(n+2)^2(n+1)}\end{aligned} \tag{23}$$

To show the joint profit variation of two firms belonging to different final markets due to a potential integration process, it is possible to perform a numerical simulation by means of an appropriate choice of demand function parameters.

3. Numerical Simulation Evidence

Figure 2 illustrates some results of the considered simulation process. More specifically, given the number of firms operating in one final market (N_A), the joint profit variations ($\Delta\Pi$) of two firms that decide to integrate are shown as a function both of carriers belonging to the other market (N_B) and of the number of the preexistent vertically integrated carriers (n).

The maximum convenience of the integration is obtained by starting from a small number of preexistent vertically integrated firms. In other words, the results of the vertical integration process (in terms of a joint profit increase) are better for the firms that decide to integrate earlier. More specifically, in the first case a symmetric situation (in terms of demand function parameters $A = B = 10$) is considered.

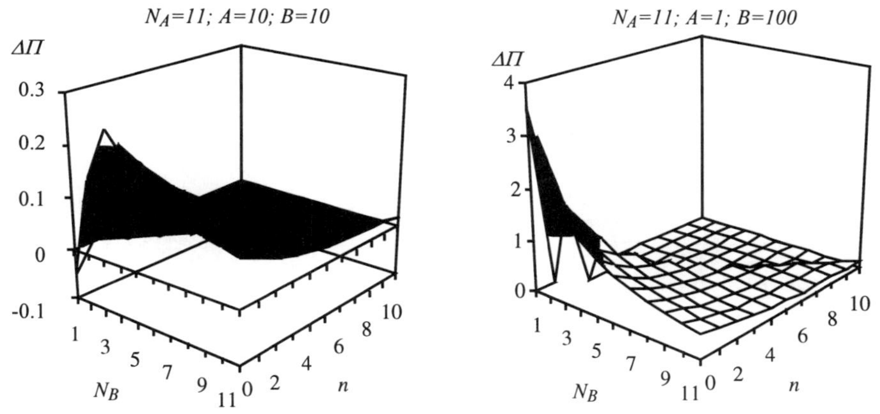

Figure 2. Total Joint Profit Variations

Notice that the presence of a monopolistic situation in the second market ($N_B = 1; n = 0$) determines the vertical integration unprofitableness ($\Delta\Pi < 0$). The profit loss of the intermediate market cannot be compensated by the profit increase in the final market. Moreover, once the market dimensions are fixed, with increasing *n* the convenience of integration decreases. In the second case, we illustrate an asymmetric situation (*A = 1; B = 100*) in which the maximum level of convenience is obtained when there is a monopolistic situation in the second market.

Nevertheless, to evaluate the real convenience of the different strategies leaving demand function parameters out of consideration, it is useful to calculate the percentage profit variation of two firms that decide to integrate:

$$\Delta\pi\% = \frac{\Delta\pi}{\Delta\pi_i} = \frac{\Delta\pi}{\Delta\pi_{MI} + \Delta\pi_{NI}} =$$

$$= \frac{A(N_A+1)\left[2N_B(n+2)(n+1)-(N_A+1)\right]+B(N_B+1)\left[2N_A(n+2)(n+1)-(N_B+1)\right]}{(n+2)^2\left\{A\left[N_B^2(n+1)+(N_A-n)(N_A+1)\right]+B\left[N_A^2(n+1)+(N_B-n)(N_B+1)\right]\right\}} \quad (24)$$

The related results are shown in Figure 3, in which it is possible to note that the different situations arising from demand function parameter variations are now removed.

In particular, in both cases the maximum convenience of vertical integration is obtained with high values of N_B and low values of *n*. Also in such cases the presence of the monopolistic situation in the second market determines the vertical integration unprofitableness under the same demand function parameter conditions.

The presented analysis points out that it is too difficult to obtain an optimal strategy which is fine in all situations, since the different market conditions may significantly affect the profitability of the aggregation strategy [10].

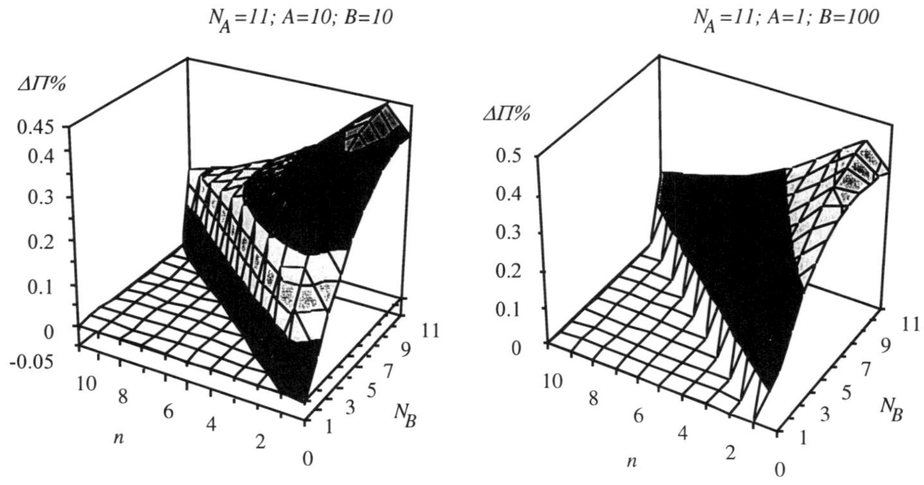

Figure 3. Total Joint Profit Variations in Percentage Terms

This is empirically evidenced by the different real strategies of the main IXCs, which are often complementary. In particular the strategies are characterized by both acquisitions and strategic alliances with firms producing complementary services. Long-distance carriers are coming into the cable TV market through agreement with specialized firms to enlarge the bundle of services and to increase the cable network quality and transport capacity. For example, SPRINT has reached an agreement with three of the main US cable TV operators: TCI, COMCAST, COX; while AT&T has joined forces with DIRECT TV, a firm operating in satellite TV. At the same time the BOCs are giving rise to a merger process among them, to achieve the critical mass necessary to compete with the direct competitors in the long-distance market.

4. Conclusions

The increasing volume of international calls and the correspondingly growing carriers' interest lead telecommunications to move from the traditional monopoly to a market with a growing competitive pressure. The paper focuses on the analysis of IXCs' growing numbers and impact belonging to the same national market. It is shown that vertical integration strategies may induce a final price reduction whereas, for profit increase, the result depends upon the preexistent market structure. Nevertheless, a growing horizontal competitive pressure among carriers involves a final price reduction. Such results are in tune with the real situation in the USA. The AT&T price decrease is by 44% of the value in 1984 (date of introduction of the Modified Final Judgement) that represents the starting point of the competition process. As a result the competition has reduced the local access network cost and, consequently, has involved a final price decrease.

Appendix

Proof of proposition 1. The following relations hold:

$$\frac{\partial t_A}{\partial n} = -\frac{e - c_B - c}{(N_A + 1)(n + 1)^2} < 0$$

$$\frac{\partial p_A}{\partial n} = -\frac{(a - c_A - c)}{(N_B + 1)(n + 1)^2} < 0$$

where $e - c_B - c > 0$ and $a - c_A - c > 0$, due to the basic assumption of profits' non-negativeness. Similar relations hold for t_B and p_B.

Proof of proposition 2. The following relations hold:

$$\frac{\partial t_A}{\partial N_A} = -\frac{e - c_B - c}{(N_A + 1)^2(n+1)} < 0; \qquad \frac{\partial t_A}{\partial N_B} = 0$$

$$\frac{\partial p_A}{\partial N_A} = -\frac{(a - c_A - c)N_B}{(N_B + 1)(N_A + 1)^2} < 0; \qquad \frac{\partial p_A}{\partial N_B} = -\frac{(a - c_A - c)(N_A - n)}{(n + 1)(N_A + 1)(N_B + 1)^2} \begin{cases} < 0 & if \;\; N_A > n \\ = 0 & if \;\; N_A = n \end{cases}$$

Similar relations hold for t_B and p_B.

Proof of proposition 4. $\Delta\pi_{MF}$ is always *>0* and the following relation holds:

$$\frac{\partial \Delta\pi_{MF}}{\partial n} = -2\left\{\frac{\frac{(a-c-c_A)^2}{b}\left[\frac{(N_A+1)^2}{(n+2)^3} + N_B\frac{N_A+1}{(n+2)^2}\right]}{(N_A+1)^2(N_B+1)^2} + \frac{\frac{(e-c-c_B)^2}{f}\left[\frac{(N_B+1)^2}{(n+2)^3} + N_A\frac{N_B+1}{(n+2)^2}\right]}{(N_A+1)^2(N_B+1)^2}\right\} < 0$$

Proof of proposition 5. $\Delta\pi_{MI}$ is always *<0* and the following relation holds:

$$\frac{\partial|\Delta\pi_{MI}|}{\partial n} = -\frac{2n+3}{(n+1)^2(n+2)^2}\left[\frac{(a-c-c_A)^2}{b(N_B+1)^2} + \frac{(e-c-c_B)^2}{f(N_A+1)^2}\right] < 0$$

References

[1] C. Antonelli, Technological Change and Multinational Growth in International Telecommunication Services, *Review of Industrial Organization* **10** (1995) 161-180.

[2] E.M. Noam, Beyond Liberalization, *Telecommunication Policy* **18** (1994) 286-294.

[3] W.J. Baumol and J.G. Sidak, Toward Competition in Local Telephony. Cambridge, MIT Press,1994.

[4] N. Economides, The Economies of Networks, *International Journal of Industrial Organization* **14**, No.2 (1996).

[5] M. Carter and J. Wright, Symbiotic Production: The Case of Telecommunication Pricing, *Review of Industrial Organization* **9** (1994) 365-378.

[6] M.A. Salinger, Vertical Mergers and Market Foreclosure, *The Quarterly Journal of Economics*, **37** (1989) 373-387.
[7] R.A Schmalensee, Note on the Theory of Vertical Integration, *Journal of Political Economy*, **81** (1973) 442-449.
[8] P. Mallela and B. Nahata, Theory of Vertical Control with Variable Proportions, *Journal of Political Economy*, **88** (1980) 1009-1025.
[9] F.M. Wetfield, Vertical Integration: Does Product Price Rise or Fall?, *American Economic Review* **71** (1981) 334-346.
[10] M. Gastaldi and N. Levialdi, Strategic Planning for Long Distance Telecommunications: A Symbiotic Production System, in press, *International Journal of Production Economics*.

Telecommunications Transformation: Technology, Strategy and Policy
E. Bohlin and S.L. Levin (Eds.)
IOS Press, 1998

CHAPTER 7

An Application of Chaos Theory to the Evolving Telecommunications/ Information System

Neal Stolleman*

Bellcore
RRC4E917, 444 Hoes Lane, Piscataway NJ 08854, USA

Abstract. This paper reviews and applies some of the concepts of chaos theory to the evolution of the telecommunications/information system. Concepts are drawn from models of population growth and the diffusion of epidemics. These concepts are modified and applied to the growth of information communities of interest and realized market shares of competing network/service providers. Regulatory constraints such as the allocation of access costs and restrictions on entry into new markets are shown to have potentially destabilizing effects on market structure, and negative consequences on long-run network operating efficiency.

1. Introduction

In 1991 theoretical physicist David Ruelle wrote: "Legislators and government officials are thus faced with the possibility that their decisions intended to produce a better equilibrium will, in fact, lead to wild and unpredictable fluctuations, with possibly quite disastrous results. The complexity of today's economics encourages such chaotic behavior, and our theoretical understanding in this domain remains very limited" [24]. Ruelle's observation, made in the context of Chaos Theory, has a direct bearing on the current transformation taking place in the telecommunications-information industry in general, and on interconnection in particular. There are two related reasons for this connection. First, the recent Telecommunications Law of 1996 will release a significant amount of pent-up energy, fueled by irresistible economic incentives. It is almost trite to say that industry and market structure will be altered in fundamental ways. Secondly, the continued injection of new wireless and broadband technologies, along with the increasing availability of Internet access and the evolution of information communities of interest, will raise the turbulence level in the industry in the same way that turning up a flame causes water to reach the boiling point faster. Increasing pressures for interconnection will stem from technological, geographical, and virtual information community forces.

* The views expressed in this paper are those of the author, and do not necessarily reflect the views of Bellcore, nor any of its past, present or future owners and clients.

In 1984 the MFJ established the parameters of an industry structure within the United States which was based, in part, on Contestability Theory. It was designed to foster competition where cost functions did not imply natural monopoly characteristics (or, more precisely, by modifying cost functions via equal-access interconnection requirements to eliminate such characteristics). The system was also designed to protect rate payers in the one sector where natural monopoly was thought to persist, the local loop. Lower prices, increased competition, new services and technology were objectives of the 1984 construct. It is difficult to argue, however, that a least-cost and optimally evolving industry structure was attained. Consider, for example: a) the combined capacity of the major interexchange carriers serving the long-distance market segment, relative to realized demand, b) the combined capacity of exchange carriers providing exchange access and telecommunications, and who want to enter the long-distance and information service segments, c) the capacity of other information and enhanced-service providers, d) the patchwork of embedded and forward-looking technologies that exists on the verge of the industry's transformation, e) the adjustment costs, in terms of hardware and software, that will be incurred as a result of (d).

It is imperative to avoid making implementation decisions today that will create a similarly unsustainable set of arrangements ten or so years from now, requiring yet another costly adjustment. The resource and economic welfare implications of embarking on such an inefficient trajectory are too high. Had different regulatory, legislative and judicial policies been implemented in 1984, the current discrete transformation may have been made somewhat smoother. Today, policy decisions made to implement 1996 legislative goals will determine: a) how rocky the ride will be towards a stable equilibrium, b) whether a stable equilibrium will in fact be attainable and c) if attained, the characteristics of the equilibrium.

This paper will draw upon some results from Chaos Theory to illustrate and suggest the complexity of the dynamics we can expect to observe in the coming period. Section 2 will provide a brief, qualitative review of some of the key aspects of Chaos Theory relevant for this analysis. Section 3 will describe specific analytical model structures. In Section 4 these models will be modified and applied to a stylized telecommunications economic system containing geographic, technological and virtual information community dimensions. The efficiency with which this system evolves will depend, in part, on the nature of interconnection arrangements, market entry policies and how those arrangements and policies are modified. The influence of regulatory constraints on the dynamics of this system will be analyzed qualitatively and through numerical simulations based on hypothetical data. Section 5 will summarize key findings and recommendations.

2. Elements of Chaos Theory

Chaos theory seeks to explain physical, biological, and, more recently, economic behavior that seems to be random but is not [2, 4-9, 12, 14-15, 17, 19, 22, 26]. Of course, observed behavior usually includes both systematic and truly stochastic components. But in certain cases, even if the purely stochastic influences are removed from observation, the resulting time series might still show an apparently random pattern – despite the fact that a relatively simple and systematic structure generates the observations. The term "Deterministic Chaos" is sometimes applied to this phenomenon because it is in fact the present state of nature that completely determines the future state – it just doesn't look that way. [3] Chaos theory examines the properties of underlying, dynamic systems that produce these kinds of observations.

One of the fundamental properties of a deterministic chaotic system is "sensitive dependence on initial conditions" [16]. Imagine two trajectories starting from essentially the same initial conditions, where the differences are infinitesimally small; for a period of time the two trajectories might even coincide. Suppose there are no purely stochastic influences present – the system is completely deterministic. Nevertheless, at some point the two trajectories will diverge exponentially [3, 16] and ultimately will be unrecognizable as ever having emanated from similar starting points.

Policy makers observe the course of events after a system is in motion. Policy recommendations are formulated based, in part, on the expected evolution of the system. If the system is characterized by deterministic chaos, it is not possible to make an accurate prediction unless the policy makers know precisely what the initial conditions were. The smallest error in measuring initial conditions can lead to wildly inaccurate predictions in systems characterized by deterministic chaos, and therefore inappropriate policy recommendations.

A second, closely related property is nonlinear feedback between the current and past states of the system [16, 17]. For example, the larger a herd of zebra in year t, the larger will be the herd in $t+1$, due to reproduction. However, a large herd at t may result in lower food supply, or a larger predator population in $t+1$ [18]. These forces will work to reduce the population. Overtime, these countervailing forces may cause the population of the herd to oscillate. The oscillation a) may settle into a stable equilibrium, b) may continue in a regular and cyclical pattern, or c) may appear to fluctuate in a more or less random, but bounded manner. Nonlinear feedback is the mechanism that enables sensitive dependence on initial conditions to exert its influence.

The term "bifurcation" is frequently used in Chaos Theory [3, 16, 25]. It refers to the fact that changing a policy or environmental control variable can cause the long-run dynamic pattern to shift. In the zebra example above (a), (b) and (c) were presented as three possible mutually exclusive long-run outcomes. Changing an environmental control variable can cause the long-run tendency of the system actually to migrate from (a) to (b) or from (b) to (c). The long-run, centralized tendency of the system's evolution as represented graphically in phase space is referred to as an "attractor". Phase space depicts different physical or economic states in an attribute coordinate system (as opposed to a time coordinate system, or shifted time coordinate system [1, 10, 21]). Essentially, an attractor is a contour map of the basins and ridges in phase space that exert gravitational and/or repulsive forces on the system's orbit [5, 16, 25]. Bifurcation changes the shape of the contour map – the orbit starts getting pulled, pushed and split in different directions simultaneously as the policy control variable is changed. An observer watching the system's evolution through time might see patterns that looked unsystematic. However, when viewed in phase space, the observer would see the changes in the attractor's contour map. A "Strange Attractor" is also a Fractal, an infinite nesting of scale-invariant, self-similar structures that arises when the policy control variable of the system reaches certain critical values [2, 3, 11].

3. Analytical Models of Chaotic Behavior

3.1 Discrete Model of Population Growth

The zebra example mentioned above is one illustration of the kind of population dynamics that has been modeled with functional forms embodying sensitive dependence on initial conditions and nonlinear feedback. Perhaps the best known functional form in this family is the logistic map [3, 16, 18, 21, 25]:

(i) $x_t = Ax_{t-1}(1-x_{t-1})$
x_t = biological population at time period t
A = real, constant parameter

The constant A is interpreted as an inherent growth coefficient, indicating how the population would grow over time if there were no environmental constraints such as food supply or a predator population. Obviously, populations cannot grow without limit, and the term in parentheses is meant to capture the restraining influence of environmental factors. When the population is small, the restraining part of the equation has little influence on the growth pattern. When the population gets above 50% of the critical value (normalized to 1) the restraining influence is very strong and the population falls in the next time period. If the decline is large enough, the population will fall below the 50% threshold, engendering positive growth in the subsequent time period.

Early in the history of chaos theory the logistic map was sometimes dismissed as being simplistic and trivial to solve, namely:

(ii) long-run steady state $x^* = (A-1)/A$

However, hidden in this apparently simple functional form is a very rich and complex structure that illustrates the main characteristics of deterministic chaos. For example, the following table illustrates the sequence of numbers obtained under two scenarios: a) the initial value of x at t=0 is equal to 0.40000 and A = 4, and b) the initial value of x at t=0 is equal to 0.40001 and A = 4 (this example is from [21]):

Observation number	**Scenario A**	**Scenario B**
0	0.40000	0.40001
1	0.96000	0.96001
2	0.15360	0.15357
.	.	.
15	0.13561	0.00180

Therefore, very slight changes in initial conditions can produce significant changes in the trajectory of the biological population, or, as will be discussed below, market shares among competing firms.

A second example of the interesting properties of the logistic map is that if the coefficient A were equal to 2, then, even if initial conditions were significantly different, say 0.35000 and 0.40000, the two sequences would converge (this example is from [21]):

Observation number	Scenario A1	Scenario B1
0	0.35000	0.40000
1	0.45000	0.48000
2	0.49595	0.49920
3	0.49997	0.50000
4	0.50000	0.50000

Depending on the choice of initial values for x and the coefficient A, a number of patterns can be generated, including:

a) convergence to a unique fixed point in steady-state equilibrium;
b) a cycle of period 2, in which x continually oscillates between the same two values;
c) a bifurcation in which each of the two fixed points in (b) splits apart into two additional fixed points; thus x oscillates between four distinct points, a cycle of period four;
d) cycles of higher order such as period eight, sixteen and as far up as period 2048 and beyond.

While these results may seem quite mechanical and dependent only on arbitrary values chosen for the coefficient A, there is something fundamental taking place. For example, the amount by which coefficient A must be increased to induce period doubling in successive cycles itself converges to a geometric sequence when the cycles are of a very high order [16, 25]. And this sequence is *exactly the same* for *all* sufficiently smooth functions f(x) that have a parabolic shape, not just the logistic map [16, 21, 25]. This geometric property is characterized by the Feigenbaum number, 4.66920166..., named after the physicist who discovered it. The Feigenbaum number has the same elevated position in mathematics as does π [21]. Period doubling has been observed in the dripping of water from a tap, the beating of the human heart and other natural phenomena including, possibly, one of the moons of Jupiter and other orbiting bodies within the solar system [13, 20, 21, 27].

How do these results connect to the topic at hand? Clearly, economic variables go through periods of adjustment as long-run steady-state solutions are attained, or approximated. The adjustment process is generally modeled as asymptotic, or at least smooth and continuous [4, 5]. The logistic map seems well suited to model the complex dynamics that can be expected as the telecommunications/information industry is transformed by increased competition, a wider array of services and applications, and changes in consumer preferences. Asymptotic adjustment in realized market shares seems inadequate in this context. Therefore, the logistic map will be modified and adapted to model the evolution of realized market shares among competing firms in the stylized telecommunications/information system described below.

3.2 Discrete Model of Epidemics

A second analytical model in chaos theory is based on a variation of the logistic map, the discrete model of epidemics [17, 23, 25]. This model is used to structure the diffusion of an epidemic through a population. Properties of the model determine whether the diffusion is complete, reaches a steady-state equilibrium at a relatively low level, oscillates in some way or collapses to zero.

There are several assumptions associated with the particular model that will be used and adapted to the telecommunications model below:

a) Each individual has the same number of contacts with other individuals in a given period of time, and has the same probability of getting infected from any other individual.
b) The total population remains constant.
c) The period during which a sick person is contagious is one unit of time.
d) After becoming infected for one unit of time, an individual again becomes a candidate for infection, i.e. there is no permanent immunity.

Given these assumptions, the percent of the population that is infected over time can be modeled according to the following process:

$$\text{(iii)} \quad X(1,t+1) = X(2,t)\,(1 - e^{-aX(1,t)})$$

$X(1,t+1)$	=	infected population at time t+1
$X(2,t)$	=	healthy population at time t
$X(1,t)$	=	infected population at time t
$X(1,t) + X(2,t)$	=	1
e^{-a}	=	the probability of one healthy individual *not* becoming infected through contact with one sick individual

Clearly, if the infected population at time t is zero, the probability of a healthy individual *not* becoming infected is 1, and infected population at time t+1 will be zero. If the size of the infected population at time t approaches the entire population, the probability of not becoming infected converges to zero and the size of the infected population at t+1 would converge to the entire population. Different evolutions are possible depending on the initial size of the infected population at t = 0, and the size of the coefficient, "a". The size of "a" itself can depend on whether the medical technology of the society is primitive ("a" is a large number) or whether an advance in medical technology increases an individual's capacity to repel an infectious disease ("a" is a small number).

In both of these models, the orbits of the state variables (size of the biological population or size of the infected population) are influenced by exogenous control variables. In the case of the logistic map, the control variables are the inherent growth coefficient, A, and maximum sustainable capacity, normalized to unity. In the epidemic model, the control variable is the probability/technology coefficient, "a". In order to adapt these models to the stylized telecommunications/information system, these exogenous control variables will be replaced by variables representing either consumer preferences, regulatory or environmental influences and endogenously determined sustainable capacities.

4. Stylized Telecommunications/Information Economic System

The analytical models described in Section 3 will be adapted and extended to a stylized telecommunications/information system. In addition, hypothetical data inputs will be used to calibrate the system and run scenarios. Because these data inputs are completely hypothetical, little attention should be paid to absolute sizes of simulation results. What is important, however, are the evolutionary patterns that are produced, and the ordinal relationships among the state variables.

4.1 Information Communities of Interest

The demand side of the system begins with three generic information communities of interest. The first community of interest can be regarded as users of traditional narrow band telephony. The second community includes users whose information needs require data communications. The third information community is of those individuals whose information requirements are satisfied with a mixture of traditional telephony and data communications, e.g. internet-based telephony. Clearly, within this framework, there is a correspondence between an "information community of interest" and an underlying technology, but the information groupings are not completely defined by technology. For example, members of the first and third groups will exchange information, as will members of the second and third groups. Information flows will cross technology boundaries.

The evolution of these information communities will have a significant effect on the evolution of the entire telecommunications/information system. Assumptions of the discrete model of epidemics discussed in Section 3 will now be used and modified to describe this evolution. In the epidemic model, each healthy individual was assumed to have the same number of contacts with infected individuals in a given period of time, and therefore had an equal chance of becoming infected with a single disease. Now, instead of becoming infected, an individual will become motivated to join a particular information community of interest. Then, after one time period, each individual will again have an equal number of contacts with all other individuals and again have a chance of being influenced to join another information community. The (unrealistic) assumption of an equal number of contacts with all other individuals is made to keep the problem tractable. Loosely speaking, however, having contact can be interpreted as being aware of the approximate "mass" of individuals that are in a particular group, and having an intuitive understanding of the benefits and content of infra-group information flows.

In the discrete epidemic model, the probability of an individual not contracting the disease is $e^{-aX(t)}$, where X(t) is the size or mass of the infected group. In the telecommunications/information model, the probability of an individual in information community 1, N(1), being influenced to join information community 2, N(2), at time period t is specified as:

$$(1) \quad Pr(1>2,t) = e^{a(1,2)N(2,t)} / [e^{a(1,1)N(1,t)} + e^{a(1,2)N(2,t)} + e^{a(1,3)N(3,t)}]$$

The probability of a member of group 1 joining group 2 at time t depends on a) the preference coefficients of group 1 members for all groups, a(1,j), and b) the sizes of all groups at time t, N(j,t). The bigger the group, the bigger its center of gravity, or tendency to pull in new members. The larger a(1,j), the more susceptible members of group 1 are to the gravitational pull of group j, or, alternatively, the externality effect of joining group j. It is assumed that in each time period, all members of a given group have the ability to transfer to another community of interest; there are no technical barriers to switching. However, depending on the preference for remaining with the current community, a(i,i), a portion of the community may remain intact.

Over time the three communities of interest will evolve according to the following process:

(2) $$N(1,t+1) = \Sigma\, N(i,t)\, Pr(i > 1,t)$$
$$N(2,t+1) = \Sigma\, N(i,t)\, Pr(i > 2,t)$$
$$N(3,t+1) = \Sigma\, N(i,t)\, Pr(i > 3,t)$$
$$i = 1 \text{ to } 3$$

Equation (2) states that the size of group 1 at time t+1 is equal to the sum of the individuals from all groups who decide to migrate into group 1. The number of such individuals is equal to the sizes of the respective communities in the previous time period multiplied by the proportion of each group inclined towards community 1.

The hypothetical numbers in Table 1 were used to calibrate the model.

Table 1. Population Preference Coefficients a(i,j) for Communities of Interest

Information Community	1	2	3
1 (initial size=74)	0	0.100	0.120
2 (initial size=15)	0.100	0.037	0.035
3 (initial size=10)	0.120	0.035	0.073

Although somewhat arbitrary, the numbers reflect preferences that are weighted more in the direction of communities 2 and 3 than towards community 1. Information community 3 also has the highest "own" coefficient. These coefficients, along with the initial value assumptions, set in motion the long-run time path shown in Figure 1:

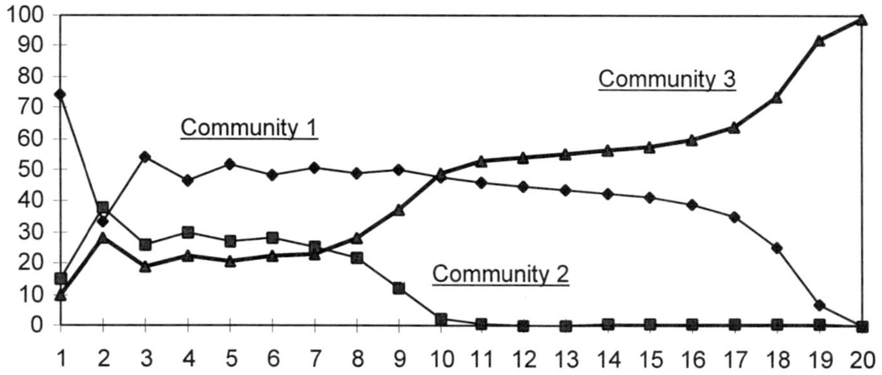

Figure 1. Evolution of Information Communities of Interest: Base Case

Community 1 initially is the largest group, followed by groups 2 and 3. In time period 2 a significant shift occurs out of group 1, which, in this scenario, turns out to be an overreaction, possibly due to intense advertising or "media hype" that induces mob psychology. In any event, a "bounceback" occurs in period 3. A relatively stable period then ensues until period 8, after which membership in group 2 decays rapidly in favor of group 3; individuals with primarily data communications information requirements expand into the hybrid voice/data domain. A steady but gradual erosion of community 1 occurs until period 16, by which time community 3 has built enough critical mass to rapidly pull in the remaining community 1 members.

Obviously, the numbers chosen in Table 1, plus the initial values for each community size, determine the evolution. Because of community 1's large initial size, a very slight increase in its "own" preference coefficient (from 0 to .001) would result in a leveling of its size after period 8; community 3 would never attain critical mass, and in fact would decay to zero, leaving the information system almost evenly split between groups 1 and 2. Thus, slight errors in measuring initial conditions, in this case consumer preferences, can lead to drastically different predictions. For comparison, the second scenario is shown in Figure 2.

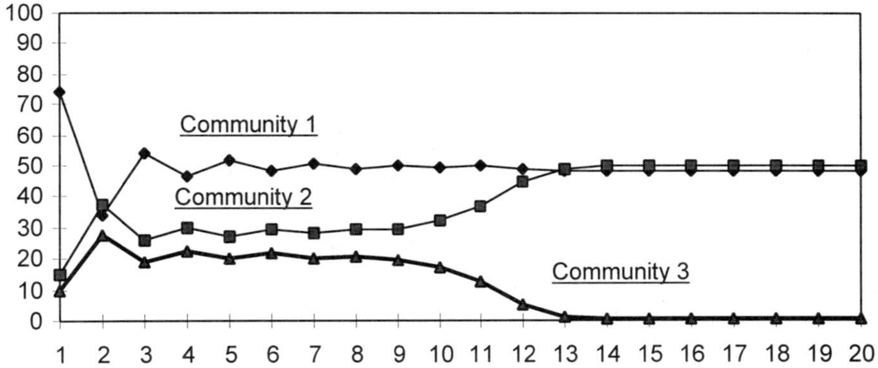

Figure 2. Evolution of Information Communities of Interest: Stronger "Own" Preference Within Community 1

4.2 Long-run Market Shares among Competing Firms

In this model, telecommunication/information services are assumed to be supplied by three generic firms. Each generic firm, in turn, can support any of the three communities of interest, but with varying degrees of efficiency. The share of each information community served by any one network/service provider will depend on both the price of obtaining access to the network/service provider and the community's preferences for different providers. Preferences for different providers reflect their respective reputations in the minds of consumers. Consumers will have different relative perceptions regarding the ability of network/service providers to deliver timely, reliable and desirable service.

Members of each community will evaluate and balance relative prices with their supplier preferences in their respective indirect utility functions. A multinomial logit specification is used to model the decision process. For example, the percentage of information community 1 members captured by network provider 2 at time t is:

(3) $M*(1 > 2,t) = e^{V(1 > 2,t)} / [e^{V(1 > 1,t)} + e^{V(1 > 2,t)} + e^{V(1 > 3,t)}]$
$V(1 > 2,t) = \alpha(1,2) + [\beta(1,2) P(1,2,t)]$
$\alpha(1,2)$ = community 1 preference for network provider 2
$\beta(1,2)$ = community 1 access price response coefficient
$P(1,2,t)$ = access price charged by network provider 2 to community 1
*denotes long-run market share

Table 2 summarizes the hypothetical preferences for each network/service provider.

Table 2. Community of Interest Preference Coefficients α (i,j) for Alternative Network/Service Providers

Information Community	Network Provider 1	Network Provider 2	Network Provider 3
1	2.5	0.8	1.1
2	0.8	1.0	1.05
3	0.8	1.0	1.5

There is a positive relationship between an information community of interest and an underlying supplier (communities 1 and 3 have clear relative preferences for suppliers 1 and 3, respectively), but the relationship is not one of fixed proportions between a technology and community of interest. Each community is willing to consider each supplier, to varying degrees.

The hypothetical price response coefficients for each community are: community 1 = -.0019, community 2 = -.0025 and community 3 = -.0028. These responses are assumed the same across network/service providers. These coefficients give rise to the access price elasticities of demand in Table 3.

Table 3. Derived Access Price Elasticities of Demand (Evaluated at Base Case)

Information Community	Network Provider 1	Network Provider 2	Network Provider 3
1	-.35	-1.0	-.95
2	-2.1	-1.87	-1.83
3	-1.85	-1.72	-1.27

Except for information community 1's elasticity of access demand for network provider 1, the elasticities tend to be on the high side. A high access demand elasticity for the non-traditional network/service providers is a reasonable assumption in an intensely competitive environment with no technical barriers to switching from one supplier to another.

A major determinant of long-run market shares is the set of access prices for each supplier. Baseline access prices are obtained by adding a 20% mark-up to the rental price of access capital. Rental prices are computed by converting hypothetical capital cost per unit access into an annuity stream as shown in Table 4.

Table 4. Computation of Baseline Access Prices

Capital cost	Discount Rate	Depreciation	Maintenance	Access Price (annual, includes 20% mark-up)
Community #1 -$1,200	15%	10%	15%	$576
Community #2 -$2,400	15%	10%	15%	$1152
Community #3 -$1,800	15%	10%	15%	$864

The capital cost figures assume economies of scope in the provision of access for information community 3, relative to separate provisioning of service for communities 1 and 2.

Finally, long-run market shares are obtained by inserting the baseline access prices into the indirect utility functions described above, and are shown in Table 5.

Table 5. Derived Long-run Market Shares (Evaluated at Baseline)

Information Community	Network Provider 1	Network Provider 2	Network Provider 3
1	70%	12.8%	17.3%
2	28.5%	34.8%	36.6%
3	23.6%	28.8%	47.5%

4.3 Short-run Market Shares among Competing Firms

The market shares shown in Table 5 are long-run shares. Long-run shares will not be attained instantly. An adjustment period will occur during which new information that affects decision-making will diffuse through the system. Once the information reaches consumers it may take an additional amount of time for consumers to react to the new data. If the initial reaction is too strong, consumers may retrench, and this process may repeat itself a number of times. The initial reaction might be too strong if consumers are made to feel that they must choose quickly, or risk being left behind in the "new information age". Traditional types of adjustment mechanisms used in economics usually give rise to a) trajectories that reach the long-run steady state asymptotically (Koyck lag), or b) inverted "U"-shaped response curves that are smooth and continuous (Polynomial Distributed Lag). [ ] The experience in the long-distance market, however, suggests that long-run equilibrium market shares may be characterized by a permanent amount of churn as consumers move from one provider to the other. In the long-distance market this activity has been primarily driven by price. Going forward, however, the telecommunications/information system will be transformed in unprecedented ways; existing firms and new entrants will carry both traditional and new services and applications. The nature of these services and applications will be modified, and consumer preferences will adapt and adjust to changing regimes. In this context, it is reasonable to assume that as consumers strive to reach a long-run equilibrium target that itself is evolving, the observed transition path towards long-run market shares will be anything but smooth and asymptotic.

One variation of the logistic map discussed in Section 3 that has been used in modeling biological populations is:

$$dP/dt = rP[\,1 - (P/L)\,] \tag{4}$$

dP/dt = rate of change in population at time t
P = population level at time t
r = inherent growth coefficient
L = maximum sustainable population

In this formulation, the population grows as long as it remains below its maximum sustainable level, L. Maximum sustainable level would be determined by food supply, for example, or the existence of a predator population. If actual population exceeds L, then shrinkage occurs, and depending on the size of the growth coefficient, r, the time path may be characterized by recurring, periodic cycles or apparently unsystematic oscillations.

Equation (4) will now be modified for application to the telecommunications/information system. The following equation shows the realized, or short-run, percentage of information community "i" captured by network/service provider "j" at time period t:

$$M(i>j,t) = M(i>j,t-1)\,[\,1 + r\,(\,1 - M(i>j,t-1) / (M^*(i>j,t) \times c)\,)\,] \tag{5}$$

$M(i>j,t)$ = realized market share of provider j capturing percent of community i
$M(i>j,t-1)$ = realized market share in last time period
$M^*(i>j,t)$ = long-run, instantaneous share at period t
r = inherent growth coefficient
c = "bias" factor

An important aspect of equation (5) is that the long-run, sustainable population variable L of the biological model has been replaced by the *endogenously determined* long-run market share of network provider j (that is, equation (3)). This change essentially introduces "predator-prey" behavior into the model. For example, if a competitor decreases price, then last year's realized market share is larger compared to the now lower long-run market share. Network provider j's long-run sustainable population falls, thereby setting an adjustment process in motion with respect to provider j's realized market share.

The variable r in equation (5) can be interpreted as an inherent growth coefficient. As its value is increased, swings in realized market share become wider and apparently more unsystematic. This kind of behavior would be expected in an environment where a) there are no technical limitations to consumers switching from one provider to another; b) consumer preferences are fluctuating in the face of evolving services and applications, the full value of which has not been ascertained; c) consumers therefore have a high propensity to switch from one supplier to another; and d) the regulatory authority encourages rapid dissemination of information and an intensely competitive environment.

Finally, the variable c is included as a "bias" factor. For example, if $c > 1$, upward adjustments in market share are exacerbated while downward adjustments are moderated. This assumption is consistent with the hypothesis that consumers are quicker to overshoot long-run equilibrium targets than they are to retrench, producing a sort of "ratchet" effect. The baseline values used in the simulations of equation (5) discussed below are: $r = 2.6$ and $c = 2$.

4.4 Usage Flows between Market Segments

Two additional aspects of the telecommunications/information systems model are the usage flows among market segments and the associated cost functions. Since there are three information communities of interest and three network/service providers, there are nine potential market segments. In order to calibrate and model usage flows and costs, certain baseline average demand and price inputs are assumed.

Table 6 summarizes baseline annual usage flows.

Members of each information community are assumed to originate usage to members of their own community, regardless of which network/service provider the receiving members utilize. Members of communities 1 and 2, however, do not communicate directly, whereas both of these groups do communicate with members of community 3. The figures in Table 6 represent average usage transmitted per individual in one market segment to each individual in the terminating market segment. Therefore, the *total* amount of usage originating in segment "a" and terminating in segment "b", for example, is equal to the product of the number of individuals in each segment, multiplied by the appropriate usage amount in Table 6.

Associated with the baseline average usage flows is a set of assumed price elasticities of demand shown in Table 7.

All of the initial elasticities are set less than unity. The elasticities tend to be larger on usage flows between communities 1 and 3 and between communities 2 and 3. Infra-community price elasticities have the lowest values.

Finally, a set of baseline usage prices is assumed as shown in Table 8.

Table 6. Baseline Average Annual Usage Flows (Market Segments = a,b,c,d,e,f,g,h,i)

	Usage flows												
							usage originating from segment X to Y at unit price						
	Originating segment=						average flow per unit in terminating segment						
segments	Information community #		**1**		**a**	**b**	**c**	**d**	**e**	**f**	**g**	**h**	**i**
a		Network provider #	1		100	100	100	0	0	0	60	60	60
b		Network provider #	2		100	100	100	0	0	0	60	60	60
c		Network provider #	3		100	100	100	0	0	0	60	60	60
	Information community #		**2**										
d		Network provider #	1		0	0	0	70	70	70	50	50	50
e		Network provider #	2		0	0	0	70	70	70	50	50	50
f		Network provider #	3		0	0	0	70	70	70	50	50	50
	Information community #		**3**										
g		Network provider #	1		30	30	30	50	50	50	200	200	200
h		Network provider #	2		30	30	30	50	50	50	200	200	200
i		Network provider #	3		30	30	30	50	50	50	200	200	200

Table 7. Baseline Price Elasticity of Demand (Market Segments = a,b,c,d,e,f,g,h,i)

	Originating segment=						**Elasticity assumptions at initial, unit**						
segments	Information community #		**1**		**a**	**b**	**c**	**d**	**e**	**f**	**g**	**h**	**i**
a		Network provider #	1		-0.20	-0.20	-0.20	-0.50	-0.50	-0.50	-0.80	-0.80	-0.80
b		Network provider #	2		-0.20	-0.20	-0.20	-0.50	-0.50	-0.50	-0.80	-0.80	-0.80
c		Network provider #	3		-0.20	-0.20	-0.20	-0.50	-0.50	-0.50	-0.80	-0.80	-0.80
	Information community #		**2**										
d		Network provider #	1		-0.50	-0.50	-0.50	-0.20	-0.20	-0.20	-0.80	-0.80	-0.80
e		Network provider #	2		-0.50	-0.50	-0.50	-0.20	-0.20	-0.20	-0.80	-0.80	-0.80
f		Network provider #	3		-0.50	-0.50	-0.50	-0.20	-0.20	-0.20	-0.80	-0.80	-0.80
	Information community #		**3**										
g		Network provider #	1		-0.80	-0.80	-0.80	-0.80	-0.80	-0.80	-0.20	-0.20	-0.20
h		Network provider #	2		-0.80	-0.80	-0.80	-0.80	-0.80	-0.80	-0.20	-0.20	-0.20
i		Network provider #	3		-0.80	-0.80	-0.80	-0.80	-0.80	-0.80	-0.20	-0.20	-0.20

Table 8. Baseline Originating Usage Prices (Market Segments = a,b,c,d,e,f,g,h,i)

					Network provider applying price to originating usage:								
	Prices for originating usage:				NP1	NP2	NP3	NP1	NP2	NP3	NP1	NP2	NP3
segments	Information community #		**1**		**a**	**b**	**c**	**d**	**e**	**f**	**g**	**h**	**i**
a		Network provider #	1		$ 0,50	$ 0,50	$ 0,50	$ 10,00	$10.00	$ 10.00	$ 0.60	$ 0.60	$ 0.60
b		Network provider #	2		$ 0,50	$ 0,50	$ 0,50	$ 10,00	$10.00	$ 10.00	$ 0.60	$ 0.60	$ 0.60
c		Network provider #	3		$ 0,50	$ 0,50	$ 0,50	$ 10,00	$10.00	$ 10.00	$ 0.60	$ 0.60	$ 0.60
	Information community #		**2**										
d		Network provider #	1		$ 10.00	$ 10.00	$ 10.00	$ 0.30	$0.30	$ 0.30	$ 0.60	$ 0.60	$ 0.60
e		Network provider #	2		$ 10.00	$ 10.00	$ 10.00	$ 0.30	$0.30	$ 0.30	$ 0.60	$ 0.60	$ 0.60
f		Network provider #	3		$ 10.00	$ 10.00	$ 10.00	$ 0.30	$0.30	$ 0.30	$ 0.60	$ 0.60	$ 0.60
	Information community #		**3**										
g		Network provider #	1		$ 0.50	$ 0.50	$ 0.50	$ 0.30	$ 0.30	$ 0.30	$ 0.60	$ 0.60	$0.60
h		Network provider #	2		$ 0.50	$ 0.50	$ 0.50	$ 0.30	$ 0.30	$ 0.30	$ 0.60	$ 0.60	$0.60
i		Network provider #	3		$ 0.50	$ 0.50	$ 0.50	$ 0.30	$ 0.30	$ 0.30	$ 0.60	$ 0.60	$0.60

(Note that a price of $10.00 for communication between communities 1 and 2 is used only because usage flows between those groups is set to zero.)

The average usage flows, prices and baseline elasticity assumptions are combined to calibrate a set of linear demand curves for originating usage in each of the nine market segments that terminate in each of the nine market segments.

(6) $Q(K>L,t) = A(K>L) - B(K>L)P(K>L,t)$

$Q(K>L,t)$ = originating average usage per member of market segment K communicating to each member of market segment L at time t

$A(K>L)$ = derived intercept for market segment (K) originating usage terminating in market segment L

$B(K>L)$ = derived slope for market segment (K) originating usage terminating in market segment L

$P(K>L,t)$ = price per unit usage originating in market segment K and terminating in market segment L at time t

K = market segments a,b,c,d,e,f,g,h,i

L = market segments a,b,c,d,e,f,g,h,i

Each market segment is the intersection of an information community of interest and a network/service provider

4.5 Cost Structure

The cost structure for each network/service provider is divided into switching, termination and transport segments (in addition to access costs discussed above). The cost functions for each of these network segments have the generic form:

(7) $\text{Cost}(i,j,t) = F(i,j) + D(i,j)\,[\,\text{Usage}(i,j,t)^{\varepsilon(i,j)}\,]$

$D(i,j) = 1$, all i and j

$\text{Usage}(i,j,t)$ = usage category appropriate to given network cost segment

= ALL originating and terminating usage, by market segment, for switching cost

= terminating usage, by market segment, for terminating costs

= originating usage from network provider X to Y, by market segment, transport cost

i = information community of interest 1, 2 or 3

j = network/service provider 1, 2 or 3

The functional form in (7) was chosen because it highlights the significant fixed cost component in network cost functions, and it also allows usage-related economies of scale to be reflected through the term $\text{Usage}(i,j,t)^{\varepsilon(i,j)}$. If $\varepsilon(i,j)$ is 0 there are no economies of scale (all costs are fixed). If $\varepsilon(i,j)$ is less than 1 there are global economies of scale (average cost per unit usage always declines), and if $\varepsilon(i,j)$ exceeds 1 average usage cost curves are "U"-shaped. Although all network/service providers are assumed to have the same cost structure as in (7), the magnitudes of the coefficients differ, reflecting differences in comparative advantage in handling different types of traffic. These differences are summarized in Table 9:

Table 9. Cost Function Parameters

Network Segment	Network/service provider	Market segment	Intercept F(i,j)	Exponent E(i,j)
Switching	Network provider 1	Info community 1	5000	0.950
		Info community 2	10000	1.001
		Info community 3	15000	0.950
	Network provider 2	Info community 1	10000	0.788
		Info community 2	20000	0.848
		Info community 3	25000	0.813
	Network provider 3	Info community 1	15800	0.479
		Info community 2	31200	0.599
		Info community 3	39502	0.524
Terminating	Network provider 1	Info community 1	1950	0.659
		Info community 2	3900	0.850
		Info community 3	5850	0.900
	Network provider 2	Info community 1	4063	0.807
		Info community 2	8126	0.924
		Info community 3	6095	0.795
	Network provider 3	Info community 1	6420	0.453
		Info community 2	12839	0.538
		Info community 3	9629	0.464
Transport	Network provider 1	Info community 1	5000	0.775
	(from 1 to 2)	Info community 2	10000	0.900
		Info community 3	8000	0.950
	Network provider 1	Info community 1	7000	0.771
	(from 1 to 3)	Info community 2	12000	0.861
		Info community 3	10000	0.850
	Network provider 2	Info community 1	5000	0.735
	(from 2 to 1)	Info community 2	10000	0.840
		Info community 3	8000	0.731
	Network provider 2	Info community 1	7000	0.771
	(from 2 to 3)	Info community 2	12000	0.861
		Info community 3	10000	0.753
	Network provider 3	Info community 1	8295	0.495
	(from 3 to 1)	Info community 2	14220	0.574
		Info community 3	11850	0.496
	Network provider 3	Info community 1	8295	0.495
	(from 3 to 2)	Info community 2	14220	0.574
		Info community 3	11850	0.496

A basic characteristic of these input assumptions is that network provider 3 tends to have higher fixed cost components than providers 1 and 2, and, in addition, a lower variable cost component as reflected in its exponent values. Network provider 1 tends to be more efficient in handling the usage generated by information community 1 than does network provider 2, evaluated at base-case volumes. In general, the fixed cost components associated with information community 3 usage are less than the sum of the fixed components for communities 1 and 2. This characteristic applies to all network providers, reflecting economies of scope in the provision of community 3 usage.

4.6 Usage Pricing Rule and Inter-firm Payments

The baseline usage pricing rule followed by all providers is assumed to be traditional profit maximization, where the previous period marginal cost of usage is used to determine current period price. The (lagged) marginal cost used by a network provider to compute prices for its originating usage consists of switching, transport and terminating marginal cost components. Further, the originating network provider is assumed to reimburse the receiving network provider for all terminating costs associated with the usage flowing from the former to the latter.

4.7 Regulatory Policy Initiatives

The model allows four types of regulatory policy initiatives to be introduced.

First, network provider 1 may be constrained to allocate a portion of its termination costs for information community 1 traffic to its termination costs for information community 2 and community 3 traffic. A regulator might promote this kind of initiative to make it easier for new entrants to terminate traditional, community 1 type traffic into the incumbent's local market. At the same time, this policy would make it more expensive for newer service applications to terminate in the incumbent's local market.

A second initiative is the allocation of a portion of provider 1's access costs to its terminating cost functions for community 2 and community 3 usage. The regulatory motivation here is to decrease local access prices and the effect, again, is to raise the cost of terminating new forms of information in the incumbent's market. On the other hand, as competition intensifies for all forms of customers, artificially lowering access prices gives the incumbent an artificial and inefficient access price advantage.

Third, the regulator can delay the incumbent's entry in the provision of new forms of telecommunications and information services, i.e., community 2 and community 3 usage. The motivation of the regulator presumably is to give other firms time to build critical mass to sustainable levels, and to prevent the incumbent from using its access revenue to cross-subsidize entry into newer markets.

The fourth type of policy control considered is forcing the incumbent's usage prices to either to a) equal marginal costs, or to b) maximize revenue instead of profit. Either way, price restrictions will force the incumbent into sub-optimal financial performance.

The effects of these regulatory initiatives will be evaluated in terms of operating network efficiency and realized market shares. Operating network efficiency is defined as the ratio of non-access costs divided by aggregate usage.

4.8 Simulations – Termination and Access Cost Allocation Rules Implemented

The following chart shows the realized access market shares for information community 1 when policy initiatives one and two are implemented simultaneously. For comparison, the level base-case market shares are also shown in Figure 3.

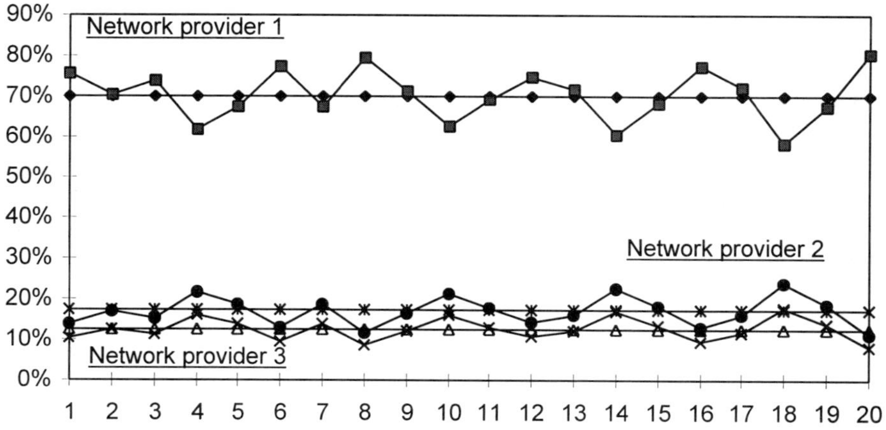

Figure 3. Realized Market Shares – Information Community 1, All Network Providers

Policy initiative one consists of the allocation of 50% of network provider 1's termination costs (for community 1 type traffic) to the terminating cost functions of community 2 and community 3 type traffic. The 50% figure is decreased in 5% increments per year until it becomes zero in year 10. Policy initiative two consists of allocating 25% of network provider 1's access costs (again, for information community 1 only) to the terminating cost functions for type 2 and type 3 traffic, with the percent decreasing 5% per year until zero is reached. In both cases, the relative shares of community 2 and community 3 traffic determine the total and marginal cost allocations to these respective segments.

The main determinant of long-run market share is access price; in the first time period, the allocation of access costs artificially lowers provider 1's access price, boosting its long-run, sustainable market share. Thereafter, as the cost allocation is phased out, network provider 1's access price steadily rises and its long-run share correspondingly decreases. However, given the dynamic market share adjustment embodied by equation (5), realized market share oscillates around, but never actually converges to the long-term target created by the regulatory policy. Figure 4 explicitly shows the relation between network provider 1's short-run and long-run market shares.

Figure 5 compares network provider 1's operating efficiency in the base case with the implementation of regulatory policies one and two (solid line denotes base case).

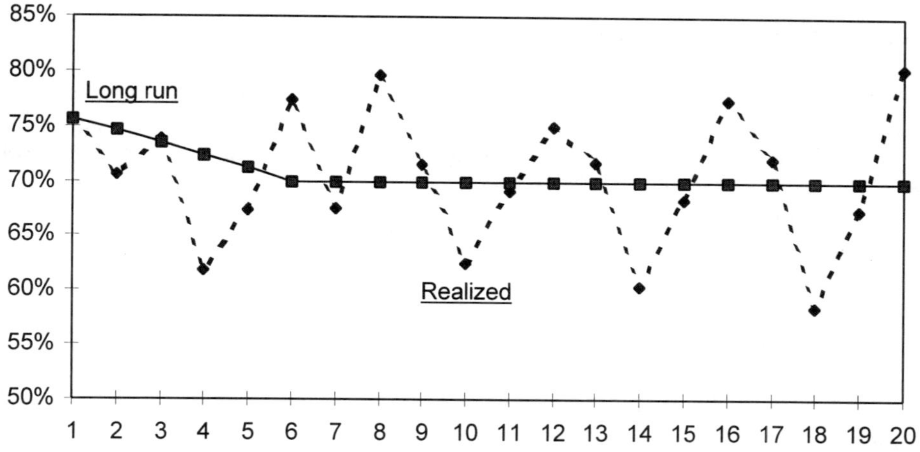

Figure 4. Realized Market Shares vs. Long-run Market Share:
Network Provider 1 – Information Community 1

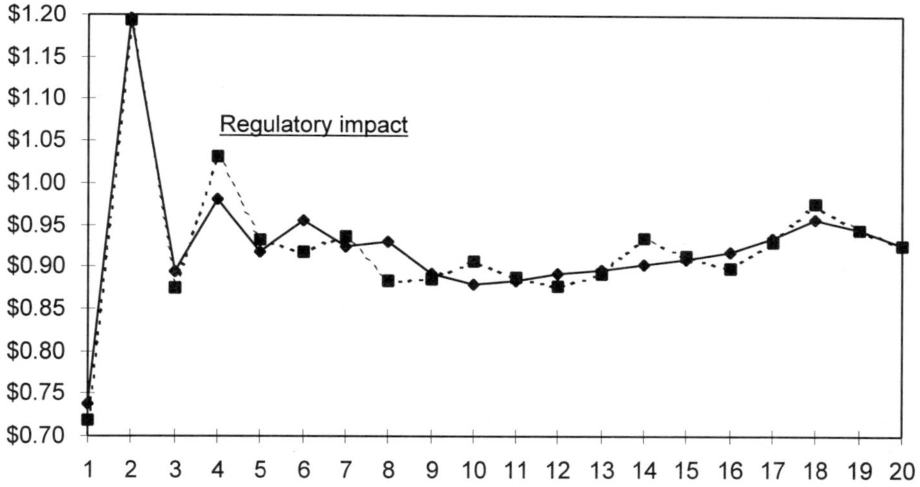

Figure 5. Network Provider 1 – Operating Efficiency:
Base Case vs. Regulatory Policies 1 & 2

The oscillation in efficiency corresponds to oscillations in realized market shares and associated fluctuations in usage. The spike in unit cost in time period 2 corresponds to the global shift in the population from information community 1 to communities 2 and 3 in that time period, as shown in Figure 1. The improvement in time period 3 efficiency corresponds to the "bounceback" of individuals to information community 1. The gradual tendency towards improved efficiency up to time period 10 reflects the disappearance of community type 2 usage, a sector in which network/service provider 1 has a comparative disadvantage. The gradual decrease in operating efficiency after time period 10 reflects the continued and accelerated migration of individuals from information community 1 to community 3, in which network provider 1 also has a comparative disadvantage.

Figure 6 compares total operating efficiency for network provider 3 under the same two scenarios.

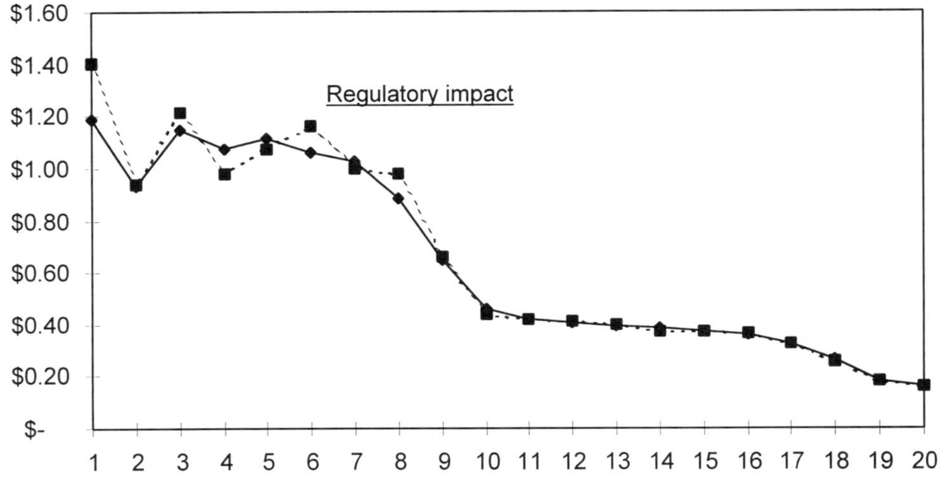

Figure 6. Network Provider 3 – Operating Efficiency:
Base Case vs. Regulatory Policies 1 & 2

The initial oscillations in network provider 3's operating efficiency reflect market share fluctuations discussed above. Also, the sharp drop in average cost until period 10 reflects the rapid migration from community 2 to community 3 shown in Figure 1. However, over time additional gains in efficiency are moderated as the population migrates out of information community 1 to community 3 at a more gradual rate. Eventually, there is no difference between the baseline and regulatory policy scenario as far as network provider 3 is concerned because, globally, information community 1 essentially vanishes.

4.9 Simulations – Restrictions on Network Provider 1's Market Entry

Assume now that the regulator keeps policies 1 and 2 intact, and then adds a market entry restriction on network provider 1 as shown in Table 10.

Table 10. Entry Restrictions Placed on Network Provider 1

			1	2	3	4
Information community #	2		0%	50%	100%	100%
Information community #	3		0%	33%	67%	100%

Table 10 indicates that network provider 1 is prevented from serving information communities 2 and 3 in time period 1. However, these restrictions are gradually lifted. Network provider 1's entry into community 2 is allowed to proceed faster than its entry into community 3. (The percentages indicate what portion of the indirect utility functions network provider 1 is allowed to satisfy for consumers in the restricted segments.)

Realized market shares for information community 2 are shown in Figure 7.

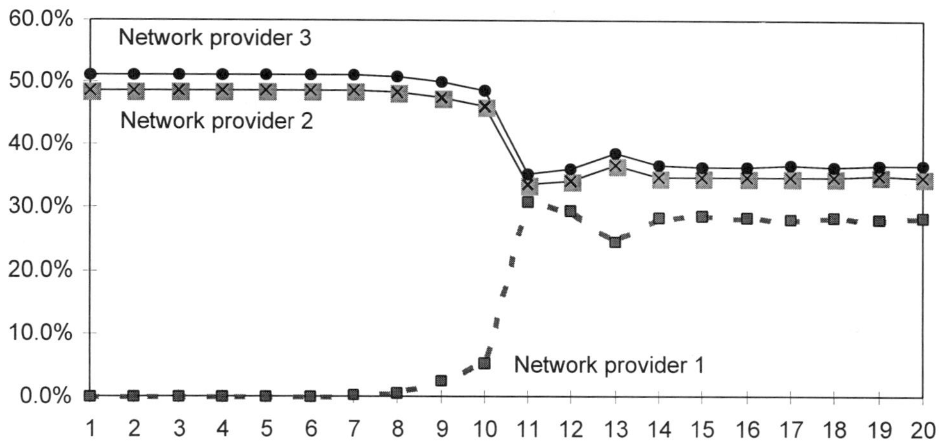

Figure 7. Market Shares for Information Community 2:
All Network Providers – Entry Restriction Placed on NP-1

Restricting network provider 1's entry into this market is analogous to pulling on a violin string; once it is let go the string vibrates above and below its long-run equilibrium value. In this instance, the initial "tension" on the violin string is relatively low, hence the vibrations converge quickly to long-term market shares. Although individuals in community 2 overreact to network provider 1's late entry, the similarity in community preferences across the three providers (see Table 2) leads to a stable outcome.

The dynamics related to information community 3 are shown in Figure 8. In this instance, the vibrating violin string does not settle back into long-run equilibrium. This result is probably due to the initial consumer overreaction to provider 1's late entry into this market (in periods 11 and 12), in combination with a very strong long-term preference of community 3 individuals for network provider 3 (see Table 2). The "tug of war" between these two forces is exacerbated by the "ratchet" effect mentioned earlier, which tends to prolong consumer overreactions.

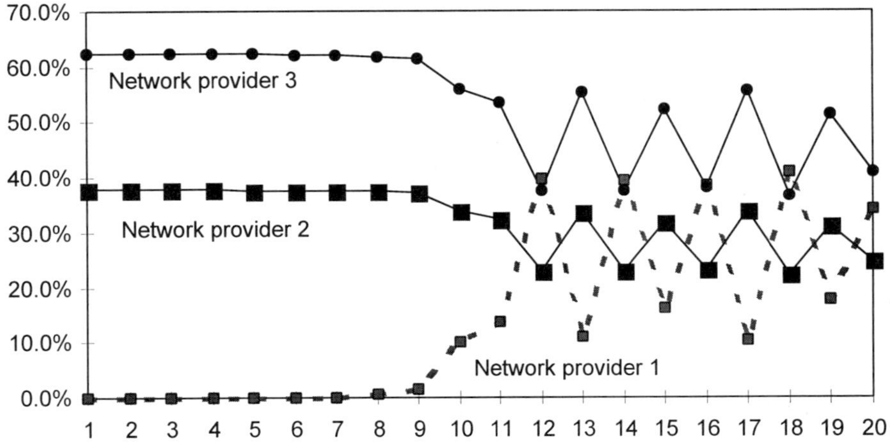

Figure 8. Market Shares for Information Community 3:
All Network Providers – Entry Restriction placed on NP-1

The effects of market entry restriction on overall operating efficiency (dashed line), relative to the base case (solid line), are shown in the following charts. Figure 9 shows the combined effect for all network providers taken together.

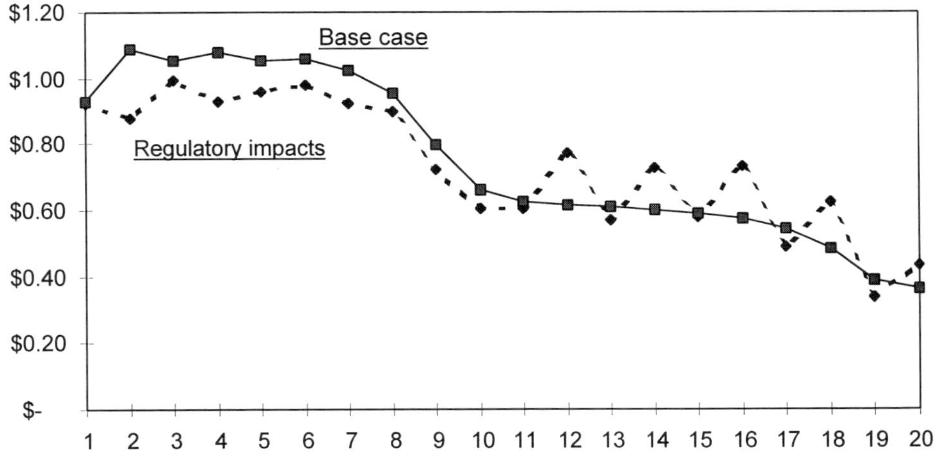

Figure 9. Operating Efficiency: All Network Providers Combined
– Entry Restriction Policy vs. Base Case

In the early time period, the entry restriction on provider 1 lowers overall costs per unit of usage. The traffic streams of information communities 2 and 3 are diverted to those networks which have comparative cost advantages. In the longer run, however, entry restrictions are relaxed and market share dynamics cause a "bounceback" towards network provider 1; overall efficiency declines, and is more volatile relative to the base case. The decline in efficiency persists because the "ratchet" effect biases fluctuations in market shares in favor of provider 1. On a global basis, the shift of demand from information community 1 to 3 results in a long-term secular decline in unit costs, regardless of the entry restriction.

The specific operating behavior of network provider 1 is shown in Figure 10.

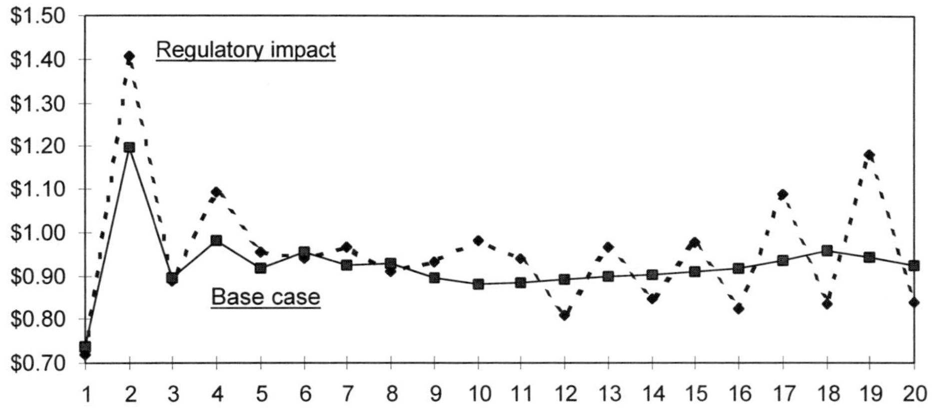

Figure 10. Operating Efficiency: Network Provider 1
– Entry Restriction Policy vs. Base Case

As discussed above, operating efficiency oscillates because of market share fluctuations. The initial traffic shift towards communities 2 and 3 in the early years is somewhat offset by the artificial regulatory-induced decrease in access prices that increase provider 1's share of community 1 traffic, in which it has a comparative cost advantage. However, as the mix of information communities inexorably changes, and provider 1 "bounces" back into markets 2 and 3 after entry restrictions are relaxed, its unit costs start to turn upward. Market entry restriction leads to increased volatility of provider 1's operating efficiency vis-à-vis the effects of the first two regulatory policies (see Figure 5)

The operating behavior of network provider 2 is shown in Figure 11.

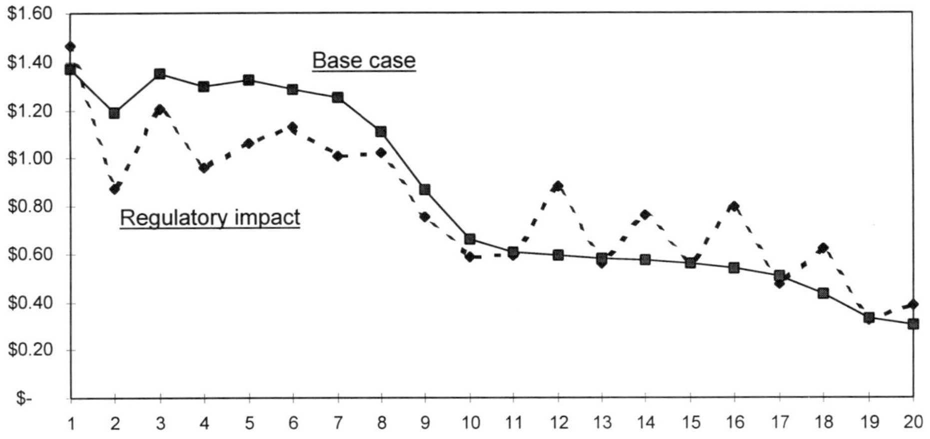

Figure 11. Operating Efficiency: Network Provider 2
– Entry Restriction Policy vs. Base Case

Entry restrictions on provider 1 improve network provider 2's performance until the "bounce-back" effect siphons traffic from provider 2 to provider 1, relative to the base case. The global change in the mix of information communities forces a downtrend in unit costs because of the provider's comparative cost advantage. However, the volatility in demand introduced by the regulatory policy is reflected in provider 2's operating efficiency.

The performance of provider 3 is shown in Figure 12.

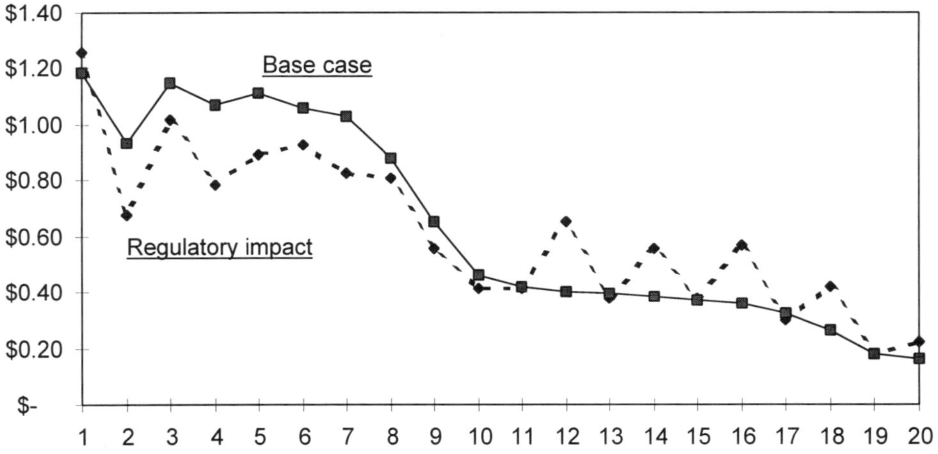

Figure 12. Operating Efficiency: Network Provider 3
– Entry Restriction Policy vs. Base Case

Network provider 3 has essentially the same story as provider 2. The entry restriction helps its performance in the short run, but hurts it in the long run, as compared to the base case. The damage is in terms not only of level of average cost per unit of usage, but also of volatility, which makes it difficult to plan effectively for additional network capacity.

Figure 13 compares the operating efficiency of the three network/service providers.

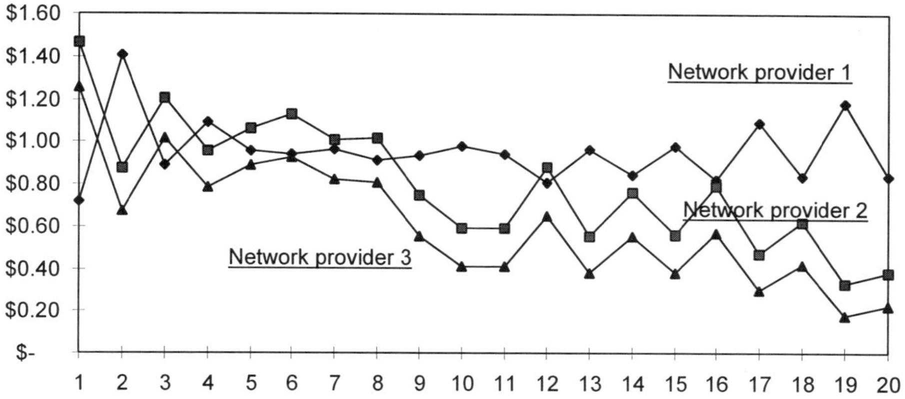

Figure 13. Operating Efficiency: Network Providers 1, 2 & 3 – Entry Restriction Policy

Network provider 3 enjoys the greatest efficiency gains in the longer run because of its comparative advantage in the market segment that is growing the fastest.

Finally, Figure 14 shows network provider 1's operating efficiency under the entry restriction policy vs. the base case if it is forced to set usage prices for community 1 traffic equal to revenue-maximizing levels as opposed to profit-maximizing levels.

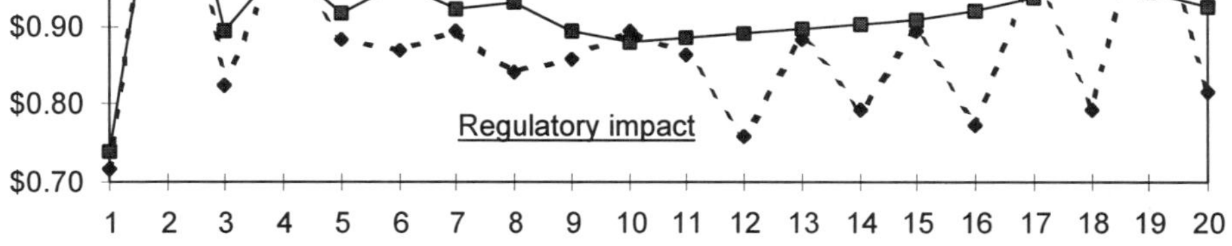

Figure 14. Operating Efficiency, Network Provider 1 – Entry Restriction Policy vs. Base Case: Usage Prices Maximize Revenue on Community 1 Traffic

Clearly, lowering usage prices stimulates demand and generally leads to improved efficiency relative to the base case (of course, profitability deteriorates). However, the extreme volatility in realized efficiency persists.

4.10 Simulations – Initial Conditions, Nonlinear Feedback and Alternative Outcomes to Policy Initiatives

The operating efficiency and market share effects of implementing cost allocation and market entry policies will depend on the interactions of various dynamic processes taking place in the relatively simple telecommunications/information system being modeled. Uncertainty regarding initial conditions or the sizes of certain critical coefficients can render policy decisions ineffective if not totally counterproductive. For example, the market entry restriction discussed above caused network provider 1's market share of community 3 to slowly rise from zero, and then enter a pattern of oscillation that at least appeared to be bounded. Figure 15 shows different market share trajectories that can be induced by exactly the same regulatory policy with selected market share adjustment coefficients.

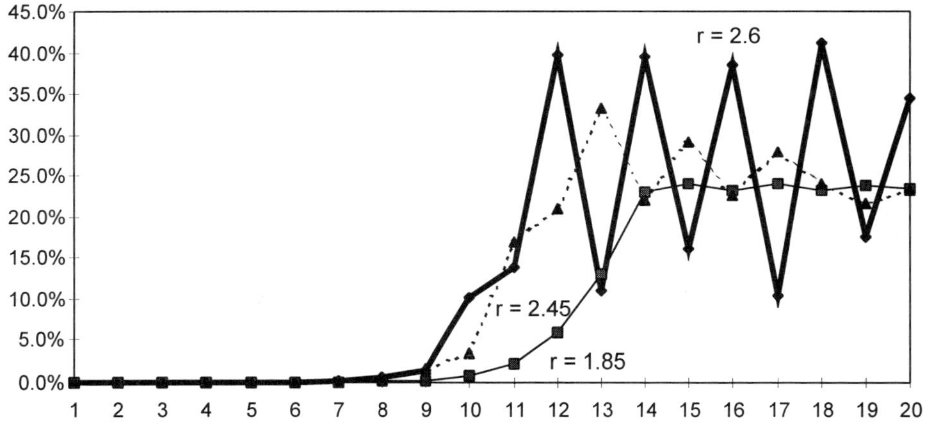

Figure 15. Network Provider 1 – Market Share of Community 3, Entry Restriction Policy: Alternative Coefficients in Market Share Adjustment

Recall that a coefficient of r = 2.6 was used as the baseline value. A coefficient of 2.45 leads to less volatility and eventual convergence. A coefficient of r = 1.85 results in a more traditional evolution of market share, similar to, but not the same as, the well-known Koyck model. The market share orbit for r = 1.85 in phase space is shown below in Figure 16. The slope of the orbit is less than 1 where it crosses the 45° line; hence the intersection is an "attractive" fixed point.

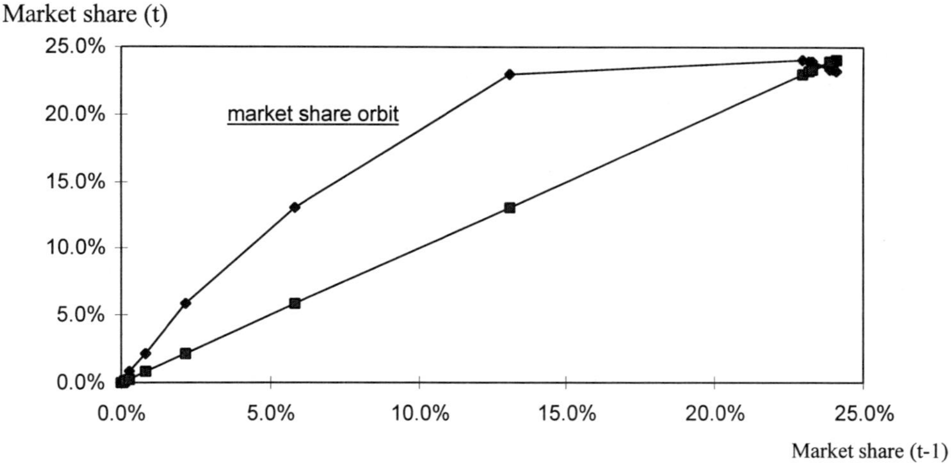

Figure 16. Network Provider 1 – Market Share of Community 3,
Entry Restriction Policy, r = 1.85

The practical implication is that if consumers' energy level is not too high, if they are not stirred up into over-reacting, then the system will gradually adjust to delayed market entry and converge to a stable solution.

A coefficient of r = 2.45 leads to an intermediate situation where delayed market entry results in initial volatility, but eventually the oscillations vanish. Figure 17 shows the market share's evolution in phase space.

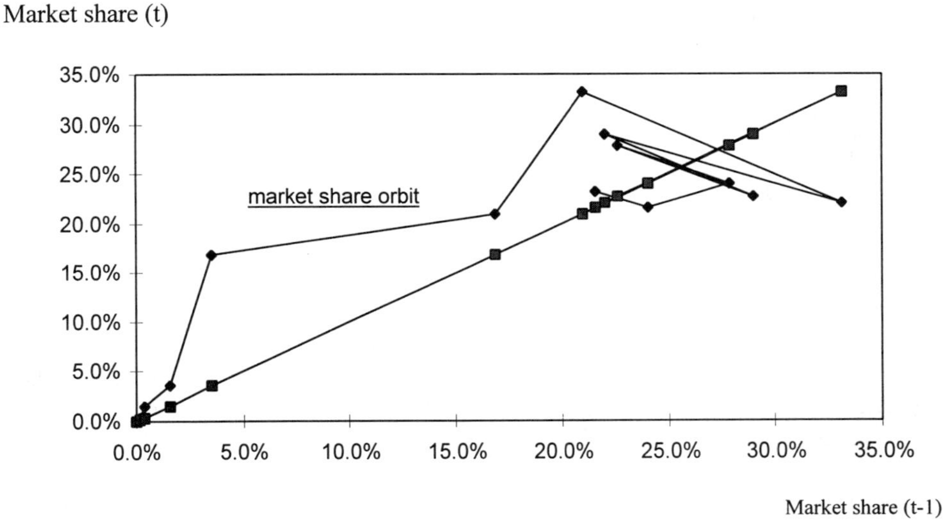

Figure 17. Network Provider 1 – Market Share of Community 3,
Entry Restriction Policy, r = 2.45

Note that the market share orbit still retains remnants of the parabolic shape shown in Figure 16.

For comparison, Figure 18 below summarizes the baseline scenario that used a coefficient r = 2.6. Most of the parabolic shape has disappeared; the large "hump" corresponds to the eruption of market share in time period 12. It is not clear whether the oscillations will weaken and die out under the pressure of community 3's strong preference for network provider 3, or whether the "ratchet" effect will cause the tug of war to persist indefinitely.

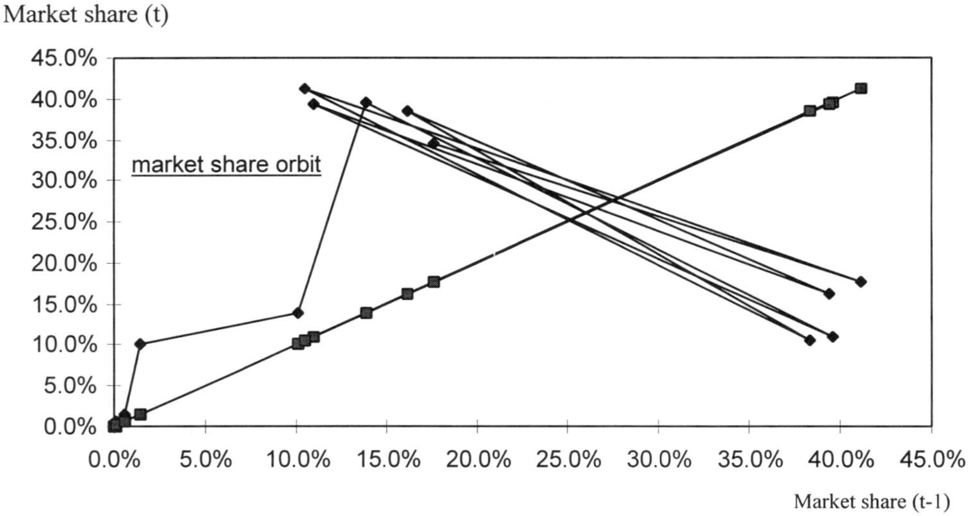

Figure 18. Network Provider 1 – Market Share of Community 3,
Entry Restriction Policy, r = 2.6

Because of sensitive dependence on initial conditions and nonlinear feedback, it does not take large deviations in parameters to produce large deviations in trajectories. In Figure 19 below, two market share evolutions are shown for exactly the same regulatory policy. One curve is based on the base-case market share adjustment coefficient of r = 2.60 used above, and the other is based on a coefficient of r = 2.63. Both coefficients appear to lead to essentially the same pattern. However, by slightly changing the size of the coefficient, the "spread" between the peaks and valleys is increased at r = 2.63.

Figure 20 below is the phase diagram for a coefficient value of r = 2.63. It is similar to Figure 18, having the initial hump and therefore a similar eruption of market share. In this case, however, the orbit is not nearly as compact as in Figure 18, and therefore its future course is even more difficult to assess than when r = 2.6. Compared to Figure 18, the market share orbit at r = 2.63 appears to be unraveling. Eventually, it could completely decay as the strong preference of information community 3 subscribers for network/service supplier 3 exerts a dominating influence.

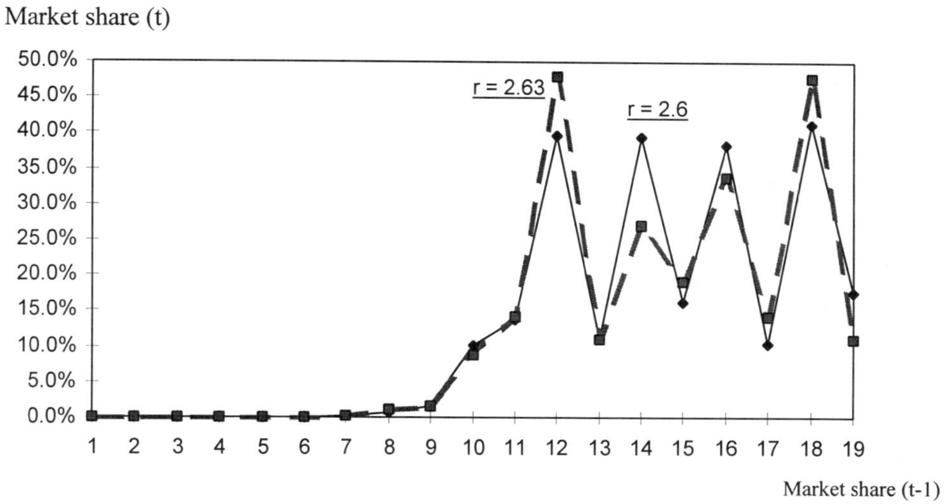

Figure 19. Network Provider 1 – Market Share of Community 3, Entry Restriction Policy, r = 2.60 and 2.63

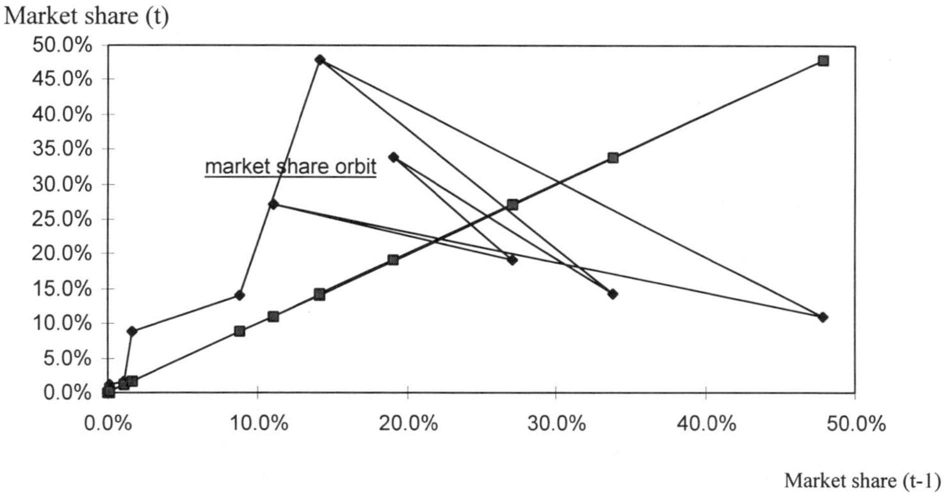

Figure 20. Network Provider 1 – Market Share of Community 3, Entry Restriction Policy, r = 2.63

The market share adjustment coefficient, r, can take on any number of values depending on the aggressiveness of the regulatory authority and the degree to which subscribers are induced into taking action. In order to give a comprehensive view of the possibilities, Table 11 below summarizes simulations of network provider 1's market share under alternative values of "r".

Table 11. Network Provider 1's Market Share of Community 3: Alternative values of "r" (time period 9-20)

r-coefficient	Time period										
	9	10	11	12	13	14	15	16	17	18	19	20
0.80	0.0%	0.0%	0.0%	0.0%	0.1%	0.1%	0.2%	0.4%	0.8%	1.4%	2.4%	4.1%
1.00	0.0%	0.0%	0.1%	0.1%	0.3%	0.5%	1.0%	2.0%	3.9%	7.0%	11.6%	17.2%
1.20	0.0%	0.1%	0.2%	0.4%	0.8%	1.8%	3.7%	7.3%	13.0%	19.5%	23.4%	23.7%
1.40	0.1%	0.2%	0.4%	1.0%	2.3%	5.0%	10.3%	17.7%	23.5%	23.7%	23.6%	23.6%
1.60	0.1%	0.4%	0.9%	2.3%	5.5%	11.7%	20.3%	24.6%	22.9%	24.0%	23.4%	23.7%
1.85	0.3%	0.8%	2.2%	5.8%	13.1%	23.0%	24.1%	23.2%	24.0%	23.3%	23.9%	23.4%
2.00	0.4%	1.4%	3.3%	10.2%	18.2%	27.6%	19.6%	27.2%	20.1%	26.9%	20.4%	26.7%
2.20	0.6%	2.9%	5.6%	19.9%	23.1%	25.1%	23.4%	24.3%	23.6%	23.8%	23.6%	23.6%
2.30	0.9%	3.0%	7.8%	21.0%	24.7%	20.7%	24.0%	22.2%	23.6%	23.7%	23.6%	23.6%
2.32	1.0%	2.9%	9.1%	19.1%	28.2%	16.8%	29.9%	15.4%	32.2%	15.1%	34.8%	15.0%
2.35	1.2%	2.9%	12.0%	18.3%	34.7%	16.5%	35.6%	14.7%	35.6%	14.7%	35.6%	14.7%
2.38	1.4%	2.9%	14.3%	19.0%	34.7%	17.7%	35.5%	15.4%	36.4%	14.1%	35.9%	14.7%
2.40	1.5%	3.1%	16.0%	19.9%	33.4%	18.9%	35.3%	17.4%	35.9%	14.9%	36.9%	13.9%
2.45	1.5%	3.6%	16.9%	21.0%	33.1%	22.0%	29.0%	22.7%	27.9%	24.1%	21.6%	23.2%
2.50	1.2%	7.8%	11.4%	35.6%	16.7%	39.4%	13.2%	36.6%	13.6%	40.0%	13.2%	35.8%
2.60	1.5%	10.1%	13.8%	39.6%	11.0%	39.4%	16.2%	38.4%	10.4%	41.1%	17.6%	34.4%
2.63	1.7%	8.8%	14.1%	47.9%	11.0%	27.1%	19.1%	33.9%	14.1%	47.8%	11.0%	27.1%
2.70	2.1%	7.7%	18.0%	31.5%	9.5%	37.9%	18.4%	40.7%	14.4%	41.1%	8.4%	38.9%
2.80	4.5%	7.2%	46.7%	26.8%	17.6%	32.7%	6.9%	40.0%	25.1%	18.2%	24.3%	17.3%

Two broad categories of results can be discerned over the space covered by the simulations. The first is in the domain from r = .8 to r = 2.3. As r is increased over this interval the market share trajectory tends to unfold into a series of almost nested "s"- shaped diffusion curves. At r = 2 a cyclical pattern resembling a bifurcation appears, and then subsides, as shown in Figure 21.

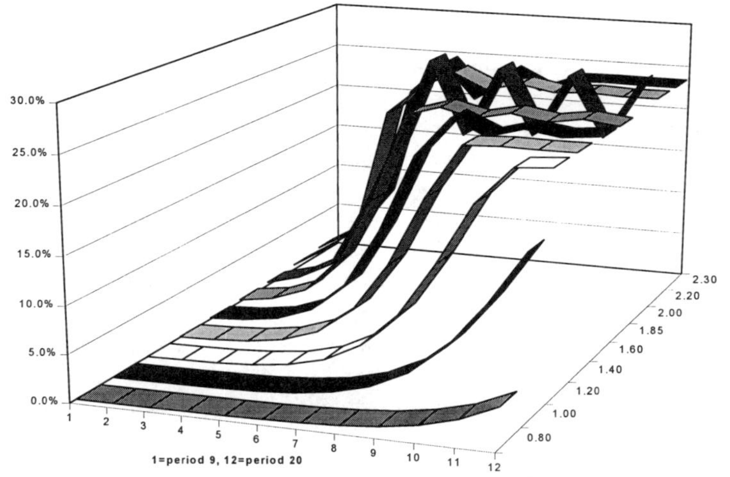

Figure 21. Network Provider 1's Market Share of Community 3: Alternative values of "r" (from .8 to 2.3)

Over the range r = 2.3 to r = 2.8, realized market share attains higher levels while oscillations become more pronounced; the highest shares are reached in time periods 12 and 18, when r = 2.63, as shown in Figure 22.

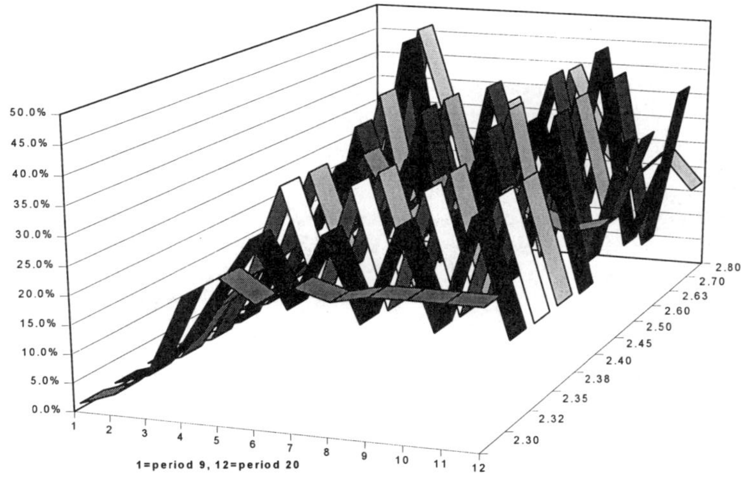

Figure 22. Network Provider 1's Market Share of Community 3: Alternative values of "r" (from 2.3 to 2.8)

5. Summary

Regulators need to be cautious in formulating and implementing policy initiatives, especially in an environment as dynamic as the telecommunications/information industry. There are just too many interacting forces at work to be able to know with sufficient certainty what the ultimate outcome of a policy decision will be. For example, unduly restraining one part of the system in the short run may cause unexpected repercussions in the long run, as shown by the model simulations. Policy makers should at least have an intuitive appreciation of nonlinear feedback. [26] It is true that the simulations are sensitive to the values chosen for certain structural parameters, but of course that is the main point. The coefficients are simply abstractions of consumer decision-making attributes: preferences for different suppliers, willingness to take risks and propensity to react to changes in the feasible choice set. Regulatory actions, in concert with technical innovations, can and will influence these attributes, especially when regulators try to impose linear constraints on a fundamentally nonlinear system. Regulatory policy should be characterized by balance, caution and a long-run perspective. And, when it becomes clear that the complexities of the environment are beyond the scope of the regulatory system, the regulatory apparatus should gracefully disengage.

References

[1] A. Aczel, Fermat's Last Theorem. ISBN: 1 56858 077 0. Four Walls Eight Windows, New York, 1996.

[2] M. Asmussen, Regular and Chaotic Cycling in Models from Population and Ecological Genetics. In: M. Barnsley and S. Demko (eds.), Chaotic Dynamics and Fractals. ISBN: 0 12 079060 2. Academic Press, Inc., New York, 1986.

[3] G. Baker and J. Gollub, Chaotic Dynamics, an Introduction. ISBN: 0 521 38897 X. Cambridge University Press, Cambridge, 1990.

[4] W. Baumol and E. Wolff, Feedback Between R&D and Productivity Growth: A Chaos Model. In: J. Benhabib (ed.), Cycles and Chaos in Economic Equilibrium. ISBN: 0 691 00392 0. Princeton University Press, New Jersey, 1992.

[5] W. Barnett and P. Chen, An Econometric Application of Mathematical Chaos. In: W. Barnett, E. Berndt and H. White (eds.), Dynamic Econometric Modeling. ISBN: 0 521 33395 4. Cambridge University Press, Cambridge, 1988.

[6] H. Berk, Nonlinear Dynamics of a Driven Mode near Marginal Stability, *Physical Review Letters* **76,** Issue 8 (1996) 1256-1259.

[7] M. Boldrin and R. Deneckere, Sources of Complex Dynamics in Two-Sector Growth Models. In: J. Benhabib (ed.), Cycles and Chaos in Economic Equilibrium. ISBN: 0 691 00392 0. Princeton University Press, New Jersey, 1992.

[8] M. Boldrin and M. Woodford, Equilibrium Models Displaying Endogenous Fluctuations and Chaos: A Survey. In: J. Benhabib (ed.), Cycles and Chaos in Economic Equilibrium. ISBN: 0 691 00392 0. Princeton University Press, New Jersey, 1992.

[9] P. Drazin, Nonlinear Systems. ISBN: 0 521 40668 4. Cambridge University Press, Cambridge, 1992.

[10] A. Einstein, Relativity, The Special and the General Theory. ISBN: 0 517 88441 0. Random House, Inc., New York, 1961.

[11] K. Falconer, Fractal Geometry. ISBN: 0 471 92287 0. John Wiley & Sons, New York, 1990.

[12] A. Gill, An Equipment Replacement Problem with Dynamic Production Planning and Capacity Considerations, *International Journal of Systems Science* **27**, Issue 10 (1996) 985-988.

[13] L. Glass, A. Goldberger, M. Courtemanche and A. Shirer, Nonlinear Dynamics, Chaos and Complex Cardiac Arrhythmias. In: M. Berry, I. Percival and N. Weiss (eds.), Proceedings of a Royal Society Discussion Meeting, February 4 and 5, 1987. ISBN: 0 521 35455 2. The Royal Society, London, 1987.

[14] J. Grandmont, Periodic and Aperiodic Behavior in Discrete One-Dimensional Dynamical Systems. In: J. Benhabib (ed.), Cycles and Chaos in Economic Equilibrium. ISBN: 0 691 00392 0. Princeton University Press, New Jersey, 1992.
[15] H. Haken, Synergetics. ISBN 3 540 08866 0 2. Springer-Verlag, Berlin, 1978.
[16] E. Lorenz, The Essence of Chaos. ISBN: 0 295 97514 8. University of Washington Press, Seattle, 1993.
[17] R. May, Chaos and the Dynamics of Biological Populations. In: M. Berry, I. Percival and N. Weiss (eds.), Proceedings of a Royal Society Discussion Meeting, February 4 and 5, 1987. ISBN: 0 521 35455 2. The Royal Society, London, 1987.
[18] J. MacQueen, Biological Modeling: Population Dynamics. In C. Bondi (ed.), New Applications of Mathematics, Penguin Books, London, 1991.
[19] K. Mainzer, Thinking in Complexity. ISBN: 3 540 60637 8. Springer-Verlag, Berlin, 1994.
[20] H. Niwa, Mathematical Model for the Size Distribution of Fish Schools, *Computers and Mathematics with Applications* **32**, Issue 11 (1996) 79-88.
[21] F. Vivaldi, An Experiment in Mathematics. In: N. Hall (ed.), Exploring Chaos. ISBN: 0 393 31226 7. W. W. Norton, New York, 1991.
[22] A. Robson, A Biological Basis for Expected and Non-expected Utility, *Journal of Economic Theory* **68**, Issue 2 (1996) 397-425.
[23] D. Ruelle, Diagnosis of Dynamical Systems with Fluctuating Parameters. In: M. Berry, I. Percival and N. Weiss (eds.), Proceedings of a Royal Society Discussion Meeting, February 4 and 5, 1987. ISBN: 0 521 35455 2. The Royal Society, London, 1987.
[24] D. Ruelle, Chance and Chaos. ISBN: 0 691 08574 9. Princeton University Press, 1991.
[25] J. Smital, Functions and Functional Equations. ISBN: 0 85274 418 8. IOP Publishing, Ltd., Temple Beck, 1988.
[26] N. Stolleman, Some Policy Issues Regarding Interoperability and the National Information Infrastructure, presentation at Interoperability and the Economics of the Information Infrastructure, Freedom Forum Media Studies Center, sponsored by Science, Technology and Public Policy Program, John F. Kennedy School of Government, Harvard University, 1995.
[27] J. Wisdom, Chaotic Behavior in the Solar System. In: M. Berry, I. Percival and N. Weiss (eds.), Proceedings of a Royal Society Discussion Meeting, February 4 and 5, 1987. ISBN: 0 521 35455 2. The Royal Society, London, 1987.

Telecommunications Transformation: Technology, Strategy and Policy
E. Bohlin and S.L. Levin (Eds.)
IOS Press, 1998

CHAPTER 8

Organizing Global Seamless Networks:

Contracts, Alliances and Hierarchies

Pablo Spiller
Haas School of Business, University of California
Berkeley, CA 94720, USA

Svein Ulset*
Foundation for Research in Economics and Business Administration (SNF)
Breviken 2, 5035 Bergen, Norway

Abstract. Drawing on transaction cost economics, this paper analyzes the development and utilization of global seamless networks and services. We expect the use of contract, hierarchy and hybrid forms of organization to vary with the amount of non-redeployable assets and non-tradable operation services, along with the level of network complexity and operational uncertainty. While the differences between the extreme forms of contract (represented by the Correspondent System and Internet) and hierarchy (represented by Concert, Cable & Wireless) are consistent with our hypothesis, the differences between hierarchy and hybrid and between the hybrid forms require further analysis.

1. Introduction

The purpose of this paper is to provide a transaction cost approach to the development and utilization of global seamless telecom networks (GSN). A global telecommunications services network is said to be seamless when its constituent elements work perfectly together without significant distortions and interruptions, as if the network were one huge computer or computer network. Rather than starting from scratch, such networks are built by interconnecting and transforming vendor-specific and partially incompatible national networks into global seamless structures, providing one-stop shopping of bundled, customized, uniform, end-to-end telecom services (voice, data, e-mail, video) and support (order processing, fault finding, repairs, billing, reports, customer service etc.). The first section of the paper discusses the need for global seamless networks. In the second section a transaction cost approach to the organization of global seamless network is presented, focusing on the transactional hazards and the organizational solutions associated with non-redeployable assets and non-tradable knowledge under considerable network complexity and operational un-

* This project is financed by Telenor [9].

certainty. A preliminary analysis of the current organization of global seamless networks is presented in section 3. Conclusions are presented in section 4.

2. The Need for Global Seamless Networks

Thanks to recent advances in digital technology, the range of interactive media has expanded rapidly from voice only to voice, data and video. At the same time, advances in multiplexing and high-capacity switching and transmission have drastically increased the speed and reduced the cost of communicating. This development has paralleled an even greater expansion in the number of available products and services within each media, such as voice-mail, electronic mail and video-conferencing. These are increasingly sold as integrated bundles of services, tailored to the special needs of each individual customer. In these bundles, not only different telecom services are included, but increasingly also support services such as fault findings, repair, technical maintenance, flexible billing, cost reports, telecom consulting, and systems integration. Primary telecom services consist of a series of different basic, enhanced and value-added services, e.g. of the following types (partially overlapping, see [7] and Table 1 further below):

MNS – Managed Network Services
(e.g. Frame Relay, X.25)

VPN – Virtual Private Network
(private switched data service on public lines)

BoD – Bandwidth-on-Demand
(e.g. allocated to video conferences)

VANS – Value added network services
(e.g. e-mail, electronic data interchange)

Systems integration
(protocols, computers, applications)

Outsourcing
(telecom and computer equipment and services)

While the first three service categories (MNS, VPN, BoD) indicate how different technologies for sharing common network facilities can be used as alternatives to leasing such facilities, the last three ones (VANS, systems integration, outsourcing) point more specifically to the value-added content and organization of such services. They all represent different ways of having a large number of separate and independent customers share the use and costs of common physical and human assets (switches, transmission, management, experience etc.) rather than pay for the extra costs of producing such services internally on private/leased and underutilized facilities and capabilities. If, for example, the multinational corporate clients of Concert were to operate their own private networks, these companies would have to: a) buy the necessary computer and switching hardware, b) have spares on hand, c) lease high-speed telephone transmission lines from the local phone companies in various countries, d) hire local technicians in each city, and then, e) figure out how to pay separate phone bills in different currencies in each nation. A global operator such as Concert would manage all these aspects and still save their clients money, as they share the network, service and support costs among different customers.

In a global seamless system, telecom services and support from the different national networks are no longer sold individually, but jointly as a package, bundled by one of the participating network operators or by some independent service provider or third party systems integrator. Economies of scope in assembly, distribution and marketing is the central economic rationale for such joint provision, meaning that there are excess resources or capabilities to be shared at decreasing unit cost as additional network units (with associated service and support) are added to the bundle, once the costs involved in developing the appropriate safeguarding mechanisms are taken into account. The question is then whether the contractual mechanisms used by Concert and the other global operators for connecting the different national networks provide sufficient safeguards and incentives for further investment and development.

3. Transaction Cost Approach

Transaction cost economics' main tenet [1, 2, 3], as it applies to the problem at hand, is that a network of interdependent network operators and service providers will be coordinated by *contract*, *alliance* or *hierarchy* depending on (i) the amount of *non-redeployable* equipment and *non-tradable* knowledge required for turning different networks into a global seamless communication system, and (ii) the level of network *complexity* and degree of operational *uncertainty* associated with providing the respective services. In a competitive environment and under sufficient network complexity and operational uncertainty, transaction cost economics would predict that fully integrated corporations will be chosen for providing those bundled services transactions (e.g. primary telecom services and associated support services) that require the most non-redeployable and non-tradable common physical and human assets, leaving contracts and hybrid arrangements to more tradable unbundled services supported by more redeployable assets.

Common resources such as physical facilities and human knowledge are *non-redeployable* if productive value is lost when transferred from original to new users or uses. Similarly, the operation services derived from these assets are *non-tradable* if the net benefits from using these services are substantially less when sold to external operators outside the corporation than when sold to internal operators paying the same price as external operators. One reason for such operation services being non-tradable is that the underlying technology or know-how is tacit (non-codified), and therefore significantly harder and more costly to transfer to outsiders than to insiders. The underlying technology and knowledge may also be non-tradable in the sense of being leaky (non-patentable and imitable) which means that the original inventor cannot prevent information dissemination and hence cannot, to a large extent, appropriate the private benefits of his own innovations. Normally, the probability of leakage is significantly higher when private knowledge or technology is transferred to outside users than when transferred to inside users. The less redeployable or tradable the distinctive resources and capabilities shared by each network (e.g. non-codified competence in interconnecting networks and synchronizing global service provision), the higher the transaction costs, and the stronger the safeguarding mechanisms needed to accomplish cost-efficient sharing of those assets (ranging from simple market discipline to fully integrated operator companies).

While some of the assets needed for global service provision are redeployable and easily tradable, others are not, and must be replaced or rebuilt each time interconnecting networks and major alliance partners are replaced. Consequently, if one network operator fails or defects, the others may suffer great losses, especially when *non-redeployable* assets and *non-tradable* knowledge are combined with high degree of *network complexity* and *operational uncertainty*. In particular, the larger the number of technically diverse networks and culturally and linguistically diverse partners, the higher the network complexity, and the higher the probability for misunderstandings, failures or defections. Similarly, the higher the performance standards in terms of data security, customization, uniform service provision and real-time interactivity for complex bundled services, and the more frequent the changes in market and technology conditions, the harder to comply with such standards and changing conditions, and the higher the level of operational uncertainty. To minimize the associated hazards, various contractual safeguards can be applied, ranging from simple contract clauses (under least hazards), to cross-ownership, mutual board representation, joint venture and full integration (under most hazards). In this paper we discriminate between the following three: (i) bilateral simple contracts (represented by the old Correspondent System and the rapidly growing Internet), (ii) alliances and joint venture (represented by World-Partners, Unisource and Global One) and (iii) fully integrated companies (represented by Cable & Wireless and Concert, besides corporate networks of multinational enterprises).

3.1 Hazards and Opportunities

To construct and operate a global network in a seamless and profitable way, considerable investments in network systems, training programs and competence building have to be undertaken. For example, to make different vendor-specific networks interoperable the participating network operators must invest in internetworking equipment such as protocol converters, gateways, bridges and routers.[1] Some of these investments are sunk or non-redeployable, such as most of the specialized internetworking equipment and installation work spent on integrating existent vendor-specific networks.[2] If the contract in this case should be prematurely terminated, and the previous network partner replaced with a new one, sunk resources and specialized equipment have to scrapped or resold at a loss, and new equipment purchased and installed.[3] Other assets are redeployable, but not easily tradable, such as the specialized and non-codified knowledge needed to provide global bundles of customized services of uniform high quality to satisfy their most demanding global customers. Not only is the market for such knowledge very thin, but it is also difficult to transfer without also transferring the personnel that has acquired it. Considerable interdependence and contractual hazards are thereby created between interacting partners (and between these

[1] Most of these are developed, produced, installed and maintained, not by the network operators, but by external equipment suppliers.

[2] E.g. VPN, X.25, Frame Relay, ATM etc.

[3] To make the BT-MCI network truly seamless and fully scaleable, and thus complete the integration of their Virtual Private Networks into a unified, dual-vendor intelligent network based on Northern Telecom equipment in North America and Ericsson in Europe, considerable amounts of costly and specialized development works have to be undertaken (Johnson et al., 1994).

and their major global customers). If one participating firm fails or defects, the others may suffer significant losses. Furthermore, as customers require increasingly more integrated bundles of customized services of uniform and high quality on a global scale rather than piecemeal offerings of random quality on a local scale, even more non-codified knowledge has to be co-developed and deployed, creating even higher levels of operational interdependence and need for global synchronized behavior, for which the chosen service organization should provide the appropriate governance structure (administrative controls and economic incentives).

However, all these hazards associated with specialized equipment and non-codified knowledge will decrease as manufacturers are increasingly able to develop intelligent networks that may support all the different vendor-specific protocols [4]. Also the level of non-tradability will normally decrease as technology and industries mature. With the introduction of advanced and more standardized software, services relying on non-codified and less tradable technical competence are increasingly being replaced with services relying more on codified and tradable software intelligence.

Indeed, digitization has made telecommunications technology a lot more decomposable, redeployable and tradable. With open standards, application software made by one firm is guaranteed to run on operating systems and communication protocols developed by a second firm, as well as manage the processing of coded information sent over physical network structures developed by a third firm. Instead of having one single big firm developing and producing all the software and hardware needed, several competing firms may do the same, sufficiently coordinated by common technology standards. Similar for service provision, where value-added services such as e-mail, electronic data interchange and teleconferencing are carried out, not only by telephone companies, but increasingly also by computer companies such as Microsoft and IBM, and specialist on-line service providers such as America Online.

Since competing standards and protocols still exist, though, interconnection equipment has to be developed and installed that may be partially specialized to the respective networks, depending on the networks' uniqueness and technological diversity. However, knowing how to exploit the business potentials associated with such software technology may still be unobservable or non-codified, and as such, constitute a competitive advantage.[4]

3.1.1 Stage One – Construction and Development

Assume, first, that there exists a sufficient number of sub-networks which are accessible through some kind of contractual arrangement. To interconnect these sub-networks and make them interoperable (seamless), bridges, routers and gateways have to be installed or incompatible network elements replaced with fully compatible ones. Likewise, operation and application softwares that do not work well together have to be modified, converted or replaced so that signals and messages can be displayed at the receiver's end precisely as intended by the sender. To achieve this, the interacting operators must first decide on a set

[4] The commentaries given by Franz Wolff, leader of a project group within Telenor, assessing alternative ordering-billing-systems after having chosen a US system called SUMMIT, are apposite (freely translated from Åpen linje, 1994, No. 5: 7): "*Being accessible to all, it is how we use technology that will determine 'who is best in class'. By exploiting the latent functionality and flexibility in our ordering-billing-system, Telenor may achieve a competitive advantage in a very competitive telecommunications market*".

of technical specifications that eventually will turn a non-seamless network system into a seamless one.

While the responsibility for carrying out the practical work is delegated to the local access operators and systems integrators, a central authority may be needed to manage the worldwide construction and development work (i.e., to monitor the worldwide construction and development tasks and resolve conflicts that may accompany such a complex undertaking). Before starting full-scale operation, a series of tests can then be run to clarify whether the different sub-networks have been designed and constructed as specified. Since the income of local operators depends on the volume of traffic received and sent to other networks, every local operator has an incentive to construct and upgrade its part of the global network as specified. This incentive, however, is not necessarily perfect, as the local operators do not capture all the surplus arising from the local traffic. Indeed, since some of that resides on the receiving end, these transactions are laden with externalities, creating potential disagreements over quality and investment levels. These problems may be aggravated when associate members (i.e., local distributors) are asked by the major partners to install specialized equipment and start network-specific training programs that are not easily redeployable or useful in other settings, thus locking themselves to the chosen partners or network alliances for a longer period. Furthermore, when changes in technology and demands call for non-trivial changes in network equipment, operation policy, obligations and remuneration policies, conflicting views and opinions will arise between captive and unequal partners. Understanding the potential conflicts that will naturally arise, and the potential for opportunistic behavior by the local operators, some type of hierarchical structure may be put in place ahead of time.

While replacement costs are small in the case of simple voice service, they may be considerable in the case of advanced data communications where the amount of proprietary components and non-redeployable physical and human assets may be larger and the potential loss in productive value from shifting a network alliance partner may be significant. Potential losses from non-redeployable assets may be larger for the major partners of global networks (e.g. BT or MCI, partners in Concert) than for the local distributors of such global services (e.g. Telenor, the local distributor of Concert services in Norway). Should, for example, MCI be forced to exit the Concert alliance, creating a new alliance with some other partner or group of partners may take years since most of the likely candidates have either signed long-term contracts with others, or are simply unavailable due to anti-trust restrictions against the ensuing market dominance (if for example MCI should propose to join Unisource, Sprint or AT&T). This delay will limit its ability to provide this type of service, making its assets specific to the non-redeployable product. Smaller and less committed partners, however, may change alliances easier (e.g. the change-over from Unisource to Concert by Spanish Telefónica, discussed below). Also affiliated local distributors, will normally have less non-redeployable assets invested in the relation, more alternatives to choose from, and consequently less need for a fully integrated governance structure.

3.1.2 Stage Two – Network Operations and Service Provision

Assume, next, that a global seamless network is successfully constructed and eventually ready for providing global communication services to a larger number of potential customers. While the technical functionality and service performance of the network system may satisfy some customers, others may not be satisfied, especially those having subsidiaries not easily accessed by the main network or not satisfied by the current menu of telecom services and available customer support services. Because of extra requirements concerning technical performance, customization and data security, some global customers may prefer to install most of the intelligence in their own switches and computers and transport most of the data on private/leased lines. This group of demanding global customers is also by most operators regarded as the most challenging and potentially most profitable market. To capture a greater part of this market, global operators must require intimate knowledge of the customer's core business and perform significantly better than the global customer's own telecom department. In other words, global operators have to turn their networks and service organizations into both specialized and flexible units, well-suited for providing customized services to a large number of customers, achieving close to a quasi-integrated relation with their most entrusting and demanding *outsourcing* customers. Similar investment in specialized technical and human assets may be needed on the customer side, creating a bilateral dependency relation similar to, but usually weaker than, the one among partners and between partners and their associate members mentioned above. How far into the customer's information system should telecom operators be allowed to integrate? It depends on how closely related the telecom function is to the customer's core business. For fear of information leakage and maladjustment, customers that use telecommunications as a critical and integral part of their core business will be more reluctant to outsource larger parts of their communication systems to external telecom operators that also will be serving competing companies.

Global customers whose main business is not communication, but something else like car production or construction, may benefit from having outside firms running their telecom business and computing service. On the other side, global customers whose main business is closely related to telecommunications, such as banking and transport, will probably be more skeptical to large-scale outsourcing. In the latter case, outsourcing major parts of the company's telecom services presupposes simultaneous transfer of firm-specific knowledge from the customer's core business to its closely related telecom business, which may be quite difficult without also transferring the people possessing the knowledge, thus increasing the chances of losing both sensitive information and key personnel if outsourcing is chosen.

Nevertheless, cost-savings from sharing expensive telecom facilities with other customers on so-called virtual private networks are too large to be ignored. Normally the price advantage of using global common carriers (due to larger scale and scope economies) will only come at the cost of less customized, less advanced and less protected information. New technology and more efficient service organizations are increasingly turning this tradeoff net positive for an increasing number of corporate customers. To the degree that such outsourcing allows the customers to focus and improve their core competences, and consequently the functionality and quality of their main products or services, such outsourcing may also contribute, albeit indirectly, to the development of more superior and higher-priced products.[5]

---

[5] Recently, outsourcing has dramatically increased in many areas of business services, due to changes in both the facilitating technology, regulatory regimes and governance forms. For example, potential scale and scope resources that previously were treated as an integrated part of firms' core assets are increasingly outsourced to better positioned and endowed specialist firms, thus enabling both upstream and downstream firms to spe-

Indeed, capacity sharing, customization and data security on public switched lines (so-called virtual private networks) are now rapidly improving, reducing the need for leased lines and private networks. Besides, global operators may still be able to price their services lower than private networks due to larger economies of scale and scope, although capacity resale on private/leased lines will eliminate some of the difference. Furthermore, dependence on their chosen external service provider can be reduced by selecting technology platforms and interface standards for their internal systems that interact smoothly with several external systems, making external suppliers more easily interchangeable and the need for quasi-integration less profound.

3.1.3 Stage Three – Revenue Appropriation

Finally, let us assume our global network operators have been in business for a while. What distinctive assets and organizational safeguards may explain differences in their economic performances? To achieve above-normal rates of return, telecom operators, like all businesses, must first develop *superior capabilities* or a *strategic position* that make them perform better than their competitors. Now, in order to appropriate a larger share of the future income generated by superior performances, the underlying capabilities must be protected by the appropriate *legal, contractual or strategic means*, or be *self-protected* due to their non-codified and tacit character [3]. If successful, such safeguards will also provide incentives for future development of superior capabilities and performance.

For example, a technical innovation in network technology that creates value for the users will only create extra profit for the original innovator to the degree (1) competing firms are prevented from copying or imitating the basic underlying value-enhancing knowledge or technology, and (2) monopoly suppliers of complementary products and services are prevented from capturing the lion's share of added value by charging higher prices on their products and services. If leaky, superior assets will soon also be used by rival firms, after which the accompanying profits will be competed away and transferred to downstream customers.[6]

In anticipating such appropriability problems, both telecom operators and service providers may try to capture a larger part of added value from their systems and service innovations by so-called *proprietary bundling*, meaning that services are intimately and inseparably bundled with the enabling, proprietary technology (software and hardware). For proprietary bundling to be successful, however, it must create value for the users. Users may find proprietary bundling beneficial if it simplifies the operation of complex systems. To

---

cialize within a smaller core of competences and activities (the AT&T divestiture of its equipment division being the most dramatic recent example). This may even happen in sectors characterized by substantial non-redeployable assets or non-tradable services, such as in many areas of customized business services and specialized technology development. As the firm's competitive core competences change, so should also its range of accompanying core products. Analyzing the effects of changing technology, regulation and governance on the optimal size and scope of telecom operators is one of the core issues of this paper.

[6] Investment in non-protected leaky assets should only be undertaken to the degree such assets are needed to access important factor or product markets, or to the degree the firm in question is especially well positioned or equipped to lead the future development of such leaky assets (e.g. due to some monopoly position or first-mover advantage). Conversely, services that external suppliers are better positioned or better endowed to carry out than internal suppliers, including the extra transaction costs from external contracting, should be outsourced rather than insourced.

keep the underlying product and bundling technology proprietary and prevent it from leaking out to competitors, most firms would normally prefer to do most of the technology development and production themselves. Due to converging and rapidly developing technology, however, many firms are forced to access some of the needed technology by contract and co-produce the bundled good with alliance partners. As a consequence the bundling technology (interface standards) becomes more open and less proprietary, although the core technology of the constituent products may still be kept proprietary by the participating firms. If the underlying technology and bundling techniques are *not* kept proprietary, operators will compete in providing the products within the same bundle. Companies will then compete on how fast they can provide essentially comparable *bundles*. Bundling, however, will adjust to the needs and sophistication of the customers. Customers with well-developed in-house management information systems (MIS) may prefer less bundling than customers with less sophisticated capabilities. As a consequence, operators will either specialize across customer types or will provide a variety of bundled services.

To summarize, advanced intelligent network (AIN) equipment supplied by different vendors will normally not work well together.[7] Although global network operators cannot derive any competitive advantage from different network systems of similar price and quality, they can derive such advantage from integrating these into more seamless (interoperable) structures and services (bundled, customized and mass-produced). The need for hierarchically organized global operators derives from (i) the need to undertake investments in specialized network equipment and in developing and deploying non-codified knowledge under (ii) high levels of network complexity and operational uncertainty associated with providing customized bundles of seamless service and support at uniform high quality on a global scale. Serving customers effectively depends, therefore, not only on the local access networks and service capabilities, but on the whole interconnected global network and service organization. The *firm-specific* parts (tacit knowledge) of the interconnected network may give the operator a competitive advantage over differently organized global operators, including private networks, to the degree the facilitating competence and technology are *superior*. Most probably, the above appropriability problem will be larger for upstream equipment suppliers and software houses whose competitive advantage is based on proprietary and partially leaky technology, than for downstream telecom operators and service providers whose competitive advantage is more based on tacit (less codified) competence in network operation and service provision. Further downstream, those corporate customers benefiting most from outsourcing their telecom and computer business are also those likely to benefit most from such upstream technology leakage.

According to the above predictions, efficient/profitable alignments of governance forms with transaction attributes are hypothesized as in Table 1, displaying different organizations for building and operating global seamless networks, depending on the attributes of the respective telecom services as indicated on the bottom line. Here, the amount of non-redeployable assets invested by telecom operators in global networks is assumed to be larger for partners (organized as equity alliance) than for associate members (organized as distribution contracts).

[7] Neither did the older and less advanced equipment inside national systems, since most of these were tailored to the perceived need of national telecom operators.

Table 1. Relationship between Attributes of Global Service Transactions and the Network Organization Chosen for Providing such Service Transactions

	I		II		III
Organization	*Contracting*		*Alliance*		*Hierarchy*
	Implicit	*Explicit*	*Loose*	*Tight*	
	Internet	Correpondent system	WorldPartners	UniSource, Global One SITA, Swift	Cable & Wireless, Concert, MNE networks
Level of non-redeployable & non-tradable assets, given high degree of network complexity & operational uncertainty	Low		Medium		High

4. Global Seamless Networks

As noted above, global networks are built by upgrading, harmonizing and interconnecting increasingly larger numbers of local and national networks, gradually achieving higher levels of global coverage and operational seamlessness. These «networks of networks» could be organized as corporations, joint ventures or contractual agreements, depending on the transactional difficulties involved in developing and operating such complex networks. The major question is whether hybrid forms such as WorldPartners and Unisource are sufficiently sustainable or will be replaced by either fully integrated corporations (Cable & Wireless, Concert, Global One) or simpler contractual arrangements (the Internet). We start answering this question by describing and analyzing the preliminary results and still evolving structures of these different global network operators in somewhat greater detail (see Table 2).

Being the most recently established organizations, Concert and Global One also have the highest global coverage. Since local access is monopolized, regulated and protected against acquisition by foreigners in most countries, and since the dominant local access providers can normally be distributors of only one global carrier, local access will be restricted in most countries. Until open interconnection is enforced locally, global seamless networks cannot be fully accomplished. The problems of local access are also exemplified and confirmed in a study by Forrester [6], emphasizing not only technical, but also the many regulatory and attitudinal, problems in contracting with the traditional monopoly operators for local access.

Table 2. Global Networks (1996)

	Internet	*WorldPartners*	*Unisource*	*Global One*	*Concert**	*Cable & Wireless*
Type of organization	C	JV	JV	JV	H	H
Date of foundation	-	1993	1992	1996	1994	1929
Turnover	-	-	$0.65 b	$0.8 b	-	$8.8 b
Staff	-	100	2230	2900	700	41 000
Carriers' carrier	Yes	Yes	Yes	Yes	Yes	Yes
Access provision	Yes	Yes	Yes	Yes	Yes	Yes
System integration and outsourcing	Yes	Some	Yes	Yes	Yes	Some
Value-added services and solutions	Yes	No	Some	Some	Yes	Some
VPN #	-	21	17	15	15	-
Private circuits#	-	25	17	-	48	13
Frame relay#	-	27	17	27	35	24
Packet data#	Yes	-	17	51	48	38
Conferencing#		-	17	35	15	-
Calling cards#		-	17	103	15	-
Internet/intranet service#	Yes	-	17	19	9	-
Managed bandwidth#	Yes	-	17	18	15	13

C: Contractual agreement, JV: Joint venture, H: Hierarchy (integrated company)
* Pre-merger data, b: billion USD
# Number of countries
- indicates either that the service is not offered or no information was supplied
Source: [5]; the turnover and staff data from Unisource are 1995 data

4.1 Implicit and Arm's-length Contracting

4.1.1 The Correspondent System

We have two different types of contractual global networks, the old Correspondent System and the much younger Internet. The Correspondent System represents the traditional way national monopoly operators have interconnected their telecoms networks and settled their payment for international communication. To divide revenue between origin and destination country, a dual price system is still used whereby, for a single call, one price is charged to users by the originating operator (the collection charge) and a second price is charged by the terminating operator to the originating operator (the accounting rate). If there is an imbalance in the volume of incoming and outgoing traffic, then the originating operator which generates more traffic should pay a certain fee to compensate the terminating operator (the settlement payment, usually half the accounting rate).[8] Currently, there are several pricing structures (combination of fixed and variable charges), especially for data communication, dependent on (i) access and transmission technology being used such as analog versus digital connections, fixed versus mobile, switched links versus dedicated or leased lines,

[8] From an economic point of view, the current accounting rate system is inefficient because it rewards those public telecom operators that keep accounting rates high and/or those that generate less traffic. Although prices have come down over the last years, international calls are still overpriced in order to subsidize local and residential service, provided by over-staffed and inefficient local monopoly operators.

and (ii) the type of service being offered such as virtual private network, frame relay, packet data, conferencing, calling cards, internet/intranet service, managed bandwidth etc. (see Table 2).

The main achievement of the Correspondent System is the operation of a global seamless network for voice telephony, and increasingly also for data communication over X.25 data networks.[9] From a given telephone a user may reach almost any other telephone on the planet.[10] Within networks employing vendor-specific and proprietary technology, equipment from one vendor can seldom be interconnected with equipment from competing vendors without considerable difficulties and extra costs. Between such networks (as well as between many networks and connecting user equipment), the use of common interface technology, standardized by International Telecommunication Union (ITU), has made such interconnections relatively easy at least for circuit voice switching and simple data communications, such as the interface for packet-switched X.25 data network. To switch and transmit more complex messages and binary information structures, advanced intelligent networks (AINs) will often be used that seldom are completely standardized and compatible when produced by different suppliers, creating severe interoperability problems in many instances. To solve these interoperability problems and create the needed international connectivity, more centralized coordination mechanisms are needed than the Corresponding System, with its traditional emphasis on bilateral contracting and lengthy standardization processes.

4.1.2 The Internet

Internet was born about 25 years ago out of an effort to connect together a US Defense Department network called the ARPAnet and various other radio and satellite networks. The objective was to build networks that could withstand partial outage (like bomb attacks) and still work. In the ARPAnet model, communications occur by having computers talk to each other and ensure that the communication is accomplished. The network itself was assumed to be intrinsically unreliable as any part of it could disappear at any moment. To send a message, the computer simply had to put its data in an envelope, called an Internet Protocol (IP) packet, and address the packets correctly. The demand for networking then spread quickly, and Internet developers from US, UK and Scandinavia, responding to market pressure, began to put IP software in every conceivable type of computer. By then the International Standards Organization (ISO) had already spent years designing the ultimate standard for computer networking without much headway. Users, however, adopted the IP instead. So did companies that developed workstations for local area computer networks (LANs), allowing all computers on such LANs to access ARPAnet facilities. One of those newer networks was NSFnet, commissioned by the National Science Foundation, with the objective of connecting computers of major universities. Due to bureaucratic and staffing problems, NSF decided to build its own network based on the ARPAnet's IP technology.

Demand grew rapidly until the computers controlling the network and the telephone lines connecting them were overloaded. The network needed upgrading and professional management, and the contract was awarded to Merit Network Inc., which ran Michigan's

---

[9] An X.25 network is any network that implements the internationally accepted ITU-T standard governing the operation of packet-switching networks.

[10] Albeit still plagued with lags, noise and bad lines in many parts of the world.

educational network, in partnership with IBM and MCI. The number of connecting networks kept growing, recently also including non-IP-based networks connected by special *gateway* technologies.

The ultimate authority for where the Internet is going rests with the Internet Society, or ISOC, a voluntary membership organization whose purpose is to promote global information exchange through Internet technology. ISOC appoints a council of elders, called the Internet Architecture Board, or the IAB, who meets regularly to bless standards and allocate resources, such as addresses. It decides when a standard is needed and what it shall be. When a standard is required, it considers the problem, adopts a standard, and announces it via the network. The Internet Engineering Task Force (IETF) is another volunteer organization. It operates through working groups which anyone can join. These groups make different recommendations that either are made available to anyone or sent to IAB to be declared a standard. These standards make computers from different vendors communicate, favoring no one in particular, whether IBM, Sun or Macintosh.

The participating networks financed by governments or private users connect without a charge, only by adopting the open IP technology or some non-IP-technology with gateway to the Internet. The modern web servers and web browsers have spawned further growth. Although the Internet is mostly used for information exchange and marketing, new communication software is allowing the Internet to carry interactive voice traffic as well. By mid-96 the Internet reached nearly 5 million host computers and 20 million users.

As alternatives to the Internet and the Correspondent System, global customers may contract for a more integrated bundle of services with some global network operator, either Cable & Wireless or one of the recently established global carriers, depending on the customers' specific need for global coverage, scope of service and level of performance.

4.2 Alliances

4.2.1 AT&T WorldPartners Association

The AT&T WorldPartners Association is a two-tier equity (partners) and non-equity (associate members) alliance, offering global business communication services, named WorldSource, based on common technical standards. It was established in June 1993 by AT&T, KDD (Japan) and Singapore Telecom. An operating organization called WorldPartners Company (WPC), with headquarters in New York, was created to function as a global support center for WorldPartners members offering WorldSource services. Later this group was joined by Unisource and four associate members (distributors) from the AT&T's previous Pacific Partners project: Telstra (Australia), Telecom New Zealand, Korea Telecom and Hong Kong Telecom (57.5%-owned by Cable & Wireless). Unisource (including the Spanish operator Telefónica with its South America operations) and AT&T then established a new company, *UniWorld*, that provided telecommunications services to multinationals operating in Europe. The new company (assets $200 million) in which AT&T participated with 40% and Unisource with 60% was located in the Netherlands.

Although AT&T WorldPartners is a significant global network alliance, it provides only a limited service portfolio of WorldSource Services to the users. The individual partners and distributors themselves have the responsibility of interfacing their existing network so that discontinuity is eliminated and end-to-end connectivity is achieved. Procurement is coordinated to ensure sufficient compatibility, but each partner retains a high degree of autonomy over its network infrastructure [7]. To become a full service provider, WorldSource Services will have to widen its portfolio significantly compared to Concert. Encompassing so many potential competing companies, however, the areas of close cooperation cannot be many, nor can they be closely related to their respective core competences.

4.2.2 Unisource

Although strictly speaking a European regional service organization, Unisource has a global reach both through its membership in WorldPartners and through a service agreement with Infonet's World Network. Unisource was first started in 1992 as a joint venture between *PTT Telecom* (Netherlands) and *Telia* (Sweden) to supply international telecom services. Joined by *Swiss PTT* and *Telefónica* (Spain) in 1993 and 1995, Unisource consisted for a period of four equity partners, each having a 25% stake.

Unisource started out by providing business support and service in 12 European countries and several countries outside Europe including Japan, Singapore and the US. Gradually, also the Unisource profile has been lowered. While Unisource initially was represented with the name Unisource in every country it served, each selected carrier now has the responsibility of representing Unisource with its own name in each country. Telia is for example serving Unisource in Denmark, Finland, Norway, UK and the Baltics. The name Unisource will not be visible and is used only as a brand name for the Unisource products, quality and prices. At least for some part of their global portfolio, the central unit of Unisource specifies the network architecture and describes how the infrastructure of each of the partners is supposed to perform. The partners procure equipment together, but own and operate it separately. Global branding and uniform quality are the result of contracts and mutual self-interest, tied to equity holdings in the joint venture [7].

In December 1994, Unisource and AT&T Corp. confirmed an agreement to establish a joint venture, UniWorld, combining their European data and voice services. In May 1996, Unisource and AT&T decided to abandon Uniworld as a marketing entity and rather merge most of their European operations into AT&T-Unisource Services (owned 40% by AT&T and 60% by Unisource). Then in spring 1997, Telefónica announced that it would leave Unisource to join Concert. With the addition of Telefónica to the MCI/BT family, Concert plc gets a strong foothold in Central and South America, where Telefónica already has multiple interests. This deal undoubtedly will have ramifications with Telefónica's Unisource partners, probably seeking retribution if Telefónica pulls out of the venture. In fact, Unisource officials have said that they would petition the European Union, claiming that a BT/MCI/Telefónica triumvirate would give Concert an unfair competitive advantage.

Lacking cross-ownership, Unisource can be regarded as a somewhat looser alliance than the original Concert, similar to AT&T WorldPartners. The joint venture into AT&T-Unisource has expanded the global reach of both organizations, but also created a rather complex and difficult governance structure combining four companies (including AT&T-Unisource) with different backgrounds, circumstances and priorities, of which two are still government-owned in largely monopoly markets. The split between services offered by AT&T-Unisource and those retained by AT&T also presents a rather confusing picture to the market place. Subject to a series of ongoing organizational upheavals and changes to its market face and branding, Unisource still has to prove that it can provide sufficient stability and direction from its multiple and diverse partners to compete with the more unified Concert and the upcoming Global One.

4.2.3 Global One

Global One, founded on February 1, 1996, is a worldwide telecom service provider for the businesss, carrier and consumer markets. It is a joint venture between France Telecom, Deutsche Telekom and Sprint, headquartered in Brussels and divided into three operating divisions with headquarters in Brussels (Europe, Global Services) and Reston (World). Although a private joint-venture company, Global One has more of an alliance or committee-based management structure than Concert, as overall control is vested in the Global Venture Board, which consists of the three chairmen of the parent companies. Services are sold by Global One's sales force in the partners' home countries (France, Germany and USA), and through local business units, national affiliates, partnerships and distribution arrangements in other countries.

Like the other global service companies, Global One aims at obtaining maximum global coverage and seamless connectivity. The company has faced a long and arduous process of regulatory approval in both the USA and the EU because of the monopoly position of its two European partners, and due to this, the company will not be able to assume its intended full form until after the January 1998 EU telecoms liberalization. Being the most recently formed major global network organization, its most immediate task is to develop significant market presence in order to reduce the lead that Cable & Wireless, Concert, Unisource and WorldPartners have developed from their earlier dates of entry into the market. Despite its late entry, Global One claims to have a sizable business: about 2500 employees, 1200 points of presence, offices in over 50 countries, and expected first-year revenues of over US $800 million.

4.3 Corporations

4.3.1 Cable & Wireless

The *Cable & Wireless group*, headquartered in London, is the world's oldest international telecommunications operator with a history stretching back some 125 years to support the business interests for The British Government.[11] After privatization in 1981 Cable & Wireless has expanded into a great number of activities around the world.

Cable & Wireless (C&W) is a federal structure, combining specialized business units (such as C&W Business Unit), with a first tier of fully owned federal operators for selected parts of the world, and a second tier of part-owned telecom operators for the rest of the world. All its multinational customers are, however, served by one company, C&W Business Network, providing global branding and uniform quality. It has operations in over 50 countries on five continents. The operations are centered around three regional hubs, namely Asia, the Caribbean and North America, and Europe.

C&W has two operational units. One of these is C&W Business Network, which aims to meet the telecommunications needs of the groups of international customers with creative

[11] Until 1934 C&W was named Eastern and Associated Telegraph Company.

voice and data solutions. These services utilize the Global Digital Highway (GDH), C&W's huge global infrastructure. The GDH is controlled by a series of Network Management Centers (NMC) located in US, Japan, Hong Kong and the UK, which has overall responsibility for managing GDH, as well as the Global Managed Data Service (GMDS) and the European Data Network. The other is C&W Marine, which is one of the leading international subsea installation and maintenance contractors, operating the largest commercial fleet of cable-laying and support vessels in the world. It continues to grow, both by acquiring other companies and by building new vessels. The company plans to invest £1 billion over five years in intelligent network services in an effort to become a major global player. This would allow customers to design their own communication systems wherever C&W has operations. Traditionally C&W has had a stronger position in Asia than in Europe and the US. Through a series of recent efforts to ally with service and network companies in Germany, Italy, UK and US, Cable and Wireless may soon be able to correct this imbalance. In the UK, however, the C&W company Mercury has been losing both business and profit margins, reflecting the problems both of competing with the incumbent monopoly operator, and of turning their own telecom activities into profitable businesses [8].

A larger ownership share in participating companies provides C&W with more influence and a stronger ability to take quick and important strategic decisions such as in implementing new standards, developing and utilizing common assets, and accessing the experience and customers that each company has in its own country. Although a more hierarchical organization may constitute a competitive advantage, smaller size and global imbalance may still represent a liability.

4.3.2 Concert

Concert was established in 1994 as a joint venture between *British Telecom* (71.5%), the largest operator in the UK, and *MCI Communication Corporation* (24.9%), the second largest carrier of long-distance telecommunication services in the USA. Concert has market and distribution agreements with a large number of companies in many countries. By internalizing the production process, Concert is probably better positioned than WorldPartners and Unisource to assure global branding and uniform quality. In November 1996 the two companies proposed to merge their activities and established Concert plc.

Concert aims to provide multinational businesses with seamless telecom solutions. Drawing on the combined strengths of its parents, it is assembling a portfolio of advanced, value-added networking products and services with worldwide availability and performance. The alliance is based on the fact that 50% of the world's multinational corporations have their headquarters either in the UK or in the US. MCI has, through its experience in the US market, a large marketing expertise, and has used a number of clever marketing initiatives to build market share. BT has for some years, together with AT&T and C&W, been an international player and therefore has global presence and experience from different markets. These will be important assets as Concert expands into other markets around the world. Concert will seek maximum control over its global network. This is done by building its own facilities, as it is doing in Sweden, and by entering into alliance/distribution agreements, either exclusive or non-exclusive, as in the Scandinavian countries.[12] Concert will have access to BT and MCI research, and will be responsible for the development of new products and services for multinationals.

One of Concert's core strengths is its integrated hierarchical structure with only two core partners of approximately the same size, facilitating decision-making as compared to the other, more loosely organized network alliances. Having established an efficient corporate structure, built the world's most advanced voice and data communications network, and cleared regulatory hurdles in both Europe and the US, Concert is now the leading provider of global services to multinational enterprises.

The recent Telefónica change-over from Unisource to Concert has strengthened Concert's lead and market position even more. Currently, Telefónica operates communications companies in Argentina, Brazil, Chile, Peru and Puerto Rico. That, along with Avantel, MCI's partnership in Mexico with Banamex, pretty much locks up Latin America for the consortium. Additionally, MCI's deal with Stentor in Canada gives it access to most of the Western Hemisphere. The deal will form Telefónica Panamerica-MCI and will be a 50/50 venture managed by Telefónica's international division. Additionally, Telefónica will become an exclusive agent of Concert Communications Services in Spain and will be able to invest in MCI's Avantel project.

4.4 Global Customer Alliances & Joint Ventures

Two of the most successful customer alliances are SWIFT (Society for Worldwide Interbanking Financial Telecommunications) and SITA (Societé Internationale de Télécommunications Aéronautiques), both set up to satisfy the special needs of their respective industry sectors. These alliances were also the first of the global carriers. SWIFT was created in 1973 by a consortium of European and American banks to provide the telecom links needed to enable the clearance of financial transactions between banks at international level. It went on line in 1977 and now has over 300 member banks. SITA is a co-operative organization established already in 1949 by 11 of the world's leading airlines. Today it has over 600 members worldwide and is the major provider of communications and information services to the global air transport industry. By building and expanding on their inherent knowledge of their industries, both these organizations have succeeded in supplying customized applications and services highly valued by their members (even at a time when the prices charged by the monopoly providers were extraordinarily high). Recently, both organizations restructured in order to increase their emphasis on serving the global communication needs of multinationals in a wider range of industries.

4.5 Global Networks of Multinational Enterprises (MNE Networks)

Only the largest multinational enterprises (MNE) can afford to interconnect local area networks into more global seamless networks on their own private/leased lines. By running their own telephone system and data networks, MNEs could better control the use of technology to secure sufficient interoperability, customized solutions and data security. For

---

[12] Through the agreements with the Nordic carriers in Denmark, Finland and Norway, Concert has access to an interesting test market, as the Nordic countries have the world's highest penetration of mobile communication per inhabitant.

some customers, the most critical aspect is the level of data security. The most sensitive information is either not sent at all, or only sent over private networks accessible to specially authorized people. As encryption and similar security techniques improve, a larger share of sensitive information may be sent over the public network operated by service providers, which in addition to telecom services also may run part of the customer's computer system. Such service outsourcing parallels and amplifies the outsourcing trend further upstream in the value-chain. As telecom technology gets increasingly standardized and decomposable, equipment, facilities and service units that previously were part of the customer's business are handed over to telecom operators or further upstream to equipment suppliers and software firms. From these upstream firms intermittent products and services may subsequently be sold back at a lower price to a larger number of competing local telecom operators and service providers, from which basic and value-added services are delivered to corporate and residential customers at a lower price and higher quality than before divestiture and disintegration.

As a consequence of advances in telecom technologies and worldwide service organization, traffic has started to migrate from corporate networks back to common carrier networks. The old trade-off between higher-priced, but more customized, advanced, and secure data communication on private lines and the lower-priced, less customized, less advanced and less secure data communication on public lines is still essential. For reasons mentioned above, this trade-off has now changed in favor of common carriage, which seems increasingly able to combine the best of both worlds.

5. Conclusions

Although the conclusions we may arrive at from comparative analysis of our rather preliminary case data are more suggestive than conclusive, several observations may be interpreted as supportive of our main hypothesis. These relate both to (i) the observed performance differences between the contractually organized Internet, the Correspondent System and the more hierarchically organized group of global network operators and to (ii) the evolving differences between the differently organized global service providers within the group of global operators. Concerning the former differences, the following tentative conclusion can be reached. By implicit contracting within the framework of one mandatory standard (TCP/IP) the Internet Society has created the kind of global seamless data network that the traditional telephone monopolies did not accomplish under the old Correspondent System, but may accomplish through the emerging global network organizations, thus partially bypassing their own and still working Correspondent System. As the Internet demonstrates, simple contracting may be the most cost-efficient and creative way of participating in the development and deployment of global networks, given that (i) the underlying critical assets (i.e., Internet-compatible network technology) are easily redeployable and tradable, and (ii) network complexity and operational uncertainty are at least moderate, or customers' performance requirements correspondingly relaxed. Furthermore, as demonstrated by Concert and the other global network operators, to achieve a sufficient level of worldwide coordination among somewhat less redeployable, but structurally more complex, networks and operationally more uncertain services for qualitatively more demanding customers, associations and non-equity joint ventures may suffice, whereas between the least redeployable, most complex networks and operationally most uncertain services for the most demanding customers, more hierarchical solutions will be needed.

Also the different development histories and performance levels of differently organized service providers turn out to be rather supportive of the above hypothesis. The less troublesome and seemingly more successful development of a broader menu of global services by Concert than by Unisource, WorldPartners and the late entrant Global One speaks for the effectiveness of using a more hierarchical governance structure in these instances compared with a looser alliance or contracting approach. It also demonstrates the difficulty of using a looser alliance or contracting structure in stabilizing and directing a larger number of diverse and sometimes competing individual companies. Global reach is important, but the multi-tier alliance structures are probably better suited for reaching globally than for providing synchronized bundles of customized service and support. For example, AT&T-Unisource seems better suited for providing global reach than C&W, which seems better suited for providing synchronized bundles of branded and customized services, while Concert (and recently Global One) represent a more balanced approach. Moreover, the more successful development of Concert compared with Cable and Wireless and Global One demonstrates the advantage of being a first mover in crafting regulatory-acceptable alliances with complementary partners, especially in a rapidly evolving and highly monopolized industry such as the telecom services industry.

Although Concert so far appears to be the most successful in developing global services, drawing any final conclusion as to the most efficient network organization would be premature. Neither can we exclude the possibility that, as internetworking technology improves and becomes more standardized, as previously bundled operations and services become less bundled, and as customer sophistication increases, most of the future coordination of global service provision will be taken care of by specialists in competitive access provision, systems integration, outsourcing and software applications, rather than by vertically integrated telephone companies.

References

[1] O.E. Williamson, The Economic Institutions of Capitalism. Firms, Market and Relational Contracting. New York. The Free Press, 1985.

[2] O.E. Williamson, Comparative Economic Organization: The Analysis of Discrete Structural Alternatives, *Administrative Science Quarterly* **36** (1991) 269-296.

[3] D.J. Teece, Profiting from Technological Innovation: Implications for Integrations, Collaboration, Licensing and Public Policy. In David J. Teece (ed.), The Competitive Challenge. Strategies for Industrial Innovation and Renewal. Cambridge, Mass. Ballinger Publishing Company, 1987, pp. 185-219.

[4] L. Guidoux, Intelligent Solutions for Data Communications in the Public Network, *Telecommunications*, October 1995.

[5] T. Hills and Graham Johnson, Telecoms Multinationals: The Future of Global Service Alliances? Analysys Publication, 1996.

[6] Forrester Research, Inc., 1996, Coping With Global Telecom, http://www.forrester.com.

[7] G. Johnson, Michael Denmead and Susan Ablett, Strategic Alliances Between Telecom operators. Analysys Publication, 1994.

[8] P. Kyle, *Wall Street Journal*, Sept. 27 (1995).

[9] P.T. Spiller, S. Ulset and B. Zelner, Strategy, Structure and Global Networks. Studies of the Telecommunications Sector. SNF-report, Bergen, Norway, 1996.

Telecommunications Transformation: Technology, Strategy and Policy
E. Bohlin and S.L. Levin (Eds.)
IOS Press, 1998

CHAPTER 9

Capabilities in the Multimedia Regime: The Case of Telecommunications Operators[1]

Massimo G. Colombo
Università di Pavia and CIRET – Politecnico di Milano

Paola Garrone
CIRET – Politecnico di Milano
Piazza Leonardo da Vinci 32, 20133 Milan, Italy

Abstract. In this paper, we use a competence-based approach to analyze the strategic behavior of telecommunications operators in multimedia. We claim that on-line interactive multimedia (OIM) services represent a new technological paradigm in the communication industry. While recognizing that the supply of OIM services requires a wide collection of resources relating to the various stages of the multimedia value chain (information contents, service provision, network, software, communication equipment, terminals), we highlight the key interface role of the service provision function. We argue that operators of telecommunications networks which target the service provision stage can leverage some peculiar capabilities, in the technological and managerial spheres. In order to develop the lacking assets and competencies, the set-up of a network of cross-industry alliances is crucial. We present data on the behavior of telecommunication operators in multimedia in the 1990s which have been obtained from the MOSAICO database. MOSAICO surveys news published by the international financial press in the years 1990-95 relating to the communication industry.

1. Introduction

The emergence of on-line interactive multimedia (OIM) services can be described as a new technological paradigm [1], defined by a class of new technoeconomic problems, a new pool of technologies (techniques, competencies and rules), and a set of shared assumptions. The paper claims that the core of such a major shift in the evolution of information and communications services is the service provision function, even if the supply of a OIM service needs a wide collection of assets and capabilities pertaining also to information contents, network infrastructure, software, communication equipment, and terminals. A service provider may be defined as a systems integrator, that is, one who acquires ready-to-use applications or raw blocks of information from content providers, presents the informa-

[1] The financial support of the Università di Pavia 1996 FAR funds and the CNR project "Gestione strategica dell'innovazione" is gratefully acknowledged. Although the paper is the result of the joint effort of the authors, Sections 1, 2, and 3 are directly attributable to Massimo Colombo, and the remaining sections to Paola Garrone.

tion to consumers, decides pricing strategies, allows browsing and choice of applications of interest, selects network resources and collects payment.

According to the evolutionary and competence-based literature, material and immaterial resources that firms have cumulated along their traditional growth path are likely to drop in value when a new technological paradigm arises. Accordingly, organizations aiming to exploit the opportunities offered by OIM services have to build new capabilities and reformulate their traditional pattern of success. By zooming on the operators of telecommunications networks (common carriers or telecoms), it is shown that though leveraging a few peculiar capabilities in the technological and managerial spheres, they are trying to develop lacking assets and competencies through the set-up of a network of collaborative relations with firms in converging industries (mainly software producers, service providers, broadcasters, and media firms).

Data used in the empirical section of the paper are extracted from the MOSAICO database, developed by Centro Studi San Salvador and CIRET-Politecnico di Milano. MOSAICO surveys news published by the international financial press (Financial Times, Wall Street Journal US edition, Wall Street Journal Asian edition) in the years 1990-95 relating to the communication industry. News are classified in a standardized format suitable to statistical elaboration. The information provided by MOSAICO on the recorded initiatives includes the type of coordination mechanism to which firms resort (autonomous investment, acquisition, joint collaboration, and so on), the functional content (R&D, production, marketing, etc.), the identity, the home country and the main sector of activity of the firms involved by the initiative.

The paper is organized as follows. Section 2 provides some definitions and illustrates the value chain for the supply of OIM services; the questions whether and why interactive multimedia is a new technological paradigm are also discussed. In Section 3 we propose a capability-based approach to the analysis of the entry strategies of common carriers into OIM services. In Section 4 the data set is described. Section 5 is devoted to the findings of the empirical analysis. The main conclusions of the paper are summarized in Section 6.

2. On-line Interactive Multimedia: The Key Role of the Service Provision

Multimedia services comprise a set of information and communication services, for both private and business users, characterized by the contextual presence of a variety of information types, that is, data, voice, sounds, moving images and graphics.[2] According to such a definition, one could question whether multimedia services are something of a novelty, since services as TV or movies are by definition characterized by the contextual presence of different information types. A finer definition, and a possible classification, of multimedia services is obtained by adding two more dimensions in the characteristic space of information services. First, whether users and information providers are linked by a communication

[2] According to [2] "multimedia relates to the information type presented. Working with multimedia systems enables the user to create, edit, transmit, receive, store, retrieve, compute and delete two or multiple types of information, such as data, text, vector graphics, pixel-oriented images, video signals (moving pictures), and audio signals (voice and sound)".

network or not is a distinction worth being made; second, whether users can interact with the source of information or not is also relevant.

One should therefore distinguish between *on-line* and *off-line* multimedia services (e.g. TV broadcasting vs. movie theaters, or electronic shopping vs. educational CD-ROMs), and between diffusive and interactive multimedia services (e.g. TV broadcasting vs. electronic shopping, or movie theaters vs. educational CD-ROMs). In this paper we focus on *on-line interactive multimedia (OIM) services*. In what follows we will set forward the argument that while diffusive multimedia services (notably digital TV) belong to the technological trajectory historically followed by the media industry, OIM services do represent a change of technological paradigm, in accordance with the definitions proposed by [1].[3] Consequently, the strategic behavior of firms and the modes for organizing economic activity will differ according to the type (diffusive or interactive) of multimedia considered. Before addressing such issues, let us briefly present the organization of the multimedia value chain.

Experts in communication technologies generally agree that the multimedia value chain can be broken down into at least five main macro stages, corresponding to different technological functions. It is commonly held that in order for a multimedia telecommunication service to be provided, one would require at least: (1) a source of information (telephone conversations, teleworking, databases, film archives, electronic newspapers, etc.), (2) an access, storage, presentation and distribution point for the information content, (3) a network to transmit the information to the vicinity of the end-user, (4) an access network and (5) a user terminal. Two auxiliary stages also have to be introduced, as software and hardware components are inputs used in most of the five main stages.

For the sake of synthesis, we prefer to address only the stage in which definitions remain open, that is, the service provision.[4] In general and abstract terms, the service provider designs and administers interfaces allowing users to access to multimedia applications [6]. More specifically, service providers acquire applications or blocks of information from content providers, present the information to consumers, decide pricing strategies, allow browsing and the selection of applications of interest, and collect payment [7].

Collins et al. [8] identify various activities (called *packaging*) similar to those described above, but emphasize that service provision can also be linked to other functions termed *manipulation* and *storage*. These latter functions are software-based and (i) guarantee that the connections user-user or user-information source meet the quality, security and reliability demands of different users, (ii) provide by way of "agents", which may be common to the system or user-specific, the range of applications demanded by various users,[5] and (iii) store and maintain information. It should be noted that if technological solutions prevail in which information manipulation and storage intelligence reside on the public network, inte-

[3] A technological paradigm is "a model and a pattern of solution of selected technological problems, based on selected principles derived from natural sciences and on selected natural technologies" [1]. It is "an outlook, a definition of the relevant problems, and a pattern of enquiry" [3]. A technological trajectory is referred to as describing "the activity of technological process along the economic and technological trade-offs defined by a paradigm". See also [4].

[4] For a more detailed analysis of the multimedia value chain, see [5].

[5] Agents are modules of the multimedia communication system responsible for guiding the user between the sources of information while disclosing the storage and processing technologies as little as possible. Their design includes artificial intelligence functions, in order to learn individual perceptions and preferences. See again [2].

gration between service provision and network operation will be possible [2]. In addition, it is well known that the computer industry is working on technological solutions in which the terminals will be partly responsible for manipulation [8]; the transformation of raw content into applications is also to be included in the manipulation process, resulting in a degree of overlap with the content supply segment.

A useful metaphor for service provision is that of an *information supermarket*, suggested, among others, by [9]. This type of supermarket stores both plug-and-play applications and pick-and-mix information, offers its own and other brand products, provides easy access and one-stop billing and, finally, is seen as a reliable supplier (in contrast to the variable performances of other solutions, such as Internet, which, extending the metaphor, are defined as a bazaar).

Elsewhere [10] we argue that OIM services are to be considered as a new technological paradigm in the communication industry. An in-depth analysis of such argument is beyond the purpose of the present paper. Suffice it here to mention that OIM services take advantage of new material technologies and provide new opportunities for fulfilling latent and ill-answered needs. As regards the latter aspect, three characteristics of OIM services are worth being mentioned. First, interactivity allows users to customize the contents of the service in accordance with their own peculiar preferences; this substantially increases the variety of the available consumption choices even for marginal users. Second, the combination of different types of information, which is typical of multimedia services, improves the quality of communication; this makes them useful in supporting activities which previously required direct human contact.[6] Third, on-line connections nullify space and time distances among users and between these and information sources; this provides the opportunity of creating new families of services such as e-mail, virtual communities, virtual museums and other cultural and educational services. As to technology, OIM services take advantage of a series of radical innovations in converging fields, such as the digitization of signals, satellite and fiber-optic based transmission systems, algorithms for signal compression and control, switching and storage devices, and others, whose combination has a supra-additive synergistic effect. Nonetheless, it is the authors' opinion that what makes OIM services peculiar with respect to the traditional technological trajectory of the communication industry is the software-based, system integration technology specific to the service provision stage. Such technology is crucial in supporting the interface activities of the service provision which were previously described, and is characterized by a set of heuristics concerning both the definition of relevant problems and the pattern of inquiry for new technological solutions which differs considerably from the traditional one of the telecommunication industry (largely shaped by network engineering technology).

3. Entry of Common Carriers into Multimedia: A Capability-based Approach

As was highlighted earlier in the paper, the multimedia service provider may be partly responsible for managing network intelligence, develops and manages the software interfaces between users, content providers and the technological infrastructure, gains access to and packages contents and retails services.

[6]An example is offered by teleworking.

What are the distinctive competencies required to be successful in the service provision field? How can they be put together?[7] A capability-based approach to the firm and its boundaries seems to effectively capture the observed distinctive features of firms' entry behavior (see [12, 13]).

There is a number of different firms (common carriers, CATV network operators, TV broadcasters, media firms, firms in the electronic and data processing industries, and so on) which do possess some key assets and competencies necessary to effectively perform the service provision function. In fact, some of the assets and distinctive competencies they have built in their core business are largely applicable to such activity. Since the marginal cost of using such assets for service provision is rather low compared with the expected returns, entering into the service provision business would allow them to improve the allocation of their resources.

Nonetheless, in order for entry to be successful, the development of new distinctive capabilities is also required. In this perspective, we contend that cooperative agreements (equity and non-equity) are in most instances far more efficient for this purpose than both arm's-length arrangements and hierarchical integration.

For one thing, it is very difficult for other firms to decode a firm's core competencies and reproduce them in a different context by means of internal investments. Obstacles to acquisitions are also numerous; they include information asymmetries between the acquirer and the target firms, and the non-separability of the assets and capabilities the acquirer is looking for from the remaining ones of the target firm (see [14]). Further, market coordination mechanisms are likely to perform rather poorly according to both static and dynamic efficiency criteria. First, provision of OIM services suffers from externality problems (interdependent values of the required assets, direct and indirect network externalities, technological spillovers in research, development and testing activities), which are a source of market failures. Second, in a world of substantial technological and market uncertainty, contracts governing relations among firms would be highly incomplete: this would lead to moral hazard problems. In addition, in order to coordinate the behavior of agents which are very likely to have divergent expectations and different priorities, the need for considerable relation-specific investments (especially in human capital) arises; under such circumstances, market relations suffer from high dynamic transaction costs [15, 16]. Lastly, the short-term horizon of arm's-length arrangements is a serious obstacle to interactive learning; in order to capture cross-firm economies of cognitive scope [17] which are instrumental to building new distinctive capabilities, more robust collaborative arrangements are required.

In this paper the focus is on operators of telecommunications networks. We contend that common carriers aiming to exploit the opportunities offered by interactive multimedia face a fundamental strategic choice. On the one hand, they can decide to concentrate on the network stage of the multimedia value chain by exploiting their traditional distinctive capabilities in network management: they would sell a commodity-type input (i.e., the transportation and distribution of digital signals) to other firms responsible for retailing multimedia services to final consumers.

[7] On this issue, see also [5] and [11].

On the other hand, common carriers can also integrate downstream by internalizing the service provision function. Such a strategic move is to be regarded at the same time as an adaptive *and* radical change. It is adaptive in so far as the choice of a service provision strategy is compatible with, and can efficiently hinge on, the collection of specific competencies and resources cumulatively developed by common carriers in their core business, relating to both the technological (i.e., management of network intelligence) and managerial spheres (especially the administrative capabilities to manage a large user base, pricing, billing, customer support services). Nevertheless, weaknesses are likely to include software technology (e.g. user-friendly interfaces and navigation tools), selection and packaging of applications, and the ability to segment final markets and sell differentiated services to (both business and private) customers. In the light of such considerations, in order to enter into multimedia service provision, telecommunications operators should improve and extend their distinctive capabilities. For this purpose, the ability to design, set up and manage a network of alliances aimed at interactive learning with firms in a variety of converging industries (software producers, specialized service providers, broadcasters, media firms) will be crucial in determining competitive success or failure.

4. The Data

Data concerning common carriers' behavior in multimedia are obtained from the MOSAICO database. We assumed multimedia to include video services, videotelephony, integrated services, multimedia interactive services, multimedia applications and contents, ATM switching nodes and technology, broadband access networks and local loops, cable and satellite TV networks, broadband SDH/SONET, video headend, non-voice terminals (including set-top boxes and PC work-stations), and software.[8] The surveyed initiatives comprise a) greenfield investments, b) M&As, c) acquisitions of non-controlling shareholdings, d) joint ventures and equity consortia, and e) non-equity agreements. This latter category includes on the one hand joint collaborative agreements, on the other licenses, technology transfer agreements, customer-supplier relations, and other commercial agreements. In what follows, this latter category will be referred to as non-equity quasi-market agreements.

The events included in the data set were classified according to the stage of the multimedia value chain to which they refer; a distinction was made between a) terminals, b) communication equipment, c) network, d) software, e) video services, f) integrated services, g) multimedia interactive services, and h) content and applications. They concern research, development, standardization activities, field trials, production, installation, commercialization, maintenance, and post-sale assistance of a multimedia system or service.[9] In addition, the identity, the home country, and the main sector of activity of the firms involved, were also recorded.

As far as common carriers are concerned, MOSAICO surveys 2,674 initiatives in the period under scrutiny. Non-equity agreements constitute more than half of the recorded initiatives, with the largest share (40.7%) being quasi-market arrangements. Autonomous in-

[8] Unfortunately, MOSAICO presently does not provide a more detailed classification of software.

[9] It is worth noticing that multiple combination of stages was allowed. The same holds true for functional activities.

vestments account for a 16.2% share, while joint ventures are 12.5% of the total number of initiatives. M&As and acquisitions of minority interest represent 10.1% and 7.3% respectively. Of the surveyed initiatives, 760 (that is to say, almost one out of three) refer to the multimedia value chain.

5. The Empirical Evidence

Should carriers limit themselves to network operation, a concentration of initiatives in the network and communication equipment stages of the multimedia value chain would result. As regards the initiative typology, one would expect internal investments in network infrastructures, acquisitions of other network operators (both common carriers and CATV operators), and arm's-length or non-equity quasi-market transactions aimed at selling a commodity-type service to other firms, to prevail. Renewal investments should also trigger quasi-market (i.e. subcontracting) agreements with equipment suppliers. In addition, the desire to control indirect network externalities may push carriers to enter multi-partner standardization agreements.

Table 1. Number of Initiatives of Telecommunications Operators (TOs) in the Stages of the Multimedia Value Chain, 1990–95

STAGE OF THE MULTIMEDIA VALUE CHAIN	*No.*	*%*
Terminals	73	2.7
Communication equipment	103	3.9
Network	133	5.0
Video services	74	2.8
Integrated services	22	0.8
Multimedia interactive services	277	10.4
Software	108	4.0
Contents	32	1.2
TOTAL IN MULTIMEDIA	760	28.4
TOTAL IN THE COMMUNICATION INDUSTRY	2,674	100.0

Source: Database MOSAICO, Centro Studi San Salvador-Telecom Italia and CIRET-Politecnico di Milano.

By contrast, the evidence provided by Tables 1 and 2 seems to support the argument that common carriers are indeed targeting the multimedia service provision function. A considerable number of the initiatives undertaken by common carriers in multimedia relates to the multimedia interactive service provision and/or software stages of the value chain. In particular, in the former stage MOSAICO surveys 277 events (that is more than 10% of the surveyed initiatives in the communication industry). In addition, Table 2 highlights the rapidly growing interest of common carriers in the service provision. The ratio of events in the multimedia service provision stage to the total number of initiatives in the communication industry substantially increases in the period under scrutiny (from 2.6% in 1990 up to almost 14% in 1995).

Table 2. Evolution of the Share of Initiatives of TOs in Multimedia Service Provision Out of the Total Number of Initiatives in the Communication Industry

YEAR	*MSP (a)*[1)]	*Total in the communication industry (b)*	*(a)/(b) %*
1990	7	274	2.6
1991	12	267	4.5
1992	53	554	9.6
1993	60	520	11.5
1994	67	499	13.4
1995	78	560	13.9

Source: Database MOSAICO, Centro Studi San Salvador-Telecom Italia and CIRET-Politecnico di Milano.
[1)] Multimedia interactive service provision.

Table 3. Number of Initiatives of TOs in Multimedia Service Provision by Type of Service

TYPE OF SERVICE	*1990–92*		*1993–95*		*1990–95*	
	No.	*%*	*No.*	*%*	*No.*	*%*
Entertainment	8	11.1	65	31.7	73	26.4
Business	24	33.3	46	22.4	70	25.3
Information	29	40.3	44	21.5	73	26.4
Commercial transactions	17	23.6	35	17.1	52	18.8
Educational	0	0.0	8	3.9	8	2.9
N.e.c.[1)]	3	4.2	44	21.5	47	17.0
TOTAL	72	100.0	205	100.0	277	100.0

Source: Database MOSAICO, Centro Studi San Salvador-Telecom Italia and CIRET-Politecnico di Milano.
[1)] N.e.c.: not elsewhere classified.

As is evident from the data reported in Table 3, there seem so far to be no "killer applications": instead, common carriers are experimenting with a wide spectrum of services. Accordingly, a considerable number of initiatives (between 50 and 70) has been surveyed in all segments (entertainment, business services, information services, transaction and commercial services), with the exception of education. Nevertheless, it appears that in the more recent period increasing attention has been devoted by common carriers to entertainment services, mainly in connection with activities in interactive TV aimed at exploring alternative technological solutions and testing market demand. It should also be noted that in the period 1993–95 there has been a considerable number of events relating to the service provision stage in the "not elsewhere classified" category: the share accounted for by such category augments from 4.2% in 1990–92 to 21.5% in 1993–95. This may possibly be a consequence of the development of Internet-based services, which actually are quite difficult to classify in a specific use category. Of course, additional evidence would be required as regards this issue. Coming back to Table 1, we also notice the limited number of initiatives focused on content and applications (32). Such a result suggests that there may be constraints and difficulties largely inhibiting common carriers' downstream integration in this stage.

Furthermore, interesting evidence concerning the type of initiatives prevalently chosen by common carriers in the different stages of the value chain is offered by Table 4. If one compares the distributions by type of the surveyed events in the various stages of the multimedia value chain with each other, and with the distributions in the communication industry, it clearly turns out that there are significant biases. In other words, the types of initiatives on which firms mostly rely do depend on the characteristics of the various stages. In order to make such biases evident, we computed initiative type specialization indices; their values are illustrated in Table 5. A value greater (lower) than 1 indicates that use of a particular coordination mechanism by common carriers is more (less) frequent in a given stage of the multimedia value chain than in general in the communication industry. The main observations as to this issue can be synthesized as follows:

- In accordance with the discussion in the previous paragraphs, in the multimedia service provision stage equity and non-equity joint collaborations together account for more than 50% of the surveyed initiatives. With a 33.9% share, the importance of non-equity joint collaborations is almost three times as large as in the communication industry globally considered. In addition, while the share of this category is stable over the period taken into consideration, that of joint ventures is on the rise.[10] Joint collaborations are also slightly above the average multimedia value (and largely above the average communication industry value) in software (23.1%) and contents (25.0%).[11]
- In the software stage, non-equity agreements account for 70% of the recorded events, especially due to the relevance of quasi-market transactions (mostly licenses and subcontracting agreements). Unfortunately, due to limits in the classification presently used by MOSAICO, we are not able to distinguish relations concerning standardized, commodity-type packages, which would be obtained through quasi-market arrangements, from those regarding specialized interface, navigation and other enabling tools (i.e. specialized middleware), which play a key role for the multimedia service provision function and should call for more robust governance modes. Additional research is needed to provide further insights into this interesting issue.
- Quite unsurprisingly, non-equity quasi-market agreements prevail in the upstream stages of the value chain (especially terminals and communication equipment, in addition to software); this can no doubt be attributed to the relevance of subcontracting relations.
- Generally speaking, acquisitions of both controlling and minority interests are outnumbered by the other types of initiatives. An exception is given by video services and, more surprisingly, by the contents stage.

[10] Joint ventures account for a 13.2% share in 1990–92 and for a 17.5% share in 1993–95. Note also that in the communication industry, the share of joint ventures marginally increases (from 11.9% to 12.3%) and that of non-equity joint development agreements drops quite substantially (from 12.1% to 9.4%).

[11] In this latter case, the limited number of initiatives (31) invites us to be cautious.

- Lastly, it is worth emphasizing the quite substantial share accounted for by autonomous investments in all stages of the multimedia value chain. Quite remarkably, the values recorded in the multimedia service provision (18.1% of the total number of initiatives in those stages, respectively) is above the average multimedia value (15.4%) and also above the average value in the communication industry (16.2%). Such evidence points to the complementary nature of autonomous investments with respect to external linkages, with the former being instrumental to developing an "absorptive capacity" [18] fundamental to taking advantage of the latter.

Let us now classify initiatives undertaken by telecoms in the multimedia service provision stage as to type and functional content. Table 6 reports the number of initiatives relating to research, development and testing (RDT) activities and their percentage share out of the total. Specialization indices are also calculated. They serve the purpose of pointing out whether and to what extent a given type of initiatives in the multimedia service provision stage is biased towards RDT activities, while controlling at the same time for the general correlation between governance modes and functional contents (independently of the specific stage). A value greater (lower) than 1 indicates a positive (negative) association between the type of initiative and the functional content concerned; the higher the value the stronger the association. The data provided in the tables strongly support the explorative nature of common carriers' activity in the multimedia service provision stage. Strategic moves in this stage are clearly oriented towards research, development and testing of new services; this applies not only to initiatives such as non-equity joint collaborations which generally entail such kinds of functional content, but also to autonomous investments, joint ventures, and non-equity quasi-market agreements. All such types of initiatives are characterized by values of the functional content specialization index greater than 2.

Lastly, we earlier proposed the argument that the capability-enhancing activity in which common carriers are involved in the multimedia service provision stage implies the establishment of cooperative relations with a variety of agents, which possess different core competencies and specialized assets. For this purpose, it is worth analyzing the identity of partners of both equity and non-equity joint cooperative arrangements. For the sake of simplicity, we have limited the analysis to bilateral relations. Partners of common carriers are dispersed among a variety of industries, with the most important ones being media and software. The former account for 16 agreements out of the 84 surveyed; the latter account for 18. On the contrary, arrangements involving another common carrier are limited to a negligible share (8%). This could be compared with the considerably higher share (24%) accounted for by intra-industry (that is, within the telecommunication service industry) agreements out of the total number of bilateral joint collaborations realized by common carriers in the 1990s.

Table 4. Number of Initiatives of TOs in the Stages of the Multimedia Value Chain by Type, 1990–95

STAGE OF THE MULTIMEDIA VALUE CHAIN	*Autonomous investments*		*M&As*		*Acquisitions of minority shareholdings*		*Joint ventures and equity consortia*		*Non-equity joint collaborations*[a]		*Non-equity quasi-market agreements*[b]		*TOTAL*	
	No.	*%*	*No.*	*%*	*No.*	*%*	*No.*	*%*	*No.*	*%*	*No.*	*%*	*No.*	*%*
Terminals	14	19.2	3	4.1	2	2.7	2	2.7	10	13.7	42	57.6	73	100.0
Communication equipment	15	14.6	0	0.0	3	2.9	5	4.9	9	8.7	71	68.9	103	100.0
Network	25	18.8	2	1.5	9	6.8	25	18.8	23	17.3	49	36.8	133	100.0
Video services	4	5.4	18	24.3	17	23.0	19	25.7	5	6.7	11	14.9	74	100.0
Integrated services	4	18.2	0	0.0	1	4.5	2	9.1	10	45.5	5	22.7	22	100.0
Multimedia int. services	50	18.1	14	5.1	15	5.4	45	16.2	94	33.9	59	21.3	277	100.0
Software	11	10.2	6	5.6	9	8.3	6	5.6	25	23.1	51	47.2	108	100.0
Contents	5	15.6	7	21.9	3	9.4	5	15.6	8	25.0	4	12.5	32	100.0
TOTAL IN MULTIMEDIA	117	15.4	48	6.3	56	7.4	97	12.8	165	21.7	277	36.4	760	100.0
TOTAL IN THE COMMUNICATION INDUSTRY	432	16.2	271	10.1	194	7.3	335	12.5	352	13.2	1,090	40.7	2,674	100.0

Source: Database MOSAICO, Centro Studi San Salvador-Telecom Italia and CIRET-Politecnico di Milano.

a) Including non-equity consortia.

b) Including supply agreements, licenses, and other commercial agreements.

Table 5. Initiative Type Specialization Indices in the Stages of the Multimedia Value Chain

STAGE OF THE MULTIMEDIA VALUE CHAIN	*Autonomous investments*	*M&As*	*Acquisitions of minority shareholdings*	*Joint ventures and equity consortia*	*Non-equity joint collaborations*[a]	*Non-equity quasi-market agreements*[b]
Terminals	1.2	0.4	0.4	0.2	1.0	1.4
Communication equipment	0.9	0.0	0.4	0.4	0.7	1.7
Network	1.2	0.2	0.9	1.5	1.3	0.9
Video services	0.3	2.4	3.2	2.1	0.5	0.4
Integrated services	1.1	0.0	0.6	0.7	3.5	0.6
Multimedia int. services	1.1	0.5	0.8	1.3	2.6	0.5
Software	0.6	0.5	1.2	0.4	1.8	1.2
Contents	1.0	2.2	1.3	1.3	1.9	0.3
TOTAL IN THE COMMUNICATION INDUSTRY	1.0	1.0	1.0	1.0	1.0	1.0

[a] Including non-equity consortia.
[b] Including supply agreements, licenses, and other commercial agreements.
The initiative type specialization index ti_{ij} relating to the type i initiative and the j stage is calculated as follows.
Let us denote with i_{ij} the number of type i initiatives in the j stage and I_i the total number of type i initiatives in the communication industry. We obtain: $ti_{ij}=[\,i_{ij}/\Sigma_i\, i_{ij}]/[I_i/\Sigma_i\, I_i]$, i=1,....,6; j=1,....,8.

Table 6. Initiatives Relating to Research, Development and Testing Activities Undertaken by TOs in the Multimedia Service Provision Stage by Type, 1990–95[a]

TYPE OF INITIATIVE	*No. of initiatives*		*Specialization index*[c]
	No.	*%*[b]	
Autonomous investments	16	35.6	2.2
Joint ventures and equity consortia	13	29.5	2.4
Non-equity joint collaborations	53	57.6	1.2
Non-equity quasi-market agreements	6	10.2	3.0
TOTAL IN MULTIMEDIA INT. SERVICES	88	36.7	2.7

[a)] M&As and acquisitions of minority shareholdings are not considered, as the functional contents coincide with those of the target firm.
[b)] [No. of initiatives of type i relating to RDT in the multimedia service provision]/[total no. of initiatives of type i in the multimedia service provision].
[c)] Functional content specialization indices fc_i are calculated as follows. Let us denote with x_i and y_i the number of initiatives of type i relating to RDT activities in the multimedia service provision stage and in the communication industry, respectively. Let us also denote with X_i and Y_i the total number of initiatives of type i in the multimedia service provision stage and in the communication industry, respectively. Then $fc_i=[x_i/X_i]/[y_i/Y_i]$, i=1,2,3,4.

6. Conclusions

In this paper we have argued that the introduction of on-line interactive multimedia (OIM) services entails the emergence of a new technological paradigm in the communication industry. We have also contended that at the core of such a major technological shift lies the service provision function. Accordingly, we have highlighted the key interface role of the service provider in spite of the fact that the supply of OIM services requires a wide collection of assets relating to the various stages of the multimedia value chain (information contents, network, software, communication equipment, and terminals, besides the service provision).

Relying on a competence-based approach, we have analyzed both theoretically and empirically the behavior of telecommunications operators in multimedia. We have opposed a MSP (multimedia service provision) strategy, which implies the internalization of the key service provision function in order to offer differentiated services to final customers, to a NWO (network operation) strategy, focused on the development of the network infrastructure and the sale of a commodity-type intermediate service to other firms. We have claimed that telecommunications operators own some of the specialized resources needed to perform the service provider role, in both the technical (management of network intelligence), marketing and sale (direct access to a large base of captive potential subscribers, well-recognized brand), managerial (administrative capabilities to manage a large user base) and financial spheres. However, we have also stressed that in order for them to develop the specialized complementary assets and capabilities required to implement such a strategy, the ability to design, set up, and manage a network of alliances with firms in converging industries is crucial.

The paper has provided extensive empirical evidence on the behavior of common carriers in multimedia in the 1990s through the analysis of data from the MOSAICO database. The findings clearly show that in the period examined telecommunications operators have been targeting the service provision function: a considerable number of their initiatives in multimedia relates to the service provision. In addition, it seems that telecommunications operators are rapidly consolidating their position in such a key stage of the value chain. This supports the view that the MSP strategy is preferable for common carriers than a strategy confined to the operation of information highways, even though additional evidence on the performances of telecommunications operators in service provision would be useful. In addition, the findings of the study show that the set-up and management of a network of equity and above all non-equity alliances with firms in a variety of industries (especially software producers, specialized service providers, broadcasters and media firms) are an essential building block of the MSP strategy. Cooperative relations with such firms allow common carriers to capture cross-firm economies of cognitive scope, thus supporting the exploration of the opportunities potentially offered by the new technological paradigm. In addition, they avoid both the (static and dynamic) inefficiencies of arm's-length transactions and the pitfalls of hierarchical mechanisms. In this way, they are instrumental to building a sustainable competitive advantage in service provision.

References

[1] G. Dosi, Technological Paradigms and Technological Trajectories. A Suggested Interpretation of the Determinants and Directions of Technical Change, *Research Policy* **11** (1982) 147-162.

[2] H. Armbrüster and K. Wimmer, Broadband Multimedia Applications Using ATM Networks: High-Performance Computing, High Capacity Storage, and High-Speed Communication, *IEEE Journal on Selected Areas in Communications* **10** (1992) 1382-1396.

[3] G. Dosi, Sources, Procedures, and Microeconomic Effects of Innovation, *Journal of Economic Literature* **XXVI** (1988) 1120-1171.

[4] R.R. Nelson and S.G. Winter In Search of a Useful Theory of Innovation", *Research Policy* **6** (1977) 36-76.

[5] M.G. Colombo and P. Garrone Entry Strategies of a Telecommunications Carrier in the Multimedia Industry, *Communications & Strategies*, **19** (1995) 227-250.

[6] R. Franz and M. Scholz, Interactive Video – Services, Success Factors and Network Solutions, *XV International Switching Symposium Proceedings*, 1995, 2.

[7] DAVIC Description of Davic Functionalities. The Digital Audio Visual Council, London, 1995.

[8] D. Collis, P.W. Bane and S.P. Bradley, Winners and Losers. Industry Structure in the Converging World of Telecommunications, Computing and Television. Draft. Harvard Business School, Cambridge, Mass., 1995.

[9] I.W. Marshall and M. Bagley, The Information Services Supermarket – An Information Network Prototype, *BT Technology Journal* **13** (1995) 132-142.

[10] M.G. Colombo and P. Garrone, La rivoluzione del multimediale alla luce delle teorie economiche evolutive. Caratteristiche paradigmatiche, fattori chiave di successo e implicazioni per le condotte delle imprese. Draft. CIRET-Politecnico di Milano, 1997.

[11] M.G. Colombo and P. Garrone, Common Carriers' Entry into Multimedia Services, *Information Economics and Policy*, forthcoming (1997).

[12] E.T. Penrose, The Theory of the Growth of the Firm. Oxford: Basil Blackwell & Mott, 1959.

[13] B. Wernerfelt, A Resource-based View of the Firm, *Strategic Management Journal* **5** (1984) 171-180.

[14] N. Kay, Collaborative Strategies of Firm. Theory and Evidence. In: A. Del Monte (ed.), Recent Developments in the Theory of Industrial Organisation. London: Macmillan, 1992, pp. 201-31.

[15] R. N. Langlois, Transaction-costs Economics in Real Time, *Industrial and Corporate Change* **1** (1992) 99-127.

[16] R. Langlois and P.L. Robertson, Firms, Markets and Economic Change. London and New York: Routledge, 1995.

[17] B. Nooteboom, Towards a Dynamic Theory of Transactions, *Journal of Evolutionary Economics* **2** (1992) 281-299.

[18] W.M. Cohen and D.A. Levinthal, Innovation and Learning: The two faces of R&D, *Economic Journal* **99** (1989) 569-596.

Telecommunications Transformation: Technology, Strategy and Policy
E. Bohlin and S.L. Levin (Eds.)
IOS Press, 1998

CHAPTER 10

Electronic Network Use and Coordination Between Producers and Suppliers: A French and US Comparative Analysis[1]

Charles Steinfield
Department of Telecommunication, Michigan State University
East Lansing, Michigan 48824, USA

Laurence Caby and Christine Jaeger
Centre National d'Étude des Telecommunications
38-49 rue du General Leclerc, 92131 Issy-les-Moulineaux Cedex, France

Robert Kraut
Department of Computer Science, Carnegie Mellon University
Pittsburgh, Pennsylvania 15213, USA

Abstract. We investigate the patterns of use of electronic data networks for transactions between producers and suppliers, comparing usage in France and the United States. We first examine whether the longer experience with an ubiquitous public data network has influenced the degree to which French firms use inter-firm networks. We then explore the implications of network use in both countries using a fundamental assumption of transaction cost economics: that firms make goods and services rather than buy them to reduce costs of coordinating with their suppliers. Based on this notion, it has been proposed that the use of electronic networks would lead to greater outsourcing, through lowered costs of coordination with the market. The approach of the present paper is to identify conditions under which firms in both countries use (a) in-house production, (b) electronic networks, and (c) interpersonal networks to coordinate with suppliers of important goods and services. Findings show that, despite lower overall network use in France, the factors that influenced network use, and the impact of such use on interfirm relationships, are quite similar in both countries. The findings support transaction cost notions that firms choose an in-house strategy to avoid the threat of opportunistic behavior. However, contrary to much recent speculation, the use of electronic networks for transactions was not associated with increased outsourcing. Rather, all three modes of coordination – in-house production, electronic network use, and interpersonal relationships – were complementary. Surprisingly, only interpersonal connections were associated with better outcomes of transactions, with some indication that greater use of electronic networks was associated with lower-quality transactions.

[1] This paper presents findings from a two-year research project jointly sponsored by the US National Science Foundation, Grant IRI-9408271, and the Centre National d'Étude des Telecommunications in France. Kraut, Steinfield, Plummer, Butler, and Hoag [1] provide a detailed analysis of the US data.

1. Introduction

With the advent of EDI and now corporate inter- and intranets, information networks increasingly are used to support inter-firm transactions. Electronic transactions can promote new revenues, increase efficiencies and allow broader market reach for sellers, while affording greater convenience and access to value-added information services for buyers. Despite these advantages, and contrary to popular wisdom, the actual incidence of electronic data networks that link across firms is somewhat less than anticipated to date.

The purpose of the present research is to provide a quantitative analysis of the factors that influence firms to participate in electronic data networks that extend beyond their organizational boundaries. A France – US comparison was chosen to investigate whether or not the existing network infrastructure alters the incentives to participate in interfirm networks. In the US, corporate intranets and the internet are quickly gaining acceptance among businesses. Moreover, costs for private networks are among the lowest in the world. In France, the Teletel network has long provided one possible public interfirm data network infrastructure, while costs for traditional private networks remain somewhat higher than in the US.

The paper is organized as follows. We first outline the methods used to obtain the data on which the paper is based. We then provide a snapshot of the relative penetration of interfirm networks in France and the United States across four industries. The four industries – advertising, apparel, pharmaceutical manufacturing, and magazine publishing – were chosen to represent information vs. non-information, and goods vs. service product attributes. Using a theoretical framework adapted from transaction cost economics, we then explore whether the same set of factors in each country influences the propensity of firms to use data networks for transactions involving important inputs to each firm's production processes. Finally, we examine how different modes of coordinating with suppliers influence the outcomes of transactions.'

2. Methods

In mid-1996 and early 1997, two telephone surveys of managers were completed, first in the United States and then in France. A sample of 250 firms in the US and 256 firms in France was randomly drawn, using commercial industry databases. The sample was stratified by industry and size, by selecting an equal number of small, medium and large firms (with size relative to each industry and a minimum size of 20 employees for inclusion in the sample) in the advertising, apparel, pharmaceutical and magazine publishing industries. Firms that were producers were selected for inclusion in the study.

The telephone survey focused on the way in which the producing firm acquired key inputs from suppliers. Respondents were asked to identify their company's principal product or service, and then were asked about a specific input needed to produce this product. Inputs were chosen to vary on whether they were products or services and whether they were tangible or intangible, enabling us to randomly sample across four types of transactions for each industry. Depending on the question, respondents were asked about the way they acquired this input in general or about the way they dealt with the "the most important supplier you've worked within the past 12 months". We stressed that the suppler could be an employee or department in the respondent's firm or an outside individual or company.

The survey included items that measured features of the input itself, features of the relationship between the producing firm and its supplier for the input, the modes of coordination that producing firms used to accomplish their transactions with suppliers, and the general use of all types of electronic networks for a variety of business purposes. Each national survey contained the same questions, and the sampling strategy ensured equivalence among the two countries in terms of industry size and type. More details about the research methodology and measurement strategy can be found in Kraut et al. [1].

3. Electronic Network Use in France and the US

Since 1983, the Teletel network has been offered commercially in France. Using a national X.25 network known as Transpac, and dedicated terminals known as Minitels, this network provided France with a national data network infrastructure long before Internet commercialization. Initially the Teletel network was envisioned as a service for the home consumer market, giving consumers access to entertainment, information, and transaction services. However, internal and business-to-business uses are prevalent, and represent nearly half of all Minitel service codes [2, 3].

Nearly all of the French firms in our sample are connected to Minitel, with usage frequency averaging at the mid-point of a 5-point scale ranging from never to very often (see Table 1). Usage varies significantly across the four industries, led by magazine publishing and advertising. Apparel manufacturers are the least likely to report Minitel usage for business purposes.

Table 1. Use of the Minitel by French Firms for Business Purposes

	% connected	**Amount of Use**
Advertising	97	3.13
Apparel	91	2.27
Pharmaceutical	98	2.89
Publishing	97	3.47
	96	**2.94**

Scale ranged from 1= never to 5=all the time
Between industry differences significant at $p<.0001$ level

Table 2. Proportion of Firms Connected to the Internet for Any Business Purpose

	France	**United States**	
Advertising	72	86	79
Apparel	15	34	25
Pharmaceutical	37	63	50
Publishing	44	82	63
	42	**66**	

Between country differences significant at $p<.0001$ level
Between industry differences significant at $p<.0001$ level
Country by industry interaction not significant

Despite (or perhaps because of) the nearly fifteen years of experience with a ubiquitous public data infrastructure, firms in France are much less likely to be connected to the Internet than firms in the United States. As shown in Table 2, 66% of US firms and 42% of French firms reported being connected to the Internet for business purposes. There were similar country differences in usage frequency (Table 3). The apparel sector had the lowest Internet usage of the industries included.

Table 3. Frequency of Use of the Internet for Business Purposes

	France	United States	
Advertising	3.56	3.87	**3.72**
Apparel	2.56	3.26	**2.91**
Pharmaceutical	2.83	3.19	**3.01**
Publishing	3.30	3.58	**3.44**
	3.06	**3.48**	

Scale ranged from 1=never to 5=all the time
Between country differences significant at p<.05 level
Between industry differences significant at p<.001 level
Country by industry interaction not significant

Network use in both countries also varies by purpose. We asked respondents to indicate whether or not they used any type of network, including internal, proprietary external, EDI, and the Internet, for six different business purposes: in their production activities, for e-mail, for taking orders for products, for sending and receiving invoices, for giving customers and suppliers product information, and for electronic funds transfer.

A high percentage of US firms reported some use of any electronic networks for each of these purposes. Approximately two thirds reported using electronic networks in their production operations, for electronic mail, for taking orders, and for giving product information, although industry variation was significant (Table 4). About half of all firms in the US sample reported use of the networks for invoicing and electronic funds transfer. French firms are not as yet exploiting the networks for these applications to the same degree as their US counterparts.

Table 4. Proportion of Firms Reporting Use of Any Network For Various Purposes by Industry and Country

	Production Processes		**E-mail**		**Taking Orders**		**Invoicing**		**Giving Product**		**Electronic Funds Trans.**		
	France	**US**	**France**	**US**	**France**	**US**	**France**	**US**	**France**	**US**	**France**	**US**	
Advert.	.56	.70	.61	.81	.34	.77	.27	.63	.60	.84	.36	.63	**.59**
Apparel	.23	.53	.21	.38	.32	.52	.23	.51	.26	.52	.26	.38	**.36**
Pharma.	.24	.48	.32	.61	.22	.62	.16	.62	.31	.70	.33	.59	**.43**
Pub.	.56	.74	.38	.72	.34	.59	.17	.47	.39	.64	.19	.43	**.47**
	.40	**.61**	**.38**	**.63**	**.31**	**.63**	**.21**	**.56**	**.39**	**.68**	**.29**	**.51**	

Between country differences: p<.0001 level for all purposes
Between industry differences: p<.0001 level for production, e-mail, and prod. info., p<.01 for EFT
Country by industry interaction not significant for any purpose

This pattern of findings, graphically portrayed in Figure 1, illustrates that the Minitel does not fully compensate for the lower computer network penetration in France. Prior research [2,3] has suggested that the Minitel might substitute for some forms of business computer networking, particularly among small firms [3]. Yet, despite the fifteen years of experience with the Minitel, overall use of electronic networks for business purposes still appears to be much lower in France than in the US. The relatively limited capabilities of the Minitel, including low bandwidth, dial-up connectivity, limited resolution terminals, crude graphics, and pricing policies that discourage long sessions clearly influence the Minitel's desirability as a business networking tool.

Figure 1 illustrates that although networks are used less in France than the US, the general profile of purposes for network use is similar. That is, in both countries, the Internet is used more for electronic mail and providing product information than the other applications. Interestingly, the profile for Minitel use in France follows a similar pattern. Given that the differences across applications in overall network use are smaller, this suggests that other internal, specialized, or proprietary networks rather than the public Internet or the Minitel are used to support transactions within and between firms that involve money flows or production processes.

We next examine whether the same factors influence firms' use of electronic networks for transactions with their suppliers of important inputs. Transaction cost theory provides the theoretical framework for exploring influences on electronic network use and the effect of such network use on make vs. buy decisions.

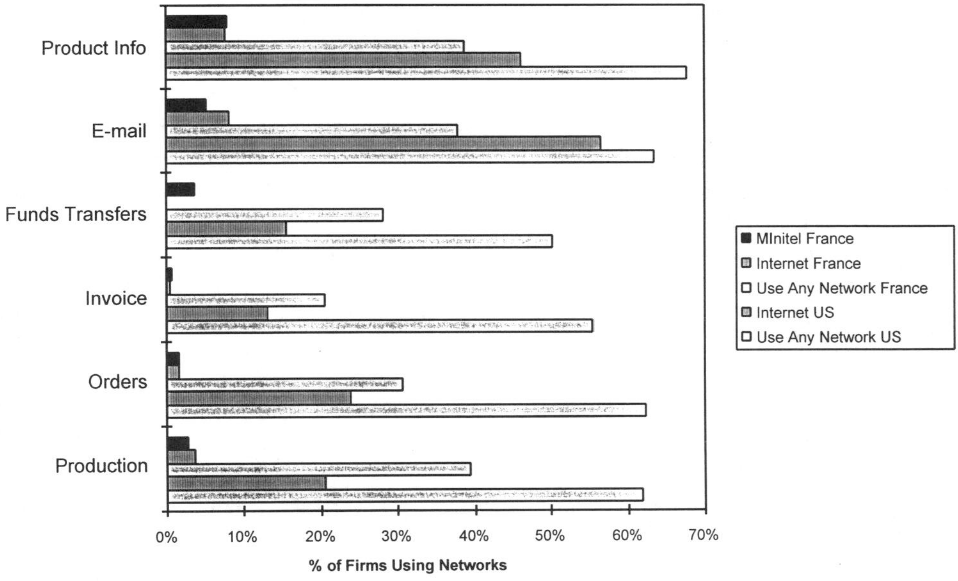

Figure 1. Using Networks in France and the US

4. Explaining Interfirm Network Use: A Transaction Cost Framework

A fundamental assumption of transaction cost economics is that firms make goods and services rather than buy them to reduce costs of coordinating with their suppliers. According to this approach, firms produce in-house in large part to reduce their vulnerability to opportunistic trading partners who might otherwise take advantage of their dependence. Most economists believe markets would generally be more efficient mechanisms for production, were it not for these costs of coordination. Indeed, according to transaction cost tenets, firms exist primarily to reduce the costs of specificity and monitoring contracts [4].

Information systems researchers have tried to make the case that modern computer and telecommunications networks sufficiently reduce the costs of coordination so that firms can get the production benefits of acquiring goods and services in the market without incurring much more coordination costs than if they produced these goods and services themselves [5]. Largely based upon a series of case studies, the expectation that telecommunications and information networks are encouraging greater outsourcing has been advanced [5, 6].

On the other hand, there appears to be a growing body of evidence that suggests that electronic networks do not necessarily encourage market-like trading patterns, even when used with external trading partners [2, 3, 7, 8, 9, 10, 11]. These studies suggest that electronic networks result in tighter integration between buyers and sellers, forming what Malone and colleagues [5] have called electronic hierarchies when suppliers are external. This suggests that it is important to examine the mechanisms that lead firms to externalize production activities, and note how these might be influenced by electronic networks.

Williamson [4] suggests a number of conditions that result in market failure and lead organizations to choose hierarchy as their coordination mechanism. Firms have bounded rationality, and cannot know all contingencies that might influence an economic transaction. Hence when uncertainty and complexity are greater, firms face an information deficit and may opt out of the market. These conditions are further exacerbated in the case where there are small numbers of suppliers to satisfy a firm's input needs, and suppliers behave opportunistically. The small-numbers problem is more likely to arise when acquiring firms need highly specific assets (or inputs) and, as a result, do not have a large number of alternative suppliers with which to do business.

According to information systems researchers who begin with the transaction cost approach, networks can reduce the costs of coordination with the market by allowing firms to inexpensively search for alternative suppliers, monitor supplier performance, and reduce the time it takes to exchange information. Hence they are proposed to reduce the effects of uncertainty, complexity and asset specificity that, Williamson felt, lead firms to choose hierarchy as a governance mechanism. On the other hand, a framework proposed by Kubicek [12] suggests that this outcome may occur in the case of "operating networks", but the opposite may hold for "strategic networks". Operating networks emphasize the use of networks to support routine transactions giving equivalent benefits to each partner. Strategic networks are developed by a focal organization for its own competitive advantage, however, and generally serve to increase dependence on the firm. Research that shows that the number of suppliers tends to decrease following EDI network implementation is consistent with this view [11, 13]. Clemons, Reddi & Row [11], for example, show that while information technologies reduce the transaction costs caused by opportunism and uncertainty regarding quality, this only occurs when the number of suppliers is reduced and long-term contracts are negotiated. Such a trading environment is inconsistent with popular views of an electronic marketplace.

Another reason to question the electronic market hypothesis is that information technology and electronic networks may have effects not only on external transaction costs, but also on internal production and coordination costs. Indeed, Coase [14] long ago speculated that new communication technologies like the telephone and telegraph reduced the spatial costs of organizations and contributed to their increased size. This implies that greater use of electronic networks should be associated with greater acquisition of inputs from in-house sources rather than external suppliers.

5. Operationalizing Variables Derived from Transaction Cost Theory

Our survey allows us to empirically examine the degree to which variables derived from transaction cost theory influence the means by which producer firms coordinate the acquisition of important inputs. It allows us to specifically examine whether greater use of electronic networks is associated with more or less outsourcing, and whether greater use of electronic networks improves coordination success. By examining these data across two national contexts, it is possible to see if cultural factors or unique national infrastructures moderate the way networks are used in producer-supplier relationships.

The variables used in our analyses are briefly described below. We created a series of multi-item scales based upon analyses of the US data [1] and replicated these measures in the French data set.

5.1 Coordination Mode

In-house production is a composite index that estimates the extent to which a firm outsources the key input or produces it in-house. If the index is high, then the firm uses a more in-house procurement strategy. It is based on three items: (1) whether the major supplier for the input was external or in-house; (2) the nature of the relation between the major supplier and the producer firm, ranging from "departments within your firm" to "owned subsidiary" to "a joint venture firm" to an "outsider" firm; and (3) the percentage of the key input which was produced in-house, ranging from 0 to 100.

Electronic links deal with the importance of using "any type of computer communication that allows you to exchange information with this supplier". We asked about the importance of using electronic links for each of six separable stages in acquisition [15, 16]. Respondents made their judgments on 5-point Likert-type scales, from "not at all important" to "very important".

Importance of interpersonal relationships in acquiring the key input for each of the six stages of acquisition was also measured. According to the definition of the term for respondents, interpersonal relations include "any type of personal connection or personal knowledge between people in your firm and people in the supplier's organization". The composite scale is highly reliable.

5.2 Product and Production Attributes

Products versus service: Transaction inputs were classified according to whether they were more "object-like" or "process-like". Products were tangible or intangible objects acquired by the producing firm (e.g., fabric or magazine articles). Services involved agreements by the producing firm to have the supplier complete one or more activities (e.g., cutting fabric).

Tangibility of an input refers to the ability to execute the transaction entirely through an electronic network with currently existing information technology. If the completion of the transaction required the transfer of a physical object between two firms then the transaction was categorized as tangible. If, on the other hand, it was technologically feasible to completely eliminate the exchange of material objects and still complete the transaction, then the input was considered intangible, regardless of whether a pair of firms actually used electronic networks.

Variability is the extent to which the item being ordered, its availability, and its price varied over time.

Predictability. Predictability is the extent to which respondents knew far in advance the specific input they would need and its quantity.

Time pressure is the respondent's judgment of the extent to which production in the firm was rushed and subject to rigid deadlines.

Order frequency measured the ordering cycle for the key input, from monthly to yearly.

Ease of description. Complexity of product description was operationalized in terms of how difficult it would be to describe the input so that it could be ordered. Respondents were asked to rate how easily the key input could be described and how quickly an expert could construct a description sufficient for ordering purposes. Surprisingly, these two items did not correlate highly with each other (r=.15) in the US data, and were entered as separate variables in the analyses.

5.3 Relationship with the Supplier

Supplier power is the strength of the major supplier in the industry, based on the respondent's estimate of the supplier's relative market share.

Trust is a two-item scale measuring the extent to which the respondent believed the prime supplier would refrain from malfeasance.

Object specificity is the degree to which the key input itself was usable only by the focal producer firm.

Knowledge specificity reflected the degree to which employees at both the focal firm and the supplier have developed skills and knowledge for dealing with each other.

5.4 Outcome Variables

Satisfaction with a supplier is an index in which respondents indicated their overall satisfaction with the supplier, their satisfaction with six aspects of the ordering process, and their satisfaction with the supplier's ability to handle exceptional orders (9 items; Alpha=.89).

Efficiency of ordering. This index indicates the average elapsed time between ordering and receiving a input, the number of different people who are involved in handling the order, and the number of communications with the supplier during a typical order cycle. The answers were reverse and standardized before averaging (3 times; Alpha=.48).

Order quality is an index representing the absence of errors in a typical order, including late arrivals, failure to meet quality standards, and containing any other error (3 items; Alpha = .62).

5.5 Control Variables

In all analyses we included variables for *industry* and *firm size*. The size of each firm was classified as small, medium, or large, relative to each industry.

6. Predicting Use of Alternative Coordination Modes and Their Relation to Make vs. Buy Decisions

We tested the influence of the various product and relationship attributes on coordination, using multiple regression analyses. Separate regressions were completed for the US and French datasets and are reported in Tables 5 and 6.

Table 5. Regression Predicting Coordination Mode Among US Firms

Predictor	**In-house Production**	**Personal Relationships**	**Electronic Networks**
Publishing vs. apparel	-.89 ***	-.12	.18
Pharmaceut. vs. apparel	-.49 **	-.31	.30 †
Advertising vs. apparel	-.79 ***	-.28	.14
Firm size	-.05	.06	.08
Ease of input descrip.	.01	.05	-.23 ***
Ease of expert descrip.	-.02	.02	.15 **
Service	-.93 *	.14	-.36
Tangibility	-2.06 ***	.15	-.86 **
Service*Tangibility	.88 ***	-.16	.25
Variability	-.01	-.05	-.06
Predictability	-.10 †	-.06	-.07
Time pressure	.01	.03	.06
Supplier power	-.12 *	-.02	.03
Order frequency	.12 *	.09	.03
Trust	.05	-.09	-.15 **
Knowledge specificity	.15 **	.35 ***	.17 **
Object specificity	.17 **	.21 ***	.18 **
N	250	250	250
Df	17	17	17
Rsq	.38	.23	.29
Adjusted Rsq	.33	.17	.24

Note: Entries are standardized beta coefficients from a multiple regression analysis.
† $p < 0.1$
* $p < 0.05$
** $p < 0.01$
*** $p < 0.001$

In general, there is some consistency in the pattern of associations across the US and French data. In-house production is strongly predicted by variables derived from transaction cost theory in both samples, even after controlling for industry and size. In particular, firms appear to internalize highly specific inputs needed often, as the theory would predict. An interesting interaction is exhibited for the product vs. service and tangibility attributes of inputs in both countries. As illustrated in Figure 2, firms in the US and France appeared to outsource tangible products, while internalizing intangible products and both types of services. These latter types of inputs are arguably more complex, which is also in keeping with Williamson's notions [4].

For US firms, personal relationships were also predicted by object and knowledge specificities. In France, knowledge specificities also appeared to lead to greater use of personal relationships, although object specificity did not. Electronic network use as a mode of coordination was also associated with object specificity in both countries.

We had expected that electronic network use would also be influenced by time pressures and order frequency, as the speed and processing advantages would suggest that benefits would be higher under these conditions. This was not the case at all for firms in the US. In France, firms did, however, use electronic networks more when order frequency was high. In both countries, it also appears that electronic networks are used more for intangible inputs, as we might expect. Contrary to the expectations derived from Malone and colleagues [5], firms in both countries used electronic networks more when inputs were difficult to describe. Hence, it appears that electronic networks are applied more for complex transactions that we might consider to result otherwise in higher transaction costs.

Table 6. Regression Predicting Coordination Mode Among French Firms

Predictor	In-house Production	Personal Relationships	Electronic Networks
Publishing vs apparel	-.40 ***	-.15 †	.05
Pharmaceut. vs apparel	-.12	-.11	.01
Advertising vs apparel	-.43 ***	-.10	-.05
Firm size	-.05	-.06	-.02
Ease of input descrip.	-.06	-.02	-.15 *
Ease of expert descrip.	.08	.04	.03
Service	-.37 **	.00	-.24 †
Tangibility	-.88 ***	.01	-.41 **
Service*Tangibility	.68 ***	.02	.29 †
Variability	.01	.12	.00
Predictability	-.04	-.02	-.10
Time pressure	.04	.11 †	.08
Supplier power	-.03	.05	.02
Order frequency	.12 *	-.10	.20 **
Trust	.03	-.06	-.08
Knowledge specificity	-.08	.17 **	-.09
Object specificity	.19 ***	.10	.14 *
N	256	256	256
Df	17	17	17
Rsq	.34	.09	.20
Adjusted Rsq	.29	.03	.14

Note: Entries are standardized beta coefficients from a multiple regression analysis.
† $p < 0.1$
* $p < 0.05$
** $p < 0.01$
*** $p < 0.001$

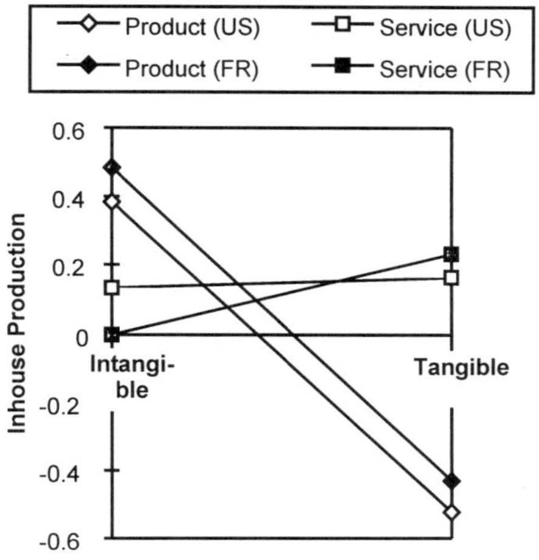

Figure 2. The Interaction of Input Tangibility and Service Interaction on In-house Production

In summary, firms appeared to increase their coordination efforts with suppliers when inputs were more complex and specific. The fact that this relationship is consistent across industries and countries lends support to the basic transaction theory framework.

We tested the effect of electronic network use on outsourcing by adding it to the list of predictors of in-house production. Table 7 illustrates that the relationship is, in fact, opposite to what would be expected if network use led to increased outsourcing. Thus, the more firms used electronic networks to acquire important inputs, the more likely they were to be acquiring those inputs from in-house sources of supply.

This should not be a surprising finding for a number of reasons. Since firms internalize complex and specific inputs in order to reduce transaction costs, it makes sense that these would be the targets for internal electronic network support as well. In fact, it is likely to be precisely because these inputs have been internalized that firms are able to exploit networks to support such transactions. Firms can take advantage of existing internal data network infrastructures and ensure standardization and compatibility between organizational units needing the input and organizational units supplying it.

A stronger test of the move-to-market hypothesis involves an analysis of the relationship between electronic network use and in-house production for those firms who report that their primary supplier is external (see also Table 7). Even for this subset of the sample, greater electronic network use was associated with more reliance on in-house sources of supply. This relationship was significant in the US sample. It was not significant in the French sample, but the coefficient was in this direction.

We conclude this section by noting that the coordination modes we examined are complementary, rather than used as substitutes for each other. In Table 8, the positive associations among the coordination modes are all significant, except for the case of the network use and interpersonal relationships link in French firms. Internalization makes it easier to build and use networks, and increases the opportunities for interpersonal interaction.

Table 7. Predicting In-house Production from Electronic Network Use for all French and US Firms and Those Whose Primary Supplier is External

Predictor	In-house Production		In-house Production	
	All firms France	Exter. Sup. France	All firms US	Exter. Sup. US
Publishing vs. apparel	-.40 ***	-.09	-.94 *	-.29 *
Pharmaceutical vs. apparel	-.12 †	.19 †	-.57 *	-.22 *
Advertising vs. apparel	-.42 ***	-.42 ***	-.83 *	-.25 *
Firm size	-.04	.03	-.07	-.02
Ease of input description	-.04	.05	.07	-.01
Ease of descrip. for expert	.08	.00	-.06	-.04
Service	-.33 **	-.01	-.83 †	-.19 †
Tangibility	-.81 ***	-.57 **	-1.83 ***	-0.54 **
Service*Tangibility	.63 ***	.03	0.81 ***	0.32 *
Variability	.01	-.14 †	0.01	-0.01
Predictability	-.02	.13 †	-0.08	-0.07 *
Time pressure	.02	.04	-0.01	0.03
Supplier power	-.03	-.01	-0.13 **	-0.01
Order frequency	.09	-.03	0.11 *	0.03
Trust	.04	.00	0.09 †	0
Knowledge specificity	-.06	-.02	0.1 †	0.01
Object specificity	.17 **	-.04	0.12 *	0.05
Electronic Network Use	.17 **	.11	0.27 ***	0.1 **
N	256	188	250	186
Df	18	18	18	18
Rsq	.36	.26	.43	.28
Adjusted Rsq	.31	.18	.39	.20

Note: Entries are standardized beta coefficients from a multiple regression analysis.
† $p < 0.1$
* $p < 0.05$
** $p < 0.01$
*** $p < 0.001$

Table 8. Correlations Among Coordination Modes for French and US Firms

	In-house Production		Network Use	
	US	France	US	France
In-house production				
Network use	.37****	.31***		
Interpersonal relationships	.16*	.13*	.20**	.08

7. Predicting Transaction Outcomes From Use of Alternative Coordination Modes

A final set of analyses focuses on the extent to which electronic network use, as well as use of other coordination modes, had positive or negative influences on firms' abilities to coordinate with their most important suppliers. Table 9 shows that among US firms, coordination through interpersonal relationships continues to contribute unique variance in explaining the coordination outcomes, once product attributes, production attributes, specificities, interfirm relationships, and use of the other coordination modes are entered into the model. Respondents are more satisfied with their suppliers and report more efficient ordering, the more they use interpersonal relationships as the basis for coordination. In contrast, neither in-house production nor use of electronic networks reliably improves coordination success on any of the outcome measures, once other variables are entered into the model. Indeed, there is a suggestion that use of electronic networks is associated with poorer order quality (i.e., more errors) ($p < .10$).

Table 9. Regressions Predicting Coordination Success from Coordination Mode Among US Firms

Predictor	**Satisfaction with Supplier**	**Efficiency in Ordering**	**Order Quality**
Publishing vs. apparel	.02	.30	.21
Pharmaceut. vs. apparel	-.05	-.03	.13
Advertising vs. apparel	-.02	.24	.38
Firm size	-.05	-.24 ***	-.03
Ease of input descrip.	.00	-.06	-.01
Ease of expert descrip.	-.13 *	-.02	.05
Service	.01	.14	-.13
Tangibility	-.24	.39	-.26
Service*Tangibility	.14	-.33	.07
Variability	.07	-.12 *	-.12 †
Predictability	-.02	-.27 ***	.06
Time pressure	.02	.01	.17 **
Supplier power	-.20 **	-.04	.10
Order frequency	.11 †	.10 †	.09
Trust	.13 *	.00	.05
Knowledge specificity	.32 ***	.03	.07
Object specificity	.00	-.08	-.19 *
In-house production	-.11	-.11	-.05
Network use	.05	-.09	-.12 †
Interpersonal relationships	.17 **	.14 *	.02
N	250	250	250
Df	17	17	17
Rsq	.38	.23	.29
Adjusted Rsq	.33	.17	.24

Note: Entries are standardized beta coefficients from a multiple regression analysis.
† $p < 0.1$
* $p < 0.05$
** $p < 0.01$
*** $p < 0.001$

Among the French firms (Table 10), reliance on interpersonal relationships for coordination was less of a factor. There was a slight tendency for greater use of interpersonal relationships to be associated with improved order quality ($p<.10$). Interestingly, use of electronic networks was negatively associated with all three outcome measures, although none of the coefficients were significant. French firms were most satisfied with those suppliers that they trusted and when they possessed greater knowledge about the supplier. To the extent that interpersonal relationships assist in the development of trust (other factors are also important – such as reputation and previous performance), these two variables may have inhibited the explanatory power of our interpersonal relationships measure.

In general, results are inconsistent with the research literature finding positive effects of EDI on order processing, which has typically examined cases in which a single, relatively sophisticated firm, such as Chrysler, uses EDI with a variety of trading partners. These results may not generalize to the larger number of unsophisticated users of electronic networks. The failure to replicate previous findings of improved outcomes with electronic network use may reflect the troubles that many firms have in mastering a complex technology.

Table 10. Regressions Predicting Coordination Success from Coordination Mode Among French Firms

Predictor	**Satisfaction with Supplier**	**Efficiency in Ordering**	**Order Quality**
Publishing vs. apparel	.11	.06	.05
Pharmaceut. vs. apparel	.11	-.20 *	.04
Advertising vs. apparel	.29 **	.06	.13
Firm size	-.02	-.15 *	.02
Ease of input descrip.	-.01	-.03	.02
Ease of expert descrip.	.13 *	.00	.12 †
Service	.20	.01	.00
Tangibility	.25	.15	.20
Service*Tangibility	-.29	-.14	-.02
Variability	.09	.15 *	.06
Predictability	-.10	-.13 †	-.08
Time pressure	.03	-.02	.01
Supplier power	.03	-.05	.13 *
Order frequency	.15 *	.05	.08
Trust	.27 ****	.07	.05
Knowledge specificity	.17 **	.08	.06
Object specificity	.04	-.11 †	-.03
In-house production	.07	-.09	-.01
Network use	-.10	-.10	-.09
Interpersonal relationships	.03	-.05	.11 †
N	256	256	256
Df	20	20	20
Rsq	.22	.21	.13
Adjusted Rsq	.15	.14	.05

Note: Entries are standardized beta coefficients from a multiple regression analysis.
† $p < 0.1$
* $p < 0.05$
** $p < 0.01$
*** $p < 0.001$

8. Conclusions

Our results indicate that, despite the head start that France may have gained with their Minitel experience, French firms use electronic networks for business-to-business transactions much less than their US counterparts. The limited technical capabilities of the Minitel terminals and the Teletel network appear to be too great, despite the externalities that such a widely diffused network has to offer. In addition, the Minitel investment may have also slowed the adoption of the Internet in France, although our data cannot be used to test this issue. Alternatively, it may be that without the Minitel, French firms would be at an even greater disadvantage. Interviews with French managers suggest that prior experiences working with Minitel-based services have made it easier to develop Internet-based services.

Although French firms exhibit lower usage rates for electronic networks, relative usage for various applications is roughly similar to that in the United States across the four industries included in our study. Public networks are used mainly for electronic mail and product information dissemination.

The factors that influence electronic network use, and the possible effects on make vs. buy decisions, appear to be similar across the two country contexts. The results reported here suggest that variables derived from transaction cost theory are predictive of when firms opt to use hierarchy over market. Interestingly, increasing input complexity and specificity not only lead to internalization, but also are associated with greater use of electronic networks to support transactions. This latter finding is not consistent with the view that electronic data network usage for transactions would lead to more market-like relations between producers and suppliers. Rather it suggests that one effect of increased network use may be more, rather than less integration between producers and suppliers. Another likely explanation is that firms first develop electronic transaction capability over internal systems, thus creating the positive relationship between electronic network use and use of internal suppliers. Our findings also suggest that network use should not be viewed as a substitute for human intervention and internalization, but rather as a complement to these modes of coordination.

Finally, electronic network use was not associated with better transaction outcomes, and there was some evidence that the contrary was true. This may be the result of the greater use of networks for more complex and intangible transactions, which would arguably be the most difficult ones. Or it may be evidence of an overall difficulty that firms are having in mastering complex technology to achieve desirable outcomes.

References

[1] R.Kraut, C. Steinfield, A. Plummer, B. Butler and A. Hoag, Coordination Modes and Producer-Supplier Integration. Paper presented to the OECD Workshop on the Economics of the Information Society, London, March 19-20, 1997.

[2] C. Steinfield, L. Caby and P. Vialle, Internationalization of the Firm and the Impacts of Videotex Networks, *Journal of Information Technology* 7 (1993) 213-222.

[3] L.A. Streeter, R.E. Kraut, H. C. Lucas and L. Caby, The Impact of National Data Networks on Firm Performance and Market Structure, *Communications of the ACM*, 1996.

[4] O. Williamson, Markets and Hierarchies: Analysis and Antitrust Implications. Free Press, New York, 1975.

[5] T. Malone, J. Yates and R. Benjamin, Electronic Markets and Electronic Hierarchies: Effects of Information Technology on Market Structure and Corporate Strategies, *Communications of the ACM* **30**, 6 (1987) 484-497.

[6] R. Wigand and R. Benjamin, Electronic Commerce: Effects on Electronic Markets. *Journal of computer Mediated Communication* **1**, 3 (1995). (Available online at http://www.usc.edu/dept/annenberg/vol1/issue3/vol1no3.html.)

[7] C. Hess and C. Kemerer, Computerized Loan Origination Systems : An Industry Case Study of the Electronic Markets Hypothesis, *MIS Quarterly*, September (1994) 251-275.

[8] C. Steinfield, R. Kraut, and A.C. Plummer, The Impact of Interorganizational Networks on Buyer-Seller Relationships, *Journal of Computer Mediated Communication*, **1**, 3 (1995) (available online at http://www.usc.edu/dept/annenberg/vol1/ issue3/vol1no3.html).

[9] E. Brousseau, Information Technologies and Inter-firm Relationships: The Spread of Inter-Organizational Telematics Systems and Its Impact on Economic Structure. Presented to the International Telecommunications Society, Venice, June, 1990.

[10] L. Caby, C. Jaeger and C. Steinfield, Explaining The Use of Inter-Firm Data Networks for Electronic Transactions: The Case of the Pharmaceutical and Advertising Industries in France. Paper presented to the International Telecommunications Society, Seville, June 1996.

[11] Clemons, E. S. Reddi and M. Row, The Impact of Information Technology on the Organization of Production: The Move to the Middle Hypothesis, *Journal of Management Information Systems* **10,** 2 (1993) 9-35.

[12] H. Kubicek, The Organization Gap in Interbranch EDI Systems. *EDI* **3**, no. 2 (1993) 105.

[13] J. Bakos and E. Brynjolfsson, From Vendors to Partners: Information Technology and Incomplete Contracts in Buyer-Supplier Relationships, *Journal of Organizational Computing* **3**, 3 (1993) 301-320.

[14] R. Coase, The Nature of the Firm. *Economica* **4** (1937) 386-405.

[15] H.R. Johnston and M.R. Vitale, Creating Competitive Advantage with Inter-Organizational Information Systems, *MIS Quarterly* (1988), 153-165.

[16] A. Kambil, Information Technology and Vertical Integration: Evidence From the Manufacturing Sector. In: M. Guerin-Calvert and Wildman, S. (eds.), Electronic Services Networks: A Business and Public Policy Challenge. Praeger, New York, 1991, pp. 22-38.

Telecommunications Transformation: Technology, Strategy and Policy
E. Bohlin and S.L. Levin (Eds.)
IOS Press, 1998

CHAPTER 11

Sustaining a Vertically Disintegrated Network Through a Bearer Service Market

Petros Kavassalis
École Polytechnique
Rue Descartes 1, 75005 Paris, France

Thomas Y. Lee and Joseph P. Bailey
Massachusetts Institute of Technology
Cambridge, MA 02139, USA

Abstract. Based upon the Internet perspective, this chapter will attempt to clarify and revise several ideas about the separation between infrastructure facilities and service offerings in digital communications networks. The key notions that we will focus on in this paper are: i) the bearer service as a technology-independent interface which exports *blind* network functionality to applications development; ii) the sustainability of an independent market for bearer service and the organizational consequences associated with such a market.

1. Introduction

During the past two years, applications like e-mail and the World Wide Web have combined with evolving network protocols to propel the Internet into the heart of a computer and communications convergence. Central to the Internet's immersion into digital convergence has been the effectiveness with which the Internet Protocol (IP) has played the role of "spanning layer"[1].

The IP abstraction enables applications to request network services independent of underlying, physical network technologies. Moreover, new underlying network technologies may either substitute for or co-exist with existing network technologies without significantly affecting the broader system. Based on this abstraction, the National Research Council recently articulated the "Open Data Network (ODN)" as an architecture for the networks of the future that generalizes the principle of separating service offerings from infrastructure facilities as demonstrated in the Internet [2]. In the same way that IP serves the Internet, the ODN relies upon a "bearer service" to function as a *technology-independent network layer* that resides above the technology substrate and enables interoperation between diverse, high-level applications and various underlying network infrastructures (Figure 1).

The NRC report describes an Open Digital Network (ODN) as a four-level layered architecture: "i) at the lowest level is an abstract bit-level service, the bearer service, which is realized out of the lines, switches, and other elements of networking technology; ii) above this level is the transport level, with functionality that transforms the basic bearer service into the proper infrastructure for higher-level applications (as is done in today's Internet by the TCP protocol) and with coding formats to support various kinds of traffic (e.g., voice, video, fax); iii) above the transport level is the middleware, with commonly used functions (e.g., file system support, privacy assurance, billing and collection, and network directory services); and iv) at the upper level are the applications with which users interact directly. This layered approach with well-defined boundaries permits fair and open competition among providers of all sorts at each of the layers" [2].

Certainly, the Internet demonstrates the technical and functional robustness of a technology-independent bearer service abstraction [3]. The bearer service is intended to support requests for service from all applications and to recognize all substrates, but as both application and infrastructure innovations turn increasingly towards user-oriented models of network architecture, technology and policy considerations related to the generalization of this abstraction should be carefully studied. Such a service blurs the boundaries of telecommunication markets.

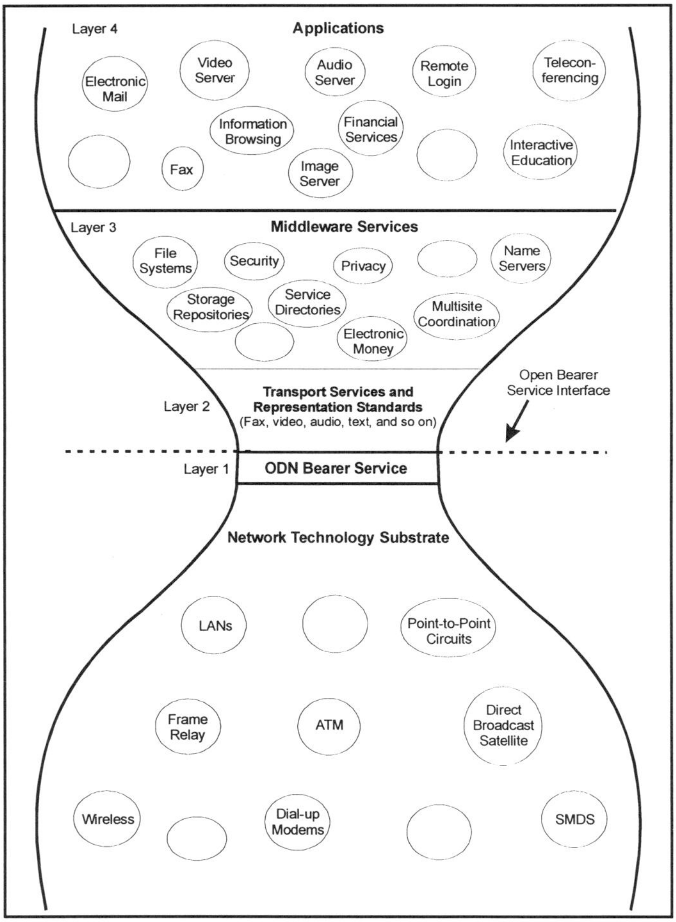

Figure 1. The Bearer Service Concept

For example, the promises of Internet telephony to combine the benefits of the public switched telephone network (PSTN) and the Internet would be possible through a bearer service – even though Internet telephony would weaken market boundaries and challenge the regulatory environment [4]. Regulators have a difficult time categorizing bearer service providers as belonging to any one existing market because such providers can offer services that cut across many existing telecommunications markets. Businesses that develop ubiquitous services or tailor applications for customers may be threatened by market entrants who have competitive products built around a new technology architecture that is able to provide great flexibility in applications design – the bearer service. Finally, customers can benefit from an integrated services environment because their data and voice communications can be transmitted across multiple telecommunications infrastructures. A new era of interoperability [5] is possible through the bearer service.

However, questions for this new market abound. Not only is the business model in question, but the technology is also in flux. In this chapter, we hope to shed some light on the bearer service issue by answering some of the questions arising from this new bearer service market. First, what does a technology-independent bearer service look like? What are the technical characteristics of a bearer service that satisfies the requirements of both elastic and real-time applications? Second, is a segregated network that separates infrastructure facilities from service offerings by a bearer service an economically viable market organization? This chapter spans both technical and economic issues to ascertain the potential of a market for bearer services. Specifically, this chapter will address the following subset of bearer service market questions:

i) What comparative advantage does a technology-independent bearer service provide over homogeneous infrastructures that support multiple service classes (e.g. ATM)? This question is especially important given the increasing interest in *Integrated Networks* solutions.
ii) In a recent paper Gong and Srinagesh [6, 7] suggest that layered network architectures may inherently engender vertically integrated markets for network services. Is this a general statement or a phenomenological one? And if the phenomenological reading is correct, what factors tend to suggest the viability of disaggregated markets for network services both at and above the spanning layer as described above?

The remainder of the chapter is organized as follows. Section 2 employs a comparative analysis to define concepts of the bearer service and the *layered network architecture* concepts. Considering IP as a *bifurcation point* in the evolution of network design, the bearer services of traditional communications infrastructures and the Internet (both the current *best-effort* and the future *Integrated Services* Internet) are surveyed to elicit design characteristics and functional differences between a technology-dependent and independent bearer service. We suggest that in a network design with a technology-independent bearer service, the communications network supports a *flexible organizational* model capable of dealing with unanticipated (applications') variability.

Section 3 associates the technology-independent bearer service with the Internet organizational model: a flexible system of regional or more extended backbones and access links (or access networks) to these backbones, managed by the Internet Service Providers (ISPs). Specifically, we address the question of whether the ISP model, which is characterized as a model where network operators exhibiting varying degrees of vertical integration compete in an open market, can sustain itself, or if one monolithic, integrated firm will emerge from mega-mergers.

To answer this question, we begin by considering the work of Gong and Srinagesh [6], who argue that a stable and sustainable equilibrium for healthy bearer service market growth might not be possible. Our analysis closely follows the definition for a bearer service formulated in section 2. Arguing that the bearer service is not a commodity product, we identify differentiable service attributes upon which an independent bearer service market could form. Through differentiation, the bearer service market can avoid a Bertrand equilibrium (i.e. pure price competition). Furthermore, we challenge assumptions about perceived trends towards vertical integration, by noting the relative independence of assets between bearer service providers and potential buyers of this service (i.e. connectivity providers who are situated at the edge of the network versus other kinds of service providers).

The chapter concludes in section 4 by underlining the role of the bearer service market in creating the potential for competition between independent and integrated providers. However, as revealed in section 3, while a market for bearer services may be economically viable, it is also likely to be fragile. The conditions for sustaining a bearer service that exhibits "spanning" capabilities and promotes efficient interoperability (as in the Internet) require an alignment of both technical and economic considerations. In addition, new institutions, not necessarily government-dependent, may be required to enable the market to form.

2. From the Railroad Gauge to Information Bitways: The Evolution of the Bearer Service Functionality

Given a pre-specified set of applications and a physical network which may include more than one substrate technology (e.g. an Ethernet-based LAN connected to a Frame relay network), the bearer service (BS) constitutes those common[1] functions which are implemented throughout the network rather than in the network end nodes, and are necessary for pairing each application's communication requirements with the performance characteristics of all components of the heterogeneous network.

For example, consider voice and fax data services transmitted via traditional, analog telephony between the United States and Canada. Neither within nor between each country do the customers need to share the consumer premises equipment (CPE). Likewise, at the level of network technologies, materials like cables, switches, and even local numbering conventions might vary. However, since all CPE and network subsystems observe a common, international numbering convention and all networks and CPE transmit, carry, and receive signals in the 4 kHz frequency spectrum, voice conversations are conducted and faxes are conveyed. For voice and fax data services, the international numbering scheme combined with the 4 kHz channel constitutes the bearer service.

[1] If every application uses the function, then it is certainly a function in common and unambiguously a component of the BS. If only one application uses the function, then perhaps it is more appropriately considered part of the application (or of the transport layer). If two or more applications utilize the function but not all applications in the set use the function, then we need to question whether the function belongs in the BS. Recall also that a separate metric for distinguishing BS functionality is whether that function can be implemented in an end node. BS functions include only those functions which cannot be implemented in an end node.

More abstractly, computer and communications technologies may be separated into three layers. The physical infrastructure (e.g. wires, switches, etc.) resides at the lowest layer. At the top lies the set of applications and service offerings supported by the underlying infrastructure. A *spanning layer*[2] bridges the two [1]. An application requests network services through the spanning layer to the substrate technologies.

In the early public switched telephone network (PSTN), telephony was tightly coupled to a specific infrastructure, so the spanning layer supported only a single application and one technology substrate (typically, the spanning layer was located *in the wires).* The development of newer applications for computers and communications, as well as advances in substrate technologies, prompted a refinement of the spanning concept. In the presence of a diverse suite of applications and a heterogeneous network, the *bearer service (BS)* constitutes a spanning layer which *escapes from the wires* and thus supports all applications over the entire network. This section uses Piore's model of organizational flexibility and production system transformation as a methodological framework for tracing the evolution of the spanning layer towards a technology-independent bearer service. Advances in shipping and the transport of physical goods are used as a metaphor for the transformation of yesterday's PSTN into tomorrow's ODN.

2.1 Production Technologies and Flexible Specialization

Piore [8, 10] describes the on-going transformation of industrial production systems towards greater variety and flexibility as a four-stage evolution[3]. The products in such systems are comprised of both independent and interdependent design features; changes in design features mark the different evolutionary stages. Independent design features "can be varied in isolation without complementary changes in other features of the design" while interdependent features "require a number of complementary adjustments" [8].

The initial stage, *mass production,* is characterized by a production system tailored to a single product. There is no room for variation. In *mass production with cosmetic variation*, product design may slightly vary existing, or may introduce new independent, design features. "The notion of cosmetic variation seems to imply a sharp dichotomy between design changes which are easy to make and those that are not" [8].

Flexible mass production extends cosmetic variation by introducing the potential for change in interdependent design features. Flexible mass production explicitly identifies, a priori, both the set of product design features which is subject to change and the set of values which each design feature may take. Therefore, the flexible mass production system represents a finite number of products, which vary in more than simple cosmetics.

Diametrically opposite mass production is *flexible specialization* where variation is virtually infinite. However, *closed flexible specialization* includes those systems where the set of design features that varies is defined a priori, but the domain over which each varying feature ranges is unknown. By contrast, *open flexible specialization* where both the set of

[2] As suggested by Clark, "a spanning layer is characterized by three key parameters that characterize the sort of interoperation it supports: i) the span of infrastructure options over which the interoperation occurs, ii) the range of higher-level applications the spanning layer supports, iii) the degree of symmetry in the services of the spanning layer and its interfaces, and in the higher layers up to the application definitions" [1].

[3] This work draws on The Second Industrial Divide [9].

variable design features and the domain over which each variable ranges is potentially infinite.

A different cognitive model applies in each stage, with flexible specialization involving a balance between "a deepening of understanding within a given cognitive frame and the pull to reintegrate back (in the production process) to a different frame in order to produce a sellable commodity" [11]. Similarly, a technology-independent bearer service offers more than a predesigned set of services. Rather, it supports an "*application-blind interface*", a functionality that enables the introduction of new applications independent of the initial strategies and service offerings of Telecommunications Operators (TOs). This functionality can be easily reintegrated to the application vendors' and users' cognitive frames, thus allowing them to introduce new applications and operate independent of the strategies and the service offerings of the Telecommunications Operators (TOs). To illustrate the continuum that spans well-defined, mass production systems and flexible, application-blind interfaces, we will discuss two contrasting metaphors, the *gauge* and the *container*.

2.2 The Gauge Metaphor

The gauge of a railroad is defined as the distance between rails or between the flanges of the wheel sets on a railroad car. The gauge determines the tracks upon which a given railroad car may travel. By extension, the gauge therefore also determines which railroad companies may exchange rolling stock and the transparency with which a customer may transport freight across boundaries between different railroad companies. Accordingly, diversity in gauge standardization implies transaction costs and other inefficiencies as customers and freight traverse rails of different gauges. Thus, for reducing technical complexity and transaction costs to internalizing mutual network externalities, railroads have been progressively converging towards a gauge standard [12, 13].

The emergence of the spanning layer concept may be derived directly from the convergence towards a rail gauge standard. Railroad tracks comprise the physical network. Differences in rail cars represent distinct applications from which a customer might choose. Gauge dimensionality is therefore a layer which resides between the physical track and the applications. Gauge standardization expanded the services that a particular rail system could offer by extending the reach of the rail network and expanding the scope of traffic (the kinds of cars) that could be carried. Standardization reflected a shift towards some flexibility where rail car design could vary (as long as it conformed to the gauge standard) and traffic could move across network boundaries.

2.3 The Spanning Layer in Traditional Communications Networks

Communications networks have traditionally been vertically bundled. Whether for telephony, radio, broadcast television, or community access television (cable), infrastructures have long been closely coupled to service provision. As Tennenhouse et al. note [14], "telephone and cable services are each carried over their own wired systems. Although radio and television share the airwaves, for practical purposes, they are discrete distribution vectors, since separate portions of the spectrum have been allocated to each type of service." With one wired network for any given application, each service resembles a mass production system. *Spanning* is trivial.

Even within a single service offering, however, the utility of a spanning layer is apparent. Telecommunications heterogeneity is subject to the same economic forces promulgating rail gauge convergence. Monopoly structures and the political environment both contributed to the rapid acceptance of the 4 kHz circuit as a worldwide standard for voice communications[4]. Subsequently, that standard emerged as a spanning layer providing transparent support for multiple applications (for instance fax and data transmission via modems) over technologically evolving analog and digital telephony[5].

The advent of digital computer and communications technologies, in addition to introducing the prospect of new applications and services, began to extend the scope of the spanning layer concept. For digital telephony, the 4 MHz channel gave way to a transfer rate of 64 Kbps using pulse code modulation (PCM). While corresponding to a single voice conversation, the data rate of 64 Kbps also introduced a new category of *value-added* applications and services (e.g. credit-card calls, file transfer, audio-conferencing, etc.).

For value-added services, however, a raw 64 Kbps bit pipe proved inadequate. Specifications for data format, routing, signaling, and many other parameters of transmission are required to enable high-speed data services. Such services include packet switching, Frame Relay, intelligent network applications, and Integrated Services Digital Network (ISDN). Consequently, value-added services led to a new network model which extended the 64 Kbps spanning layer with delay, loss, and other data characteristics. Rather than cosmetic, these additions drove the system towards flexible mass production where the network supported a discrete set of delay values, etc. to support a limited number of new, pre-specified services.

The additional switching and control functions for these new services are *logically* implemented on top of the elementary 64 Kbps spanning layer, yet *physically* implemented elsewhere than the Central Office Equipment. As examples of the distinction between the logical and the physical, consider X25 and ISDN. The transmission of packets between nodes attached to the same transmission link is ensured by a specific *connection-oriented* protocol, the X25[6]; the *data link* layer of the X25 converts an *unreliable* for packet transmission bit pipe to a packet-link, and, in collaboration with the *network* layer, supervises naming, addressing, routing and congestion control[7] [16, 17]. In a similar way, the ISDN (2B+D) provides a customer with two B channels of 64 Kbps[8] and one X25 full duplex D channel at 16 Kbps for "out-of-band" signaling; furthermore, under the *Intelligent Network* model, ISDN separates the flow of control information from user information[9] [18, 19].

---

[4] As in the case of the railroads, however, those forces may sometimes be unable to achieve complete standardization. In broadcast television, spatial lock-in contributed to the emergence of three different standards (NTSC, PAL, and SECAM).

[5] However, the development of applications other than those selected by the network operator, though technically feasible from the beginning, became effective only following a change in political climate as signaled by the FCC's Carterfone decision [15].

[6] Other technologies for connection-oriented packet switching include SNA of IBM and DECNET of Digital Equipment.

[7] We refer to OSI terminology.

[8] These channels can be indifferently used for voice and data transmission.

[9] Even though it is theoretically possible and very appealing to separate user and flow information, this separation (unbundling in the telecommunications jargon) is actually difficult to implement, for reasons relating to the particular strategic interests of the Telecommunications Operators [20, 21].

Despite the new functionality and extensions to functionality supported by the spanning layer, the spanning capability was essentially limited. While a single infrastructure may have supported more than one application, service offerings continued to be tied to the infrastructure in a largely many-to-one correspondence. Infrastructures were not application-blind. For example, changes in the X25 protocol would necessitate change both within the network and at the end nodes. Overall, the public network resembled the shape of a "patchwork" of vertically integrated, flexible mass production systems where each network supported a different set of applications. Within each of these production systems, one can still distinguish between the conception of a service and its delivery, because the spanning layer is *being embedded* in the substrate technology. Variation exists, but only the variation which has been ex-ante conceived and designed by the engineers of the infrastructure-facilities owners.

2.4 The New Metaphor: The Container

Early transportation systems provide examples of vertically integrated systems. For the purpose of transporting goods over land, dedicated technologies including railroad tracks, cars, and rail yards were created. Likewise, air and water transportation services warranted similarly single-purpose infrastructures for ships and airplanes. Over time, however, it became common for freight transporters to link different media, such as water, air, or land. Intermodal freight transportation refers to the linear combination of two or more transportation services. For example, moving a product from a warehouse in Hong Kong to a retailer in the American city of Chicago might entail trucking the product to the harbor, shipping the product to a port on the West Coast of the United States, and then conveying it by rail to Chicago.

Contrasting with vertically integrated architectures, where a dedicated infrastructure is used to realize separate applications, is the *horizontal integration* model [22], where diverse applications are developed independent of any underlying architecture. Instead, developers build to a common *abstraction*[10]. The common abstraction, a spanning layer now located *on top* of the transportation (or communication) modes, projects a virtual infrastructure which itself comprises one or more interoperable physical networks. The spanning layer pairs applications to infrastructure facilities because applications request network service[11] from the spanning layer, not the underlying infrastructure, and the spanning layer translates a request for network service to the infrastructure protocols.

For the transportation industry, horizontal integration, which emerged only in the 1950s, was made possible by the development of and agreement upon a standard freight container. Muller noted that "when it was publicly demonstrated in 1956 that standard containers could move successfully on a land-sea intermodal journey, a commercial revolution was started... *It was the container's unique role as common denominator among modes that was revolutionary* (our emphasis)" [23].

[10] Messerschmitt [22] defines abstraction as "the conscious hiding of unnecessary implementation of functional details while making visible behavioral properties that are essential and important to other modules. Abstraction helps to ensure the independence of the architectural modules, as one module cannot be dependent on the hidden properties of another module".

[11] The term "network service" indicates here the actions and responses (i.e. functionality) provided by underlying infrastructures to applications' requirements. Do not confuse this network service with what we usually call services (or service offerings), i.e. the customer services provided by a network infrastructure to final users.

2.5 The Internet Protocol: A Robust Technology-independent Spanning Layer

In the communications industry, diversity and, consequently, requirements for a common denominator among different technological media also become increasingly important. Local and metropolitan networks providing inexpensive and fast interconnections for personal computers and workstations proliferate inside and outside firms and among research institutions and universities. But successive technological waves have dispersed the computer networking environment in different architectures (LANs: Ethernet, Token Ring, Apple Talk, MANs: Frame Relay, SMDS etc.). Interconnection between these architectures, as well as connections between these private networks and the public telecommunications infrastructure, appeared as the natural priority. In the 1980s, three technologies competed to capture the interconnection market (SNA, B-ISBN, ARPA's Internet protocols), with the *Internet protocol suite* becoming more recently the solution most often adopted [24].

The Internet approach to interconnection involves the separation of infrastructure bitways from applications by defining an interface (Internet Protocol) to the basic infrastructure facilities and then exporting that interface for application development. Formally, IP provides a uniform method of addressing (which is independent of physical hardware addressing) and a variable size datagram *(i.e. a standard data container)*[12]. From this interface any number of diverse applications (e.g. the World Wide Web as an example of new, unanticipated applications) may be constructed [25]. Likewise, principles of datagram encapsulation enable any number of substrate technologies to transmit IP (e.g. Ethernet, token ring, X25, ISDN, Frame Relay, even ATM) [26].

IP takes an additional step towards supporting flexibility by dissociating *transport* from *network* functions (respectively performed by TCP and IP, TCP being over IP) [25]. As a result, IP incorporates only that functionality which must be implemented within the network as opposed to at the periphery or end-user nodes. In this way, IP minimizes the possibility that changes in applications (including transport protocols) or infrastructure components will require changes in all of the routers, hubs, and gateways throughout the network substrate.

By separating spanning layer development from both infrastructure facilities (and applications), IP characterizes a different technological trajectory[13] that stems more immediately from the container metaphor. Essential characteristics of this trajectory are that applications may work over multiple substrates (e.g. network technologies) and that these substrates do not pre-specify the development of new applications. The separation both above and below moves IP further along Piore's framework towards flexible specialization. Certainly IP offers flexibility[14] and features variability both above and below the spanning layer.

---

[12] To use Rose's terms [26], IP has no wires associated with it.

[13] In the sense of Dosi [27].

[14] Borrowing from Bar [28], "... it is the flexibility of its network resources that will determine a company's ability to experiment with telecommunications technologies, learn from the experimentation, and repeatedly reorganize itself to capture cumulative benefits. True network flexibility is not only the ability to support a range of applications over a given network configuration but also, and perhaps more importantly, the simultaneous ability to re-configure a network in order to provide various application mixes and to design new applications that take advantage of new configuration possibilities."

2.6 Enhanced IP with RSVP and ATM: Different Approaches to Horizontal Integration

The Internet protocol suite (TCP/IP) provides the foundation for the current data communications infrastructure in the United States, Northern Europe and much of the rest of the world. However, the current architecture provides only a *best-effort* delivery service. Consequently, it cannot deal with applications requiring service guarantees, such as real-time applications (telephony, video etc.). To respond to this objective, new service models are being defined in the Internet together with protocols to reserve capacity according to applications' requirements. Essentially, the goal is to bring out an architecture for Integrated Services Packet Networks – ISPN [29].

New concepts for spanning, therefore, propose to complement IP datagrams with functions to provide not only best-effort delivery but also variable Quality of Service (QoS). With the major part of the research effort devoted now to creating a new "building block" able to specify QoS parameters (such as bandwidth, delay and loss characteristics), the search for a new Internet architecture seems cognitively to approach the ODN model. Both approaches impose a narrow point in the protocol suite, "isolating the application builder from the range of underlying network facilities, and the technology builder from the range of applications" [3].

This newer approach for internetworking goes beyond network interconnection, to horizontal integration. It adopts one of the most successful architectural elements of the current Internet – the existence of a spanning layer located *outside of the wires,* providing, on top of the IP, *additional functionality* for applications supporting services with *multicast* and *bounded delay* constraints. At the time this chapter was written, the Internet Engineering Task Force (IETF) is working towards standardization of a set of extensions to the datagram functions. Essentially, their effort will standardize the protocol which will generate a *flow* from the source to the destination (i.e. a sequence of packets associated with a single application and sharing common requirements for bandwidth allocation) and will then give the network elements (routers and switches) the ability to directly recognize the existence of this sequence [30]. Stated in another way, this approach aims to define a *service model* which describes[15], in terms of Quality of Service, the set of network services offered to the applications, and then establish and maintain at each network element a *reservation state* to provide the service requested by the application.

Key mechanisms for the implementation of that scenario include [31, 32]: i) an application-initiated *flowspec* (corresponding to a set of information), signaling to the network the type of service this application requires; ii) a process of *admission control* to decide whether the network can provide the requested service; iii) a *resource reservation* process which establishes and maintains state[16] to all nodes along the path(s) of the flows, thus providing the requested service[17]; iv) a *packet service algorithm* to schedule packet transmission which lets the network switches and routers meet the flow's requirements. A specific protocol, the RSVP, has been standardized (and is now being implemented) for these purposes. Even though RSVP resides on the top of the datagram, it does not belong to the transport protocols family (usually implemented in the end nodes), since RSVP does not

[15] Shenker [29] proposes such a service model with two kinds of real-time service, several classes of best-effort services and a modified virtual leased-line service (for "bulk" virtual capacity).

[16] Clark [25] calls that a soft state, in the sense that reservation is maintained at each network element but this is periodically refreshed by the end points.

[17] This is called signaling in the telecommunications world.

transport applications data but only accommodates flow control functions. The new Internet spanning layer can therefore be defined as the *reservation paths*[18] the applications "follow", in order to operate under their own requirements in Quality-of-Service (QoS). Apparently, as RSVP over IP is only concerned with the QoS of the forwarded packets[19], this bearer service can work with any substrate technology, including ATM virtual paths.

ATM proposes a different approach to the horizontal integration of the existing patchwork of vertically integrated communications services, through a single substrate which has been designed to accommodate a diverse type mix of traffic (voice, data, video etc.) [22]. In such a model, the virtual network consists of a single technology. ATM's container, corresponding to the 4 kHz circuit or the 64 Kbps channel, is a 53-byte cell [33]. But unlike the PSTN, ATM constructs virtual paths of varying bandwidth out of 53-byte cells. As alluded to above, a second characteristic of the ATM model is the architectural integration of bitway technologies (Figure 2).

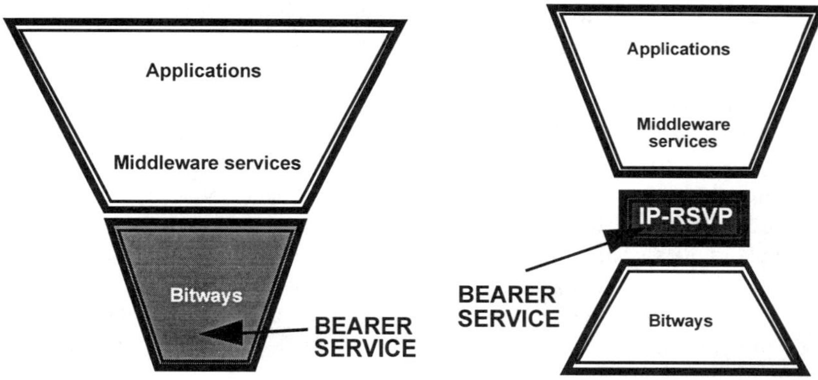

Figure 2. ATM Versus RSVP Over IP

In more general terms, the consideration of the ATM layer as the "spanning layer" raises some problems. One is that ATM is simply one technology among many, and given the rapid evolution of LAN technologies (fast Ethernet etc.), it would be hard to imagine ATM becoming the universal technology basis for applications development[20]. A second problem relates to implementation issues. Telecommunications Operators have adopted a "command and control" approach for signaling (SS7) between ATM switches, which has been criticized for promoting centralized control [22]. Consequently, if ATM should go "to the desktop", specific regulation for "unbundling" and more symmetric implementation might be defined. Still, the efficiency of such policies is open to debate, given the relative success

[18] We can imagine individual resource reservation but also reservation for groups of "calls", or multicast data distribution and multipoint-to-multipoint paths.

[19] As Braden *et al.* [31] point out, the RSVP is not itself a routing protocol: "RSVP is designed to operate with current and future unicast and multicast routing protocols. An RSVP process consults the local routing database(s) to obtain routes."

[20] *Network World*, The shrinking world of ATM, 30 June 1997.

of the Internet which is not founded on a "command and control" philosophy and is governed with minimal regulation.

2.7 From Spanning Layer to Bearer Service

This section began by considering the traditional telecommunication infrastructure and its successive evolutions in light of Piore's perception of the transformation of the industrial system. The organizational archetype is the *Bell System* design, a ubiquitous *mass production* system for telephony, where application and substrate are not only tightly coupled but in fact developed in concert. In time, this model has produced a slight differentiation, by offering a number of additional services (fax and data transmission via modems) over the initial spanning layer (i.e. a 4 kHz point-to-point channel, the telecommunication equivalent of the *railroad gauge).* With the introduction of digital technologies, we observed a more fundamental transformation in network production towards increasing variety at the level of service offerings. However, the application and service design are still designed into the system ex ante. Therefore, flexibility is *formally* developed within a "closed set with a finite number of elements" *(flexible mass production).*

The section then explored newer network models. Interconnection of heterogeneous networks (as exemplified by the Internet) and, now, horizontal integration mark the continuing shift from *cosmetic variations* in mass production to more *flexible mass production* or possibly *flexible specialization.* In these models, applications may be able to go over multiple network technologies (substrates) without the need to design applications and services in advance, as a function of particular substrate constraints.

By fitting the evolution of network architectures to Piore's framework (Figure 3), we were able to extract functions of the *spanning layer* and predict how the *bearer service* may evolve. A trivial spanning layer, embedded in the substrate technology, seems to characterize early railroads and telecommunications mass-production networks. Increasing service variety, due to digital transformation of the network, required developing extensions to the functionality supported by the spanning layer. However, the spanning layer itself remained tied to the infrastructure *wires.* It escapes from the wires only with the Internet, which suggests a network model that might be described by the *container metaphor:* a specific protocol (IP) plays the role of a common denominator among different substrate technologies. Because IP is a technology-independent spanning layer which enables independent variation of applications above it as well as substrate technologies below it, it is a layer above the bitways which enables more design flexibility and innovation.

In the near future, integrating flow control mechanisms (assuring Quality of Service) with IP should extend Internet's application variability to include services with real-time constraints. Again, rather than implementing this functionality within the bitways, an enhanced IP embeds the additional functionality into the spanning layer. In this way, a new "building block" (RSVP over IP) appears for supporting a network composed from Integrated Services Packet Networks (ISPN). These characteristics of the new Internet "building block" (the span of underlying infrastructure facilities and the range of applications that can be served) in combination with the existence of *symmetrical* properties (the type of delivered service is enforced by the end points) make it seem a sort of implementation in the Internet real world of the ODN *bearer service* concept, for two reasons. In the ODN network model, the discussion on the bearer service is in abstract terms and there is no provision for implementation. But the position of the "spanning layer" is exactly the same as the bearer service; it is close to the network technology substrate. Throughout, the whole system may scale over number of networks and different substrates. And second, the ODN bearer service layer is clearly defined as providing quantified Quality-of-Service measures for bandwidth, delay and loss characteristics. Simultaneously, this layer must also provide mechanisms for accounting at a sequence of packet levels and incorporate feedback loops to utilize accounting statistics in capacities such as evaluating QoS commitments and traffic management.

Piore transformation		Freight transport	Comm. networks	
Mass production	Tech-embedded	Early railway	Early voice	Spanning
Mass production with cosmetic variation			Fax and data 4 kHz analog	
Flexible mass production			Digital voice	
More flexible mass production			ATM substrate-bound	
Flexible specialization	Tech-indpt	Container-based multi-modal	Internet + multiple apps/subs	Bearer service

Figure 3. Trajectories of Technical Change

This approach contrasts with an "all-ATM network" strategy for horizontal integration. Though ATM uses a common container (a fixed-size cell), in at least one respect, ATM appears to evolve directly from the *railway gauge* metaphor: while ATM is application-blind, it does not offer independence from the substrate technologies. Essentially, ATM tends more towards a flexible mass production system (or a *flexible specialization with large and discontinuous change* system[21]) by pre-specifying the basic commodity products for meeting customer needs or constructing higher level applications (ATM was designed as an applications-independent platform with three general applications at its center: telephony, high-speed data communication and video delivery).

To conclude: a spanning layer lies between a set of network substrates and higher-level applications. *As a special case of a spanning layer, the bearer service is both substrate-*

[21] Piore [8] defines flexible specialization with large and discontinuous change as follows: "In any system, some moves will be incremental and some will be large and discontinuous... An open (flexible system) for which a change is large and discontinuous is in some respect similar to a system of flexible mass production".

independent and application-blind. Only those functions which cannot be implemented at the periphery of the network substrate are included in the bearer service. Because not all substrate technologies can support all application requirements, application requests must be matched to particular substrate characteristics. To preserve transparency, the bearer service conducts the pairing rather than having applications explicitly request particular underlying technologies.

3. The Economics of the Bearer Service

The evolution of communications spanning layers from homogeneous substrate/application pairings to the bearer service may change the economics of this industry. While section 2 concentrated on how spanning layers have evolved to incorporate heterogeneous applications and substrates, this section will explore the economics of this evolution and describe the eventual structure of the bearer service market. We describe why the conclusion of vertical integration described by Gong and Srinagesh [6, 7] may not be the only way to sustain a bearer service market. Essentially, we will raise questions about possible *governance structures* in networks with layered architectures.

3.1. Institutional Forms Associated with the "Container" Network Model

The argument we developed earlier built around a distinction between two different trajectories of technical change. The first derives from traditional telecommunications *mass production* organizational models and involves a pattern of innovation consisting of anticipating user needs in communications services and designing the corresponding underlying network architecture. In contrast, the Internet and newer approaches for horizontal integration of an infrastructure become increasingly heterogeneous, and separate bitways from applications (and service offerings). Technically, this is possible by defining an interface (IP) to the basic infrastructure facilities and then building on top of that interface the general functionality for applications. We argued that this particular approach to technological change leads to a new technological trajectory which we may metaphorically call *container-based interoperability.* Its essential difference from previous organizational models is that applications may work over multiple substrates (i.e. network technologies) and that these substrates do not determine the development of new applications. It is increasingly clear that this new trajectory brought up considerable flexibility in the design of applications the customers use (as exemplified by the extraordinary development of the World Wide Web). Because it "blindly" uses network resources, system production (applications and service offerings) takes the characteristics of an open-ended product line able to generate (possibly infinite) varieties.

The *container approach* appears to be different than mass production since it is associated with institutional innovations which have begun to crystallize into a *new specialization,* as is apparent from the market of *Internet Service Providers (ISPs).* To explain ISPs' sudden appearance, one may argue that the adoption of a design imposing a technology-independent bearer service introduced incentives to explore new forms of organization which had previously been rejected [24]. If IP involved a separation of the applications from the infrastructure facilities, efficient utilization of the resources would not require a multilevel switching system (and the corollary huge investments raising powerful entry barriers), as in the case of traditional telecommunications networks, but only simple forwarding of "IP packets" (i.e., the network is *not cognizant* about the applications it carries). The question then is to organize a flexible system of regional or more extended backbones and access links to these backbones. In order for this to occur, entrepreneurial, flexible, and small organizations have emerged in the Internet Services Provider market. They offer a bundle of services to their customers (such as Internet Access Providers, individual organizations, firms, and academic establishments), mainly for IP transport, but also may provide the necessary hardware and software, and customer support [34].

Although still a somewhat embryonic sector, the ISP proliferation challenges the huge classic organizational model of the Telecommunications Operators (TOs). However, a number of colleagues suggest that ISPs might not be able to stay as unintegrated firms and will move downwards, trying the acquisition of their own infrastructure facilities. And, on the other side, backbone providers (mostly, traditional Telecommunications Operators) will integrate upwards, since, according to Gong and Srinagesh [6, 7], the economics of the *layered networks* would not tolerate niche players leveraging off the investments of the infrastructure owners[22]. Essentially, as the same authors also point out, the question is to know whether ISPs and TOs with varying degrees of integration can coexist, or if the ISP model based on external provisions of network capacity is only a transitory one. In large part, the response to this question depends on the existence of an efficient market for technology-independent bearer service.

3.2. Unbundled Bearer Service?

This market may not be sustainable, according to Gong and Srinagesh [6, 7]. Their argument consists of a series of propositions which we summarize here:

i) In networks with layered architectures, competition at the bottom layer (infrastructure facilities) of the network hierarchy is unsustainable. Markets for raw transport under conditions of excess capacity, and oligopolistic competition for a homogeneous good, turn easily to "destructive competition" arising from Bertrand equilibrium[23] (prices decrease to marginal costs following the example of the leased line market). Consequently, facilities-based carriers competing for raw transport essentially on price may not cover their sunk and fixed costs and be able to afford competition with non-facilities-based resellers and ISPs.

ii) One way to avoid Bertrand competition is through bundling of bottom-layer transport with higher services, closer to final customers (vertical integration). With bundling, carriers are able to differentiate their services, segment the market, and price accordingly.

iii) As facilities-based companies integrate with others at higher layers, variable costs rise significantly along with the minimum efficient size.

iv) Given the above hypothesis, "policies promoting competition in the provision of unbundled bearer service among owners of physical networks may ultimately fall".

---

[22] As P. Srinagesh kindly reminded us, recent mergers between backbone providers and ISPs (GTE-BBN, MFS-WorldCom-Alternet) seem to give empirical evidence to this statement. We note, however, that in many industries, waves of mergers coexist with, or are punctuated by, other less centralized forms of organization.

[23] Gong and Srinagesh describe raw transport as a service that is like a commodity. A commodity is a product that is so standardized and sold by enough firms that no firm can set the price – all firms are price takers. The Bertrand competition model in economics holds for such a market. Assuming that all firms have a different marginal cost and they will exit the market if the price is set below or equal to their marginal cost, the only sustainable market price is equal to the marginal cost of the second-lowest firm.

Gong and Srinagesh extend their analysis to other forms of bundling and horizontal mergers, to demonstrate that integration may be a natural outgrowth of competition in the convergence spurred by technological advancement (layered digital architectures) and deregulation. Although increasing diversity in service offerings, *economies of scale* and *subadditivity* seem still powerful and responsible for generating growth, as in the old trajectory of *mass production*. However, counterexamples from the Internet come easily to mind. While it may still be too early to tell whether the Internet Service provision is "sustainable", it is true that there is a mix of integrated and non-integrated ISPs, with market competition rewarding more and more product differentiation[24] and competence accumulation (i.e. learning) in improving offerings[25].

In analyzing the economics of layered networks, one should probably make a distinction between economies of scale associated with fixed investments in facilities and other forms of growth (new applications generating markets and user-oriented learning). Some firms may grow even though vertically integrated suppliers representing closed systems will dominate the market. Stated in another way, the question is to know whether the classical Chandlerian hypothesis [35] indicating that in infrastructures, merger-generated megafirms yield a clear advantage in the competition, is a general statement or a phenomenological one.[26]

This chapter does not suggest answers to all these questions. It simply addresses the topic of an efficient market for bearer service which seems to be a necessary condition (but not sufficient) for allowing different organizational structures to coexist. Refocusing into our primary concern, the remainder of this section will explain the characteristics of the "bearer service" *product* (very different from a commodity product). Next, we will explore opportunities for a sustainable bearer service market under two different perspectives, i) *product differentiation* and ii) allocation of *rights of control.*

3.3 What Exactly Is the "Bearer-service" Product?

Defined as a network substrate-independent interface with Quality of Service (QoS) characteristics (section 2), the bearer service *as a product* looks like *sheer bandwidth with QoS attributes,* available over wholly owned networks as a standard (platform-independent) good with quality options (QoS^1, QoS^2, ... QoS^n). From this definition it should be inferred that bearer service is different from a commodity product and, consequently, the bearer service market would be different from raw transport markets.

A commodity product has the characteristic that the demand elasticity of substitution is infinite – if the price of the commodity goes up, then the demand goes to zero. If the price goes down, the vendor receives all the demand[27]. This particular characteristic of a commodity results in the service providers (or vendors) setting their price at the same level[28].

---

[24] Varian in [30] recognizes an evolution of the Internet with ISPs now differentiating their offerings without necessarily getting into the content business. Options for differentiation include fast-packet switching technique, proprietary routing techniques, development of tacit complementary assets for security and network management and, probably, capacity management techniques such as yield management.

[25] A quick examination of The List (http://www.thelist.com), a listing of Internet Service Providers, offers many examples of non-integrated ISPs with good local or more extended implementation.

[26] On this interesting theoretical question, see in particular Dosi [36].

[27] This is true as long as we live in a world with weak increasing returns.

[28] Assuming slight product differentiation, the market would be shared by a number of companies and an equilibrium price would be established – given different product characteristics and availability.

The leased line market may not have an infinite demand elasticity of substitution, but it is much larger than the elasticity for substitution of a bearer service product. The reason comes from the ability of the leased line vendor to fulfill the service request it receives from the customer versus the bearer service provider doing the same. In the leased line case, customers specify the bandwidth of a connection often days or weeks in advance of using that bandwidth. There may be some specification about the reliability of that link, but, in general, the leased line is always there for the customer. In a bearer service market, the customer demands service in a dynamic manner dependent upon the application's service requirements – perhaps only a few milliseconds before the service is delivered. Therefore, the bearer service provider must choose technology substrates to service application requests based upon the expected demand. This not only results in a usage-sensitive pricing policy to limit the use, but it also encourages the service provider to make the exact provision in network capacity to minimize costs (taking advantage of statistical sharing to service the users' requests). The result is a probability of blocked service requests that does not exist in the leased line market. Customers then see a noticeable difference between bearer service providers, resulting in a price versus performance decision on their part. Because one bearer service provider underprovisions its network less than its competitor, it will also increase its performance by refusing fewer service requests, thereby commanding a greater price for its service. The result is a demand elasticity of substitution that is less than the leased line market indicative of a non-commodity product.

Another commodity characteristic that leased lines have, and the bearer service product does not, is transport between two points as opposed to between a point and a shared "cloud" (Figure 4). The leased line customers specify the points they wish to connect, not only their bandwidth requirement. It is a point-to-point connection. In the Internet, however, the service between two points traverses through the cloud and the full point-to-point performance is dependent on the sender's access link *and* the recipients. Likewise, the bearer service does not specify two communication points a priori, but it only gives access to the shared cloud.

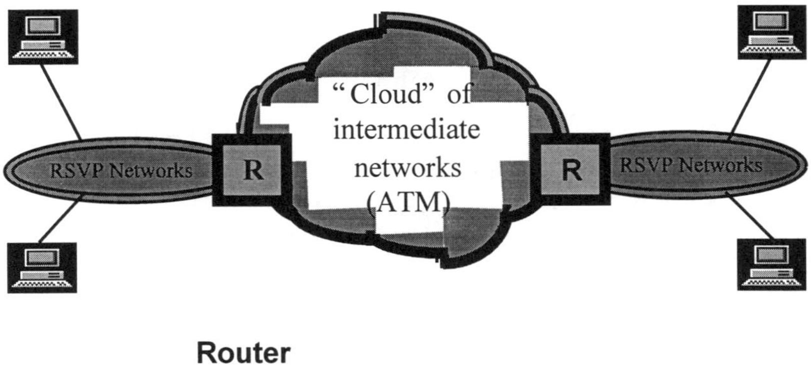

Figure 4. The Next Internet

While the bearer service cloud does more than the Internet cloud (e.g., provides a guaranteed quality of service), the ultimate functionality depends upon the access link of bearer service provider connecting the sender to the cloud and the quality of the access link connecting the recipient of the cloud. If the same bearer service provider connects both the sender and the recipient, there may be no reason to traverse the cloud and the provider can handle the transmission internally. The result is a provider that is able to offer better service because it has a large network of customers. The economics of *increasing returns* apply here, giving customers greater value if there are more customers connected to their bearer service provider's network.

The "cloud" metaphor is used to describe the statistically shared wide-area transport of the Internet. In this chapter, we argue that the "cloud" metaphor can also be used to describe the bearer service network topology. More precisely, we assume a network topology with a fast-packet (for example ATM) "cloud" and RSVP connections to Users' Networks. Overall, the leased line product does look more like a commodity than a bearer service because the offering is relatively homogeneous and static. The service demands of customers in the leased line market do not change dramatically from day-to-day because the service has been established a priori for a fixed time period (usually months or days). The bearer service provider must service requests very dynamically and it must service requests for an unknown or short period of time (perhaps seconds or minutes). Furthermore, the leased line product that one provider can offer versus another provider is not very different. In the bearer service market, there are distinguishing characteristics about the provision of bearer service that can differentiate the providers, thus making competition less susceptible to "destructive prices"[29]. In addition, bearer service provision requires complex investments in different families of assets ranging from network hardware and management tools to link capacity.

3.4 Bearer Service Differentiation

Since the bearer service is not a commodity, it is important to explore ways that it can be differentiated when provided by competitors. Independent bearer service providers can: i) choose different substrate technologies, ii) design a different network topology than their competitor and iii) design a different pricing policy for their service. Each of these three methods will be explored in detail below and compared to the leased line market.

The choice of substrate technology is the first way that bearer service providers can differentiate themselves from a competitor. As discussed earlier, while the bearer service can support all applications and all substrates, it is not possible for all application/substrate pairings to work together. But, by choosing a particular substrate technology, the application space can be constrained. For example, if the bearer service provider chooses Ethernet as its substrate technology, it cannot offer a guaranteed quality of service to the customer, while its competitor with an ATM substrate can offer this service. The subset of applications offered by the Ethernet-substrate provider is smaller than that of the ATM-substrate provider, but their cost structure is also different. By choosing Ethernet, the costs are lowered and, therefore, they can pass on some savings to the customer. In summary, the application space may be constrained by choosing a particular substrate technology.

[29] Besides, even globally decreasing prices in the leased line market are not so "destructible". Big discounts and "back of the book" tariffs for large-volume and long-term contracts seem to correspond to nonlinear pricing strategies and other pricing discrimination techniques.

The choice of network technology is another degree of differentiation between the bearer service providers. As outlined above, connection to the *cloud* is very different than connection between two points. Therefore, the way a bearer service provider connects to the cloud and how dependent the provider is on the cloud for service are differentiating factors. The better the connection to the bearer service cloud and the less dependent the provider is on the cloud, the lower the probability of a denied service request. To use a metaphor from transportation, the bearer service provider can provide transport to a "hub" or exchange point where you can hop on another link. If this exchange point is in a remote location, or experiences heavy congestion, then the service is providing you less value than if it takes you to a hub with greater service.

Pricing policies can be a powerful service differentiation mechanism. Since service requests for bearer service provision must be answered dynamically, it is likely that pricing will also be set in a dynamic manner (for example, as a function of the requested QoS). Pricing policies for the Internet to give a quality of service better than "best effort" are discussed in McKnight and Bailey [37]. How a provider chooses to make capacity provisions, and how it defines prices to maximize its revenues, should distinguish one bearer service provider from another[30].

3.5 Shared "Residual Control Rights"

Perhaps another way to evaluate the sustainability of a bearer service market is to consider the physical assets, in the post-investment stage, which make the relationship between suppliers (S) and buyers (B) of bearer service – B may correspond to a non-facilities-based ISP. Then, following Hart [38], we can assess the economic implications of different governance structures.

One critical issue for governance structure is what Grossman and Hart [39] call the *rights of residual control.* These are the rights to make decisions about the usage of an asset which is not anticipated in a contract between the parties involved (incomplete contracts). There are different ways to allocate these rights, integration (in both directions, from S to B or B to S), or non-integration, with integration involving the ownership of an asset (which goes together with the residual rights of control over the asset). The choice of the ownership structure depends upon the specifics of the relationship investments and the distribution of the information required to make decisions. Grossman and Hart [39] develop a theory of integration based on the above principles which allows them to formally evaluate the costs and the benefits of integration (or non-integration).

The allocation of the residual rights of control is obvious in the mass production model. As long as the bearer service resides *within the wires,* the rights of control over bearer service assets (i.e. network resources offered under *Open Network Architecture* provision) are naturally held by the infrastructure providers. Resellers and other *third parties* are

[30] Other ways the bearer service providers could differentiate themselves include marketing and customer service. While these differentiating factors have nothing to do with the transport of bits or the ability of an application to access the bearer service, it does influence the customers' decision when they trust their provider or achieve benefits from better customer support. Brand recognition that results from advertising and marketing can influence a decision-maker and also lead to differentiation between competing service providers. These differentiation tools are understood by the marketing literature and are beyond the scope of this chapter for adequate analysis.

bound by a contract which gives them access to network resources, but it is understood that the infrastructure owner has the right to use the asset in any way that is not inconsistent with the law. The infrastructure provider (S) would have the right to define production capacity and decide on technological investments, assuming that the initial contract is silent about these. Because the operating firm possesses the residual rights, it will receive a greater fraction of the *ex post* surplus created by its investments, so these rights work as incentives to make investments and expand the relationship with the third parties.

But with bearer service *escaping from the wires,* flexibility is increasing because the *openness* of the IP protocol makes the assets less complementary, i.e. less specific to the relationship. Under the assumption of a network with *fast-packet cloud* and RSVP routers to connect to that cloud (Figure 4), investment decisions occur at both the *cloud* and the *periphery,* so non-integration is likely to be a serious hypothesis[31]. As Piore [8] concludes in more general terms, when the rights of control are held by organizational components upstream in the production process, the organization may be more decentralized[32].

4. Conclusion

What we suggest is that the impressive diffusion of the Internet may signify a transition towards flexible models of organization in the communications industry. Flexible organization involves independence of applications from the investment cycles of the infrastructure facilities and it significantly depends upon the efficient existence of a bearer service market. Such a market, if implemented, may establish the *symbiosis* of backbone suppliers with independent ISPs.

One of the main conclusions of this chapter is that the bearer service may be sustainable in the absence of vertical integration. It is the flexibility of the bearer service itself which helps the bearer service providers to differentiate themselves from their competition and set prices based upon varying degrees of value to their customers. The differentiation of the service can be realized in many forms. The bearer service providers can choose a substrate technology to better support a set of applications, they can design a network topology to minimize the number of denied service requests, and they can formulate a pricing policy different than their competition. The different offerings can appeal to different classes of users and provide a mechanism for price discrimination and service flexibility, so both the provider and the customer benefit.

However, the ability to vertically integrate to absorb the bearer service is also possible. We realize that there are benefits to vertical integration from a substrate provider to a bearer service provider, or from a bearer service provider to a middleware provider. While this chapter does explore a rationale behind an independent bearer service market, it does not adequately explore the comparative benefits of vertical integration and disintegration (nor the conditions favoring the one or the other approach). These are important questions which would be a valuable extension to this chapter. To respond to these questions, it appears that

[31] Hart [38] considers non-integration as optimal, as long as assets are non-complementary (independent). In this case, the effect of the full control transfer to S (a backbone operator) is to reduce B's relationship-specific investments, which may reduce openness and, as a corollary, total surplus.

[32] The thoughts outlined above are at a preliminary stage and one should pursue them further to come up with firm conclusions.

one must combine elements from the theory of *contracts* [38, 39] with the *competencies-based* analysis of organizations [40].

Overall, the bearer service appears as an opportunity which might prevent "closure" and collusive price-fixing in future integrated services networks and, more generally, should reinforce Schumpeterian *creative destruction* forces. But, while a market for bearer services may be economically viable, that market is also fragile. New institutions may be required to enable the market to form. For example, to check short-term influences for horizontal and vertical market integration in a digital network with a layered architecture, regulation mechanisms, although not necessarily government intervention, might be appropriate for defining risk-sharing agreements between providers of services above and below the bearer layer. May the stock market create these mechanisms?

References

[1] D. Clark, Interoperation, Open Interfaces, and Protocol Architecture. Laboratory for Computer Science, Massachusetts Institute of Technology, mimeo, 1995.

[2] Computer Science and Telecommunications Board (CSTB), National Research Council, Realizing the Information Future: The Internet and Beyond. National Academy Press, Washington, D.C., 1994.

[3] Computer Science and Telecommunications Board (CSTB), National Research Council, The Unpredictable Certainty: Information Infrastructure Through 2000. National Academy Press, Washington, D.C., 1996.

[4] Internet Telephony Interoperability Consortium (ITC), Workplan 1997. MIT.

[5] J. Bailey et al., The Economics of Advanced Services in an Open Communications Infrastructure: Transaction Costs, Production Costs, and Network Externalities, *Information Infrastructure and Policy* **4** (1995) 255-278.

[6] J. Gong and P. Srinagesh, The Economics of Layered Networks. In: L. McKnight and J. Bailey (eds.), Internet Economics. MIT Press, Cambridge, MA, 1997, pp. 63-77.

[7] J. Gong and P. Srinagesh, Network Competition and Industry Structure, *Industrial and Corporate Change* **5** (1996) 1231-1241.

[8] M. Piore, Corporate Reform in American Manufacturing and the Challenges to Economic Theory. Department of Economics, Massachusetts Institute of Technology, WP #90-079, 1989.

[9] M. Piore and C. Sabel, The Second Industrial Divide. Basic Books, USA, 1984.

[10] M. Piore, Beyond Individualism. Harvard University Press, Cambridge, MA, 1995.

[11] M. Piore, Fragments of a Cognitive Theory of Technological Change and Organizational Structure. In: N. Nohria and R. Eccles (eds.), Networks and Organizations: Structure, Firm and Action. Harvard Business School Press, Boston, MA, 1995.

[12] C. Kindleberger, Standards as Public, Collective and Private Goods, *Kyklos* **36** (1983) 377-396.

[13] D. Puffert, The Economics of Spatial Network Externalities and the Dynamics of Railway Gauge Standardization. Stanford University, Dissertation Thesis, 1991.

[14] D. Tennenhouse et al., Virtual Infrastructure: Putting Information Infrastructure on the Technology Curve. Laboratory for Computer Science, Massachusetts Institute of Technology, TM-523, 1995.

[15] G. Brock, The Telecommunications Industry: The Dynamics of Market Structure. Harvard University Press, Cambridge, MA, 1981.

[16] J. Walrand, Communication Networks. Irvin, Boston, MA, 1991.

[17] D. Bertsekas and R. Gallager, Data Networks. Prentice-Hall, Englewood Cliffs, N.J, 1992.

[18] G. Dicenet, Le RNIS: Techniques et Atoûts. Masson, Paris, 1988.

[19] M. Elton (ed.), Integrated Broadband Networks: The Public Policy Issues. Elsevier Science Pub., North Holland, 1991.

[20] R. Mansell, The New Telecommunications. Sage Pub., London, 1994.

[21] P. Kavassalis et al., The Internet: A Paradigmatic Rupture in Cumulative Telecom Evolution, *Industrial and Corporate Change* **5** (1996) 1097-1126.

[22] D. Messerschmitt, The Convergence of Telecommunications and Computing: What are the Implications Today?, *Proceedings of the IEEE* **8** (1996) 1167-1185.
[23] G. Muller, International Freight Transportation. Eno Transportation Foundation, USA, 1995.
[24] P. Kavassalis and R. Solomon, Mr. Schumpeter on the Telephone: Patterns of Technical Change in the Telecommunications Industry Before and After the Internet, *Communications and Strategies* **26** (1997) 371-408.
[25] D. Clark, The Design Philosophy of the DARPA Internet Protocols, *Computer Communication Review* **18** (1988) 106-114.
[26] M. Rose, The Simple Book: An Introduction to Management of TCP/IP - based Internets. Prentice-Hall, Englewood Cliffs, N.J, 1991.
[27] G. Dosi, Technological Paradigms and Technological Trajectories, *Research Policy* **11** (1982) 147-162.
[28] F. Bar, Network Flexibility: a New Challenge for Telecom Policy, *Communications & Strategies* **2** (1991) 113-123.
[29] S. Shenker, Service Models and Pricing Policies for an Integrated Services Internet. In: B. Kahin and J. Abbate (eds.), Standards Policy for Information Infrastructure. MIT Press, Cambridge, MA, 1995, pp. 315-337.
[30] R. Braden et al., Resource Reservation Protocol (RSVP) - Version 1 Functional Specification. Internet draft, 1996 (ftp.isi.edu/internet-drafts/draft-ietf-rsvp-spec-14.ps).
[31] R. Braden et al., Integrated Services in the Internet Architecture: an Overview. Internet Engineering Task Force, RFC 1633, 1994.
[32] D. Mitzel et al., A Study of Reservation Dynamics in Integrated Services Packet Networks. Proceedings of Infocom, 1996 (ftp://catarina.usc.edu/pub/mitzel/Infocom96/infocom96.ps.Z).
[33] S. Downs, From Narrowband to Broadband: ATM and the Transformation of the Telephone Network. Massachusetts Institute of Technology, Master's Thesis, 1992.
[34] P. Srinagesh, Internet Cost Structures and Interconnection Agreements. In: L. McKnight and J. Bailey (eds.), Internet Economics. MIT Press, Cambridge, MA, 1997, pp. 121-154.
[35] A. Chandler, The Visible Hand. The Belknap Press of Harvard University Press, Cambridge, MA, 1977.
[36] G. Dosi, Organizational Competencies, Size and the Wealth of Nations: Some Comments From a Comparative Perspective. Presented to XI International Economic History Congress, Milan, Sept. 1994.
[37] L. McKnight and J. Bailey (eds.), Internet Economics. MIT Press, Cambridge, MA, 1997.
[38] O. Hart, Firms, Contracts and Financial Structure. Oxford University Press, Oxford, 1995.
[39] S. Grossman and O. Hart, The Costs and Benefits of Ownership: A Theory of Vertical and Lateral Integration, *Journal of Political Economy* **94** (1986) 691-719.
[40] D. Teece et al., Understanding Corporate Coherence, *Journal of Economic Behavior and Organization* **23** (1994) 1-30.

Part IV

Current Issues in Pricing

Telecommunications Transformation: Technology, Strategy and Policy
E. Bohlin and S.L. Levin (Eds.)
IOS Press, 1998

CHAPTER 12

A Yardstick Approach to Optimal Access Pricing

Dale E. Lehman
Department of Economics, Fort Lewis College
Durango, CO 81301, USA

Dennis L. Weisman
Department of Economics, Waters Hall, Kansas State University
Manhattan, Kansas 66506-4001, USA

Abstract. The issue of efficient pricing of unbundled network elements in telecommunications markets is being debated as competition is introduced into local telecommunications markets. There is currently no consensus among economists regarding the optimal pricing rule, nor does it seem likely that such a consensus will form any time soon. The problem is that no pricing rule has yet been found that addresses the myriad of complex issues surrounding the introduction of competition. These include static, technical, and dynamic efficiency, "incumbent burdens", subsidy-laden retail rate structures, discrimination, and imperfectly competitive downstream markets. In this paper, we examine a variant of the efficient component pricing rule that we term the "yardstick efficient component pricing rule" or Y-EPCR. The Y-ECPR addresses three common criticisms of the standard ECPR. First, the ECPR may not provide ideal incentives for unobservable cost-reducing innovation. Second, the ECPR "keeps the incumbent firm whole"; and third, the ECPR creates incentives for the incumbent firm to withdraw from the retail market entirely since it realizes the same level of profits from either wholesale or retail supply. In general, we find that optimal access pricing rules necessarily involve trade-offs between the various moral hazard problems.

1. Introduction

The issue of the efficient pricing of unbundled network elements in telecommunications markets is being debated around the world by economists and public policy analysts as competition is introduced into local telecommunications markets. Though many pricing rules have been proposed, there is currently no consensus among economists regarding the optimal pricing rule, nor does it seem likely that such a consensus will form any time soon. The problem is that no pricing rule has yet been found that addresses the myriad of complex issues that surround the introduction of competition into local telecommunications markets. These include static efficiency, technical efficiency, dynamic efficiency, "incumbent burdens", discrimination, subsidy-laden retail rate structures, and imperfect competition in downstream markets.

Interconnection pricing is further complicated by a number of issues during the *transition* to competition. Among these are: depreciation overhang, residual incumbent burdens, distorted retail price structures, economies of scale and scope, rapid technological change, and an intensely politicized environment of rent seeking/rent protection by entrants, incumbents, and regulators. Amidst this setting, designing interconnection prices which simultaneously address the incentive problems of all stakeholders is difficult at best.

In particular, our setting is one fraught with moral hazard problems. A moral hazard is a particular incentive problem that arises when the economic agent does not incur the full costs of a loss and, as a result, fails to put forth the efficient level of effort (which cannot be observed directly) to avoid that loss. For example, an individual may not take adequate precautions in locking the doors on a rental car or parking the rental car to avoid damage because he does not pay the full cost in the event of theft or damage. The moral hazard problem explains why most insurance policies require co-payments and/or deductibles. In the present context, interconnection pricing which ensures that incumbents will get all costs reimbursed reduces the incentive to minimize cost; pricing on the basis of unverifiable costs reduces entrants' incentives to provide realistic estimates of those costs; and pricing which protects the existing retail rate structure reduces the incentives of regulators to pursue more efficient pricing.

In this paper, we examine a variant of the efficient component pricing rule that we term the "yardstick efficient component pricing rule" or Y-EPCR. The Y-ECPR addresses three common criticisms of the standard ECPR. First, the ECPR may not provide ideal incentives for unobservable cost-reducing innovation. Second, the ECPR "keeps the incumbent firm whole"; and third, the ECPR creates incentives for the incumbent firm to withdraw from the retail market entirely since it realizes the same level of profits from either wholesale or retail supply. In general, we find that the optimal access pricing rules necessarily involves some trade-off between the various moral hazard problems.

The format for the remainder of this paper is as follows. In Section 2, we present a brief overview of access pricing issues. A discussion of the different concepts of costs as they relate to the access pricing issue is developed in Section 3. Section 4 is a discussion of the criticisms that have been leveled at the ECPR. Section 5 addresses the yardstick approach to access pricing and its underlying incentive properties. A brief policy discussion, the conclusions and directions for future research are contained in Section 6.

2. An Overview of Access Pricing Issues

One of the fundamental issues concerning the introduction of competition into local telephone service markets concerns optimal access pricing.[1] These issues are in the forefront in the US market as a result of the 1996 Act. As European telecommunications markets are opened to competition in the next several years, access pricing for new competitors in formerly monopoly markets will become a central focus of regulatory attention. In this section, we briefly summarize the debate over the ECPR, while offering some new perspec-

[1]The FCC will formally address the issue of access charge reform in the third part of its trilogy of orders implementing the 1996 Telecommunications Act. The first part of the trilogy addresses local service competition and interconnection while the second part concerns universal service. In the absence of RBOC entry into interLATA markets, setting access charges may be viewed as a different pricing problem than setting interconnection/unbundled element prices. Post entry, RBOCs and IXCs will be competing in the downstream market and access charges and interconnection pricing become similar pricing problems.

tives on the relationship between the optimal access charge and its underlying marginal production cost.

If there is one theme that rings constant throughout the debate on the optimal access charge it is the importance of establishing a "level playing field". The term "level playing field" (or LPF) is not well-defined in the sense that it means different things to different parties. For our purposes here, we define LPF to be a set of regulatory rules that ensures the least-cost provider is not precluded from being the least-price provider. With this definition, we proceed to develop a clear understanding of the rationale for various access charge pricing rules that have been advanced in the course of what has now become a rather protracted debate.

We begin with some brief notation to simplify the discussion that follows. Let w represent the access charge and c its underlying marginal cost of production. Also, let s represent the marginal cost of the self-supplied input to the production of local telephone service. (Without loss of generality, we assume that each unit of basic telephone service requires one unit of access and one unit of the self-supplied input.) Hence, the entrant's marginal cost of providing basic telephone service is $w + s$. A common proposal for access pricing, generally advanced by the interexchange carriers (IXCs), is to set the price of access equal to its marginal production cost, so that $w = c$.[2] This proposal is sometimes referred to as "cost-based" access pricing, though, as we shall see subsequently, this reference is sufficiently imprecise (there are, of course, many different measures of cost) to be very useful in this discussion.

The proposal to set $w = c$ certainly has intuitive appeal, at least on the surface. It would appear to be consistent with a LPF in that the LECs would not be able to exploit an artificial cost advantage merely because of their control over essential access facilities. In other words, the cost of access to the IXCs would be the same as the cost of access to the LECs.[3] It follows immediately that with $w = c$, the least-cost provider of local telephone service is the provider with the lowest marginal cost of the self-supplied input (s).

Upon closer examination, however, setting $w = c$ will frequently turn out to be inconsistent with a LPF. Recall that according to our definition of a LPF, a set of regulatory rules must not preclude the least-cost provider from being the least-price provider. Suppose that due to incumbent burdens (e.g., universal service obligations, rate averaging, carrier-of-last resort obligations, etc.), the LEC cannot set its retail price for local telephone service below P^L, where $P^L > c + s^L$, and where the superscript "L" denotes LEC. The IXCs, however, are not subject to these constraints. Suppose that the IXC sets a retail price of $P^I < P^L$, where the superscript "I" denotes IXC. Observe now that the following scenario is possible: $P^I < P^L$ when $c + s^L < c + s^I$, or $s^L < s^I$. That is, the least-cost provider may be precluded from being the least-price provider and the LPF property breaks down. This leads to the following conclusion. In the presence of incumbent burdens, setting $w = c$ is generally not technically efficient.

There are two other potentially serious problems with cost-based access charges. First, setting $w = c$ may provide incentives for the LECs to discriminate against the IXCs in the provision of essential access facilities. To the extent that this discrimination may be largely unobservable to the regulator, it could be difficult to protect against. Sibley and Weisman [13,14] show that the LECs do not have unequivocal incentives to discriminate against their downstream rivals when $w > c$.

---

[2]See Economides and White [4].

[3]To be precise, we are assuming here that the marginal cost of supplying access to the IXCs is the same as the marginal cost of the LEC supplying access to itself.

Second, cost-based access charges provide suboptimal incentives for the LECs to invest in unobservable cost-reducing innovation. Hence, the cost of providing exchange access may exceed its minimum possible level. When the incumbent firm begins to lose market share to rivals, as AT&T did to MCI prior to divestiture and as the LECs are doing to CAPs presently, the benefits of such unobservable investment accrue in part to new market entrants that interconnect with the incumbent firm's network. Because this investment is unobservable, and therefore cannot be rewarded directly, the incumbent firm may have insufficient incentive to invest. In such cases, it may be efficient to (i) design regulatory policies that, at a minimum, provide no artificial advantage to the entrant (and possibly even handicap the entrant);[4] and/or (ii) design interconnection rates that provide sufficient incentive for the incumbent firm (possibly through the use of price caps) to undertake unobservable network investment.[5] This leads to a second conclusion. In the presence of unobservable discrimination and/or cost-reducing innovation, cost-based access charges are generally not technically efficient.

A second proposal for access pricing is known as the efficient component pricing rule (ECPR).[6] According to the ECPR, the efficient access price is equal to the marginal cost of access *(c)* plus the opportunity cost borne by the LEC in providing wholesale access to an entrant rather than supplying the retail local service itself $(P^L - c - s^L)$. Hence, the ECPR = $(P^L - s^L)$. Note that it would also be appropriate to refer to the ECPR as cost-based access pricing in that the appropriate cost measure here includes both marginal production costs and opportunity costs.

The ECPR clearly satisfies the LPF property. To see this, note the following. Let $P^I = ECPR + s^I = (P^L - s^L) + s^I$. Hence, $P^I < P^L$ implies that $s^I - s^L < 0$ or $s^I < s^L$ .Hence, the least-cost provider is also the least-price provider. The rationale behind the ECPR is an appealing one. Namely, each market provider should be required to generate (through its prices) the same level of contribution or subsidy necessary to satisfy social equity goals of regulation (i.e., what were formerly incumbent burdens). In fact, the ECPR is sometimes referred to as the equal-tax rule or parity principle.[7]

---

[4]The idea of handicapping an entrant may seem somewhat inconsistent with the LPF property. In fact, this only underscores the possibility that there may well be trade-offs between static and dynamic efficiency. The LPF property is inherently static in nature.

[5]Laffont and Tirole [7, p. 518] examine a similar problem in the case of renewable franchises, wherein the current operator may lose the franchise when it is put up for renewal. They suggest that it may be efficient to favor the incumbent firm during the bidding process to enhance the firm's incentives to invest in unobservable cost-reducing innovation and modernization. There are significant parallels between a franchise that is lost to another operator and market share that is lost to a competitor who interconnects with the incumbent firm's network.

[6]See Baumol and Sidak [2] for an extensive discussion, and Kahn and Taylor [6] for a thought-provoking review of the relevant issues. The ECPR is actually believed to have originated with an earlier paper by Willig [17]. See Laffont and Tirole [7] for an examination of the market and technological conditions under which the ECPR is optimal, and Larson and Lehman [9] for the conditions under which the ECPR is a special case of Ramsey Pricing.

[7]See Baumol [1]. Imputation rules for pricing the LEC's retail services (e.g., IntraLATA toll) are identical conceptually to the ECPR for pricing the LEC's wholesale services (e.g., exchange access and basic network service elements). The price floor for the incumbent LEC's retail service is set equal to the marginal cost of providing the service plus the contribution foregone in supplying the retail service itself rather than the wholesale inputs necessary for a competitor to supply the retail service. An imputation rule of this type ensures that all participants in the telecommunications market in question generate (through their prices) a minimum level of subsidy or contribution per unit. Recognize that when the incumbent LEC's retail price is set equal to this price floor, the LEC will be indifferent between supplying the retail service or the wholesale service since its margin is the same in both cases.

3. Cost Measures in the Access Pricing Debate

Thus far in our discussion we have not been precise in identifying the various underlying cost measures that comprise the heart of the debate over access pricing issues. We remedy that deficiency in this section.

In the last section, we defined *w* as the access price and *c* as the marginal cost of providing access. However, the precise manner in which *c* should be measured has been the subject of intense debate in regulatory proceedings in the United States following the passage of the 1996 Telecommunications Act [20]. The LECs contend that *c* should reflect their own marginal cost of providing access. In contrast, the local market entrants, the IXCs, contend that *c* should be based on a hypothetical benchmark that reflects a telecommunications network that operates at maximal efficiency using the most current technology, but holding switch locations fixed [16], as this is the standard purportedly satisfied in a competitive marketplace. Hence, the marginal cost of access (*c*) can be classified as either (i) actual or (ii) hypothetical.

The opportunity cost borne by the LECs in providing access to rivals in the local exchange market place is a key and controversial component of the ECPR.[8] In principle, this opportunity cost is measured as the contribution loss borne by the LEC when it supplies the wholesale access input rather than the retail service. The appropriate measurement of opportunity cost has not received a great deal of attention. Suffice it to say that the measurement of opportunity cost entails a number of complex and subtle issues which we take up in later sections of this paper.

4. Standard Criticisms of the ECPR

The conventional ECPR has been widely criticized, both legitimately and otherwise, on a number of different grounds. We briefly discuss each of these criticisms in turn.

First, the ECPR has been criticized because it "keeps the LECs whole" (i.e., monopoly profit levels are maintained).[9] As a result, the LECs have an incentive to withdraw entirely from the retail market.[10] This weakness leads to a possible strength, however. Because the LEC earns the same contribution whether it provides wholesale access or the retail local service, it has no incentive to discriminate against its rivals.

Second, the ECPR has been criticized for not stimulating lower retail prices. Spulber and Sidak [15] have shown that going from a regulated average-cost price does, in fact, lead to lower retail prices in several plausible oligopoly models in which there is competitive entry. Intuitively, this occurs because an entrant with lower retail prices stimulates demand sufficiently to drive the required contribution per unit down to the point where retail price

---

[8] The ECPR framework was rejected by the FCC [18] because it was viewed as being inconsistent with a competitive marketplace. The particular sections of the FCC's interconnection order that address the pricing of unbundled network elements are currently under appeal [Iowa Utilities Board, 19] and the FCC's pricing rules have been stayed in the interim. A final decision from the Eighth Circuit Court of Appeals is expected later this year.

[9] See Economides and White [4]. A rebuttal to this paper is Larson [8].

[10] Kahn and Taylor [6] contend that because the ECPR satisfies the "indifference principle", the incumbent LEC has no incentive to discriminate. They also argue that the LEC will not have an incentive to withdraw from the retail market. We do not believe it is possible to satisfy both of these conditions simultaneously, which is why the standard ECPR may give rise to some potentially serious incentive problems.

equals average cost for the entrant calculated at a lower ECPR. To the extent that regulators set retail prices, then the ECPR appears to mitigate regulatory incentives to rebalance rates. In particular, the ECPR permits regulators to maintain pricing of some services above incremental cost as a means of providing contributions towards lowering other service prices.

Third, because it preserves contribution levels, the ECPR is faulted for not giving the incumbent LEC incentives for cost reduction. With respect to underlying incentives for cost reduction, reductions in both s and c increase the incumbent's profits, so the ECPR mechanism would appear to give the incumbent incentives for cost reduction. We provide a detailed examination of this issue in the next section.

A fourth criticism of the ECPR is that a fall in s^L (the firm's cost of the self-supplied input) has the perverse effect of raising the ECPR. This perversity, however, is more apparent than real. To see this, suppose that the IXC charges a retail price of P^I and that $s^I < s^L$, so that the IXC replaces the LEC as the retail firm. In this case, the IXC's profit per unit of output is $P^I - s^I - (P^I - s^L) = s^L - s^I$,[11] so that the IXC's profit equals its retail cost advantage. Thus, if the LEC lowers its retail cost, the IXC's profit falls, which is intuitively appealing since its cost advantage is decreased.

The next section presents our variation of the ECPR which addresses these potential deficiencies in the ECPR, albeit imperfectly. The imperfections arise because the varied efficiency characteristics of access pricing are in conflict and no access pricing rule can simultaneously address them all.

5. The Yardstick Approach

The pioneering work on the use of yardsticks in economic regulation is by Schleifer [12]. Yardsticks have also been used as a means to compensate hospitals and physicians for medical procedures [3]. Lehman and Weisman [10] examine the use of yardsticks in cost reimbursement rules for telephone pools and access pricing. The primary advantage of yardsticks is in setting up a "tournament" of sorts that allows a monopoly provider to compete against an identical twin. Because the linkage is severed between the firm's compensation and its actual costs, the firm has ideal incentives for investment in unobservable cost-reducing innovation.

We consider a general access pricing rule of the following simple form:

$$w = c + \beta[P - c - s^L], \qquad (1)$$

where $\beta \in [0,1]$ and the other terms are as previously defined.[12] We also assume that P is set by the regulator in this problem in order to focus on the particular incentive issues of interest. In Eq.1, c represents the direct cost of providing the unit of the access service and the terms inside the brackets represent the opportunity cost. We begin with the simplest case of $\beta = 0$, which corresponds to pure cost-based access pricing. As noted above, the IXCs have generally favored an access pricing rule of this type because it purportedly places them on equal footing with the incumbent LEC.

[11]Note that the ECPR is now computed using $(P^I - s^L)$ rather than $(P - s^L)$ because the market price of local service is P^I rather than P^L when the IXC replaces the LEC in the market for local telephone service.

[12] β is defined as the probability that the LEC would have provided the retail service if it were not required to supply the access service to a rival in the downstream market.

There are a number of issues that need to be examined with this particular approach, however. The IXCs claim that c should be based on some hypothetical network planning model, using the most efficient technology, but subject to the constraint that switching locations remain fixed. One obvious problem with this approach is that there is no market determination of c. A network planning model can be constructed that outputs a particular c, but there is no way of empirically verifying whether this c would ever be the outcome of some competitive market process. The LECs claim that c should be based on their own specific costs of providing network access. This proposal suffers from problems of moral hazard identical to those that arise under rate-of-return regulation. The LEC has sub-optimal incentives to invest in unobservable cost-reducing innovation when it is compensated, in whole or in part, for its actual costs.

5.1 Yardsticking Direct Cost

The yardstick approach to setting c solves the theoretical problem of moral hazard but raises the practical problem of measurement in the absence of an identical twin firm. Consider a generic yardstick pricing rule of the form, $w = \tilde{c}$, where $\tilde{c}$ is the actual access cost of the incumbent LEC's identical twin. Normalize output to unity and express the profit function for the LEC in the upstream access market as follows.

$$\pi^u = w - c(e) - \Phi(e) = \tilde{c} - c(e) - \Phi(e), \tag{2}$$

where $c(e)$ is the LEC's actual costs, e is unobservable effort, $c'(e) < 0$, $c''(e) > 0$, $\Phi(e)$ is the disutility of effort, with $\Phi'(e) > 0$ and $\Phi''(e) > 0$. To see that the moral hazard problem is solved with this access pricing rule, we note that maximizing Eq.2 with respect to e yields $-c'(e) = \Phi'(e)$, or the marginal benefit of effort is equated with the marginal cost of effort. We should also note here that in breaking the link between the incumbent LEC's compensation and its actual costs, we solve the moral hazard problem, but we may distort what the FCC refers to as the "build or buy" decision [18, §620].

The absence of an identical twin firm need not doom the yardstick approach. In particular, two substitutes for the identical twin firm are readily available. The first would substitute the average of a suitable "peer group" of firms' costs for the LEC's actual costs. Selection of a peer group is somewhat easier than selection of an identical twin firm, but still suffers from the defect that firms will be rewarded (penalized) if their costs are lower (higher) than average, and this can result from either technical (in)efficiencies or exogenous cost-causing characteristics (e.g., population density, geographical variations, etc.). In this case, a staggered yardstick can be used, where firm level costs are estimated as a function of other firms' costs and relevant exogenous characteristics (see Lehman and Weisman [10] for a more detailed description of this approach).

5.2 The ECPR

We now relax the assumption that $\beta = 0$ and consider the access pricing problem in terms of both the direct costs and the opportunity costs of providing access to a rival to serve the downstream market. Again, normalizing output to unity, we can express the upstream profit function for the incumbent LEC as

$$\pi^u = P - s^L(e) - c(e) - \Phi(e), \tag{3}$$

We have assumed that $\beta = 1$, so that the access pricing rule is given by $w = P - s^L$. Note that we have maintained the assumption that the LEC's costs for providing access to a rival are the same as when it provides access to itself. In Eq.3, the access costs (c) and the costs of the self-supplied input (s^L) are expressed as a function of cost-reducing effort, e. Assuming pure price cap regulation $(dP/dc = 0)$,[13] maximizing Eq.3 with respect to e yields $-(s^{L\prime}(e) + c'(e)) = \Phi'(e)$, or the marginal benefit of effort (now the sum of cost reductions for the two inputs necessary for downstream production) is equated with the marginal cost of effort. Recognize that under rate-of-return regulation or price cap regulation with profit-sharing, $dP/dc > 0$, which implies that $-(s^{L\prime}(e) + c'(e)) > \Phi'(e)$, or the incumbent LEC under-invests in cost-reducing effort. This, of course, is the well-known moral hazard problem that arises when the incumbent LEC's prices and actual costs are linked. Hence, pure price cap regulation appears to address one of the moral hazard problems, specifically that of preserving the incentives to invest in unobservable cost-reducing innovation.

Note also that the same inferior cost-reducing effort incentives result if $\beta < 1$, since the first-order condition for effort becomes $-\beta[s^{L\prime}(e) + c'(e)] = \Phi'(e)$ with the result that effort will again be suboptimal. Intuitively, the reason for this is that, with $\beta < 1$, the firm bears all of the unobservable costs of cost-reducing effort but only recoups the observable cost reductions. So, ratcheting down retail prices linked to cost reductions has the same qualitative effect as providing the firm with only a fraction of the opportunity cost when selling wholesale. In both cases, unobservable cost reduction will suffer since the LEC is forced to share the benefits of unobservable cost reductions (through retail price reductions in one case and the fractional opportunity cost in the other).

The second incentive problem that has been linked to the ECPR concerns whether the incumbent LEC would simply withdraw from the downstream retail market entirely because it earns the same margin in the monopoly upstream access market. (In essence, the concern is that the incumbent LEC would simply free-ride on the marketing efforts of its rivals.) Economides and White [4] claim that such a pricing rule is inconsistent with a competitive marketplace because it simply preserves the same level of monopoly profits for the incumbent. This argument is fallacious on numerous grounds. First, the ECPR does not preserve monopoly profit levels if the opportunity cost calculation is based on the prevailing market price (P). Preserving the cost-reducing incentives of the ECPR does not require that P remain constant, only that it not vary directly with the incumbent LEC's actual costs. Hence, if competition in the downstream retail market puts downward pressure on P, the margin that is foregone in the retail market or the opportunity cost is reduced accordingly, *ceteris paribus*. Hence, the ECPR does not preserve the level of monopoly profits. Second, the use of the term "monopoly profits" is a misnomer here. Regulators have traditionally used the monopoly power of the LECs to subsidize selected markets (e.g., residential rates in rural areas). Hence, these "monopoly profits" if it is even fair to characterize them as such have been used in part to satisfy social equity goals of regulation. It is unclear why all market participants in the telecommunications industry should not be required to participate in the funding of such goals, provided that the method of funding these social equity goals is competitively neutral and minimizes economic distortions.

Let us now turn to the question of whether the incumbent LEC would have an incentive to withdraw from the downstream retail market. Let $\tau(m)$ denote the incumbent LEC's downstream retail market share and m its unobservable marketing effort, with $\tau'(m) > 0$ and

[13] See Sappington and Weisman [11, p. 81].

$\tau''(m) \leq 0$. Also, let $\psi(m)$ represent the disutility of marketing effort, with $\psi'(m) > 0$ and $\psi''(m) > 0$. With output normalized to unity, the aggregate profit function for the incumbent LEC with $\beta = 1$ is given by

$$\pi = [1 - \tau(m)][P - s^L(e) - c(e)] + \tau(m)[P - s^L(e) - c(e)] - \Phi(e) - \psi(m). \tag{4}$$

Maximizing Eq.4 with respect to m and allowing for the possibility of corner solutions yields $\pi_m \leq 0$ and $m[\pi_m] = 0$. Hence, under the ECPR, $m^* = 0$ by complementary slackness and there is no incentive for the incumbent LEC to expend effort in retaining the retail business function. The intuition for this result is straightforward. Marketing effort is costly to the incumbent and serves not to expand the size of the profit margin, but merely to redistribute it from the access to the retail market segments. Hence, marketing effort confers only costs with no offsetting benefits.

5.3 The Fractional ECPR

There are two solutions to the market retention moral hazard problem that readily suggest themselves. The first is to use some discount factor to shade the access margin below the retail margin. This would provide stronger incentives for retail market retention, but there is no rigorous basis on which to establish such a discount factor. Hence, we exclude this possibility. The second solution is to use the incumbent's retail market share as a yardstick or signal of its effort in retaining the retail business function. This amounts to setting $\beta = \tau(m)$. With output normalized to unity, the aggregate profit function for the incumbent LEC is now given by

$$\pi = \tau(m)[1-\tau(m)][P - s^L(e) - c(e)] + \tau(m)[P - s^L(e) - c(e)] - \Phi(e) - \psi(m). \tag{5}$$

This fractional yardstick EPCR will provide the incumbent LEC with incentives to invest in strictly positive levels of marketing effort (for $\tau < 1$) and, at least in part, addresses the retail market retention moral hazard problem. The incumbent LEC will underinvest (respectively, overinvest) in marketing effort (relative to an identical downstream standalone rival) whenever $\tau > (<)\ ½$. However, at least one problem remains. Because $\beta < 1$, the incumbent LEC generates a higher rate of retail contribution to the social equity goals of regulation relative to its downstream rivals. In other words, the obligation for funding the social equity goals of regulation is borne disproportionately by the incumbent LEC. In order to correct for this asymmetry, it is necessary to set $\beta > \tau(m)$. Hence, simultaneously correcting for the moral hazard market retention problem and for the asymmetry of financing the social equity goals of regulation requires that $\beta \in (\tau, 1)$. Yet, when $\beta < 1$, $e < e^*$, so the incumbent LEC will invest sub-optimally in cost-reducing innovation. This analysis suggests that the optimal access pricing rule will necessarily involve some trade-off between the two moral hazard problems (and quite possibly the discrimination problem as well).

Some of these trade-offs may be formalized as follows. Express the aggregate (upstream and downstream combined) profit function for the LEC as:

$$\pi = \tau(m)[1-\tau(m)][P-s^L(e)-c(e)] + \tau(m)[P-s^L(e)-c(e)] - \Phi(e) - \psi(m) \tag{6}$$

Totally differentiating Eq.6 with respect to e and m yields the results that $de/dm < 0$ and $d^2e/dm^2 > 0$. Thus, there is a trade-off between e and m. The positive second-order derivative means that the "savings" in either form of effort decreases as effort is shifted towards the other.

We can also multiply Eq.6 by r, a regulatory regime variable defined as the fraction of economic profits that the regulator permits the firm to retain. Under pure rate of return regulation $r=0$, while under pure price caps $r=1$. Sharing plans produce $0<r<1$. Straightforward, but tedious, manipulation of the first-order conditions for the modified Eq.6 yields $de/dr > 0$ and $dm/dr > 0$. In other words, both cost reduction and marketing effort are increasing functions of r. Similarly, if we consider changes in price (P is assumed to be exogenously determined by the regulator), we find that $de/dP > 0$ and $dm/dP > 0$, so both types of effort are increasing functions of retail prices. The similarity of these two sets of results derives from the fact that both price and regulatory regime provide positive incentives for cost reduction and marketing effort under the fractional ECPR. The opportunity cost margin is increased in one case and the retained profits of the firm are increased in the other.

5.4 The Yardstick ECPR

Previous sections have revealed that the ECPR has efficient cost reduction incentive properties but poor market retention incentives. What happens if we permit access charges to be a linear combination of actual costs and the ECPR (based on yardsticked costs and actual opportunity costs)[14], i.e., $w = \beta[\tilde{c} + P - c - s^L] + (1-\beta)c$? Ignoring market retention incentives for now, the firm's profit function is given by

$$\begin{aligned}\pi &= \tau(P-c-s^L) - \Phi(e) + (1-\tau)[w] - (1-\tau)c \\ &= \tau(P-c-s^L) - \Phi(e) + \beta(1-\tau)(\tilde{c} - 2c + P - s^L)\end{aligned} \tag{7}$$

after substitution of w from above. The first-order condition for Eq. 7 is

$$\beta[1-\tau][-2c' - s^{L\prime}] + \tau[-c' - s^{L\prime}] = \Phi'(e). \tag{8}$$

Note that the correct comparison for efficient cost-reducing effort is now $-[c' + \tau s^{L\prime}] = \Phi'$. This equates the marginal cost of effort with its marginal benefits (upstream and downstream, the latter reflecting its market share). Note that Eq.8 yields efficient cost-reducing innovation for $\tau = 1$. Note also that there is overinvestment in unobservable cost-reducing innovation for $\beta = 1$, because the ECPR and cost yardsticks produce efficient effort individually, and both are applied here. A third result is that the firm underinvests in cost-reducing innovation for $\beta = 0$. So, it appears that there is a value of β for which cost-reducing innovation is efficient.

[14] It would be more straightforward to use a weighted average of yardsticked costs and the actual ECPR but, unfortunately, this formulation will not work. A fraction of the yardsticked cost is itself a yardstick measure of costs, and thus yields efficient cost reduction incentives on its own. Adding the fractional ECPR to this would yield excessive cost reduction incentives.

Manipulation of Eq.8 reveals that this efficient value is

$$\beta = \frac{c'}{2c' + s^{L'}} < \frac{1}{2} \tag{9}$$

Previously, we determined that to recognize the market retention moral hazard problems, $\beta \in (\tau, 1)$. Eq.9 implies that there exists a β that simultaneously solves both moral hazard problems only when $\tau<\frac{1}{2}$ (this is a necessary condition). More generally, there will be a trade-off between cost-reducing innovation and market retention effort. To the extent that regulatory regimes do not permit the regulated firm to be the full residual claimant for cost reductions (e.g., when sharing is used, so $r<1$), then the tendency for our yardstick ECPR (Y-ECPR) to result in overinvestment in cost-reducing innovation is mitigated.

Thus, the Y-ECPR is a general form of access pricing which embodies the following elements: direct costs are measured using a yardstick not tied to the firm's actual costs; opportunity costs are calculated on the basis of the firm's actual costs; and access prices are a fraction of the combined yardstick costs and opportunity costs. The value of this Y-ECPR lies in its ability to address both unobservable cost-reducing innovation and market retention, although generally in an imperfect manner. This should be contrasted with setting access charges equal to costs, which has good market retention incentives but poor cost reduction incentives, and the standard ECPR which has good cost reduction incentives but poor market retention incentives.

5.5 Yardsticking the Regulator?

There is one additional access price structure which is amenable to the yardstick approach. Thus far, we have focused on the incentives of the incumbent for cost reduction, market retention, and discrimination. We have mentioned the potential for inefficient incentives for competitive entry absent the ECPR. One other agent plays an important role in determining access charge properties. We have assumed that the regulator's behavior is exogenous up to this point.

Suppose we consider regulatory determination of prices and how the regulator's behavior may be influenced by benchmarking it against other regulators. The rationale for the ECPR derives mainly from the fact that some services have traditionally been priced well above cost in order to keep other services priced lower. It is these incumbent burdens that raise the LPF issue and it is the ECPR which produces efficient entry incentives *given the retail price distortions*. To the extent that the regulator sets these retail prices, a mechanism which produces regulatory incentives to reduce retail price distortions will reduce the efficiency distortions inherent in the access pricing problem.

A relatively simple approach would be to consider the ECPR with opportunity cost calculated on the basis of some "yardstick" retail price rather than the retail price set by the regulator. For example, business line prices could be an average of all other jurisdictions' prices (states or nations) excluding the regulator setting this retail price. By breaking the link between wholesale pricing under the ECPR and regulated retail pricing, regulatory incentives will change. To see this, consider a regulator using residual pricing to determine one retail price, P_R, where it can manipulate another retail price, P_B, and where the ECPR will be applied in the upstream market ($w=P_B-s^L$). For simplicity, assume that the firm's total profits = 0, so that

$$P_R = \frac{c + s^L - P_B(Q_B+1)}{Q_R} \tag{10}$$

where Q_B is the quantity of the service producing the contribution to keep down P_R. We are ignoring any own-price or cross-price elasticities here in order to facilitate the exposition. We also let c represent the total cost function for providing all of the firm's services. Now consider the ECPR using the yardstick price in place of the regulator's own price:

$$\widetilde{w} = \widetilde{P}_B - s^L, \tag{11}$$

and

$$\hat{P}_R = \frac{c + s^L - \hat{P}_B Q_B - \widetilde{P}_B}{Q_R} \tag{12}$$

The hats denote the regulatory-chosen prices under the yardsticked approach. Then, there are two scenarios to consider.

Scenario 1: The regulator keeps the same retail price for the service providing the contribution: $\hat{P}_B = P_B$. Comparison of Eq.10 and Eq.12 reveals that

$$P_R > \hat{P}_R \text{ iff } \widetilde{P}_B > P_B$$

Regulators who distort the contributing service less (more) than their yardstick and who do not wish to change the contribution level will now set the receiving service's price higher (lower) than before. In other words, regulators who distort prices "more than average" will be penalized and led to reduce the level of subsidy to the receiving service. They are forced to do so because the ECPR no longer permits them to provide the same wholesale contribution as the retail contribution. On the other hand, regulators who distort prices less than average, are rewarded by permitting them to lower P_R (assuming that they keep P_B unchanged).

Scenario 2: Suppose regulators wish to keep the price constant for the receiving service: $P_R = \hat{P}_R$. In this case, comparison of Eq.10 and Eq.12 reveals that regulators who distort the contributing service more (less) than average will be forced to distort the contributing price even more (less). This results directly from the fact that the effectiveness of receiving contributions under the ECPR is reduced if contributions are calculated based on the yardstick retail price rather than the actual retail price. While theoretically possible, this policy is not likely to be sustainable as even current levels of contribution (no less increased levels) will be difficult to sustain in a competitive environment. It does raise the possibility that regulators who insist on preserving distortions will be forced by the yardstick to use increasing distortions in order to maintain the same level of subsidized price. An adverse selection equilibrium becomes possible in which price distortions for contributing services reflect only those regulators who insist on preserving contributions rather than those who are willing to reduce price distortions when subject to the yardstick.

6. Summary and Conclusions

The access pricing issue currently occupies center stage in the economics of regulation. The issues are complex and the ensuing debate has polarized economists as well as public policy-makers. In a very real sense, the adversarial nature of the debate centers on contrasting views of both the future and the past. The past may be prologue, but it can have real efficiency consequences. To paraphrase Alfred Kahn, it is always easier to tell the other guy to forget about his sunk costs [6, p.30].

In this paper, we have focused on a number of incentive issues that have plagued the ECPR and explored the effectiveness of yardsticks in addressing these concerns. Abstracting from problems with imperfect downstream competition, we find that the optimal access pricing rule generally involves a trade-off between two moral hazard problems (and quite possibly the discrimination issue as well). In general, we find that it is not efficient to allow equal margins in both the upstream and downstream markets as the standard ECPR proposes. This equating of margins gives rise to potentially serious incentive problems, and may not be necessary to dampen incentives for discrimination on the part of the incumbent LECs or for provision of efficient cost reduction incentives.

On the other hand, we find little to recommend setting access charges equal to any particular measure of costs. This provides little incentive for cost reduction, and in the presence of incumbent burdens it creates inefficient, excessive competitive entry incentives. The fractional ECPR, combined with yardsticks for direct costs, the Y-ECPR proposed here, provides a mechanism for simultaneously addressing all these incentive issues. It solves none of them optimally, but other mechanisms are forced to solve one incentive issue at the expense of another.

It would be desirable to have a simpler pricing rule (marginal cost, and even the ECPR has a simpler structure), but it does not appear feasible in light of the myriad stakeholders and complicating transitional factors. Indeed, if the market were already competitive, no regulation of interconnection prices would be required, and in a monopoly environment, there is no essential input to price. It is precisely the environment in which we seek to facilitate competition, maintain other public interest objectives, and provide efficient incentives for all parties, in which our pricing problem emerges. Not surprisingly, our "solution" is not as simple as we might like – nor is it too complex to implement.

We are not advocating yardsticking regulators at this point. Such a policy is fraught with legal and practical issues. However, the importance of showing how such an approach might work is that yardsticks do provide a mechanism under which regulatory incentives can be subject to a type of competitive pressure. Ultimately, it is reduced regulatory distortions which will render the efficient access charge problem manageable. No other proposed access charge mechanism provides incentives to reduce such distortions.

Based on our analysis, we offer the following conclusions:

- Improved regulatory regimes, such as price caps, reduce all of the incentive problems simultaneously.
- Cost reduction incentives are efficiently provided by using yardstick measures of direct cost.
- The fractional ECPR permits market retention efforts to be improved (they are suboptimal under the pure ECPR) without necessarily sacrificing cost reduction incentives, if the yardstick approach to direct cost is maintained.
- Yardsticking regulatory behavior has additional properties which merit further study, as regulatory incentives are intertwined with incumbent and entrant incentives.

References

[1] W. Baumol, Some Subtle Issues in Railroad Regulation, *International Journal of Transport Economics* **10** (1983) 341-355.

[2] W. Baumol and J.G. Sidak, Toward Competition in Local Telephony. Cambridge MA: MIT Press, 1994.

[3] D. Dranove, Rate-Setting by Diagnosis Related Groups and Hospital Specialization, *Rand Journal of Economics* **18**, No. 3, Autumn (1987) 417-427.

[4] N. Economides and L.J. White, Access and Interconnection Pricing: How Efficient Is The 'Efficient Component Pricing Rule'? *The Antitrust Bulletin* **XL**, No. 3 (1995) 557-579.

[5] A.E. Kahn, Can Regulation and Competition Coexist? Solutions to the Stranded Cost Problem and Other Conundra, *The Electricity Journal* **7** (1994) 23-35.

[6] A.E. Kahn and W.E. Taylor, The Pricing of Inputs Sold To Competitors, *The Yale Journal on Regulation,* **11**, No. 1 (1994) 225-240.

[7] J-J. Laffont and J. Tirole. Access Pricing and Competition, *European Economic Review* **38** (9), December (1994) 1673-1710.

[8] A.C. Larson, The Efficiency of the Efficient Component-Pricing Rule: A Comment, *Antitrust Bulletin* (1997) (Forthcoming).

[9] A.C. Larson and D.E. Lehman, Essentiality, Efficiency, and the Efficient Component-Pricing Rule, *Journal of Regulatory Economics* (1997) (Forthcoming).

[10] D.E. Lehman and D.L. Weisman, Telephone Pools and Economic Incentives, *Journal of Regulatory Economics*, **10**, September (1996) 123-146.

[11] D.E.M. Sappington and D.L. Weisman, Designing Incentive Regulation For The Telecommunications Industry. Washington D.C.: American Enterprise Institute and Cambridge MA: MIT Press, 1996.

[12] A. Shleifer, A Theory of Yardstick Competition, *Rand Journal of Economics,* **16**, No. 3, Autumn (1985) 319-327.

[13] D.S. Sibley and D.L. Weisman, The Competitive Incentives of Vertically Integrated Local Exchange Carriers: An Economic and Policy Analysis, *Journal of Policy Analysis and Management* (1997) (Forthcoming).

[14] D.S. Sibley and D.L. Weisman, Raising Rivals' Costs: Entry of an Upstream Monopolist Into Downstream Markets. Kansas State University Working Paper, March 1997.

[15] D.F. Spulber and J.G. Sidak, Access Pricing and Deregulation. Northwestern University Working Paper, February, 1996.

[16] W. Boulton, F.R. Testimony on Behalf of AT&T Communications of the Southwest, in Kansas Docket No. 97-AT&T-290-ARB, December 1996.

[17] R.D. Willig, The Theory of Network Access Pricing. In: H.M. Trebing (ed.), Issues in Public Utility Regulation. East Lansing, Michigan: Michigan State University Press, 1979.

[18] Federal Communications Commission, CC Docket 96-98. In the Matter of Implementation of the Local Competition Provisions in the Telecommunications Act of 1996. First Report and Order, Released August 8, 1996.

[19] Iowa Utils. Bd. v. FCC, No. 96-3321, 1996 WL 589204, at *7 (8th Cir. Oct. 10, 1996).

[20] S.652 (The Telecommunications Reform Act of 1996) 104th Congress, 2nd Session, January 3, 1996. Pub. L. No. 104-104.

Telecommunications Transformation: Technology, Strategy and Policy
E. Bohlin and S.L. Levin (Eds.)
IOS Press, 1998

CHAPTER 13

International Settlements: A Time for Change *

James Alleman and Barbara Sorce

Interdisciplinary Telecommunications Program,
University of Colorado, Boulder, CO 80309-0530, USA

Abstract. In the late 1970s, when competition was introduced in the United States, competitive forces pushed international telephone prices down. Lower prices resulted, *inter alia*, in an increase in international calls (and minutes) from the U.S. to countries whose prices were high relative to U.S. prices. This increase in minute volume in turn resulted in a significant increase in settlement payments outflow from the U.S.

The U.S. is not the only country to experience settlement deficit. Several other countries, including Sweden, Australia and Japan, have growing deficits on settlement payments.

Three issues are involved in settlements – the "correctness" of the procedure from the perspective of:

- public policy
- balance of payments and
- the operating companies.

The effects of call-back and/or resale business on the settlement process are addressed. The paper, then, examines alternative settlement policies: first-best pricing, bill-and-keep, benchmark and cost-based pricing. The modelling demonstrates that: a) call-back does not necessarily force lower accounting rates on monopoly carriers, indeed, call-back can exacerbate the settlement problem; b) cost-based pricing is the preferred approach to settlements; c) bill-and-keep and the fifty-fifty split of the accounting rate are both inappropriate on efficiency grounds; and d) settlement rates are set at monopoly prices under certain market structures.

1. Overview

1.1 Background and Issues

In 1865, 20 European nations came together to form what eventually became the International Telecommunication Union (ITU). One reason this Union was established was to develop a means of dividing the revenues from international services among origin, destina-

* Partial funding for this research was provided by Hagler Bailly Consulting, Inc. This review has benefited from discussions with Vince Snowberger and Kenneth Stanley. Ms. Barbara Miller offered excellent support and editorial assistance. The usual disclaimer applies.

tion and transit countries. The original methodology developed, i.e., the accounting rate system, is still with us today, albeit in a format that has been progressively modified [1].

The original accounting rate system worked well as long as the following criteria were met: international services were jointly provided by monopoly partners; the prices charged for a call were approximately equal in different directions (the principle of symmetry); incoming and outgoing traffic was approximately in balance for each bilateral relationship between countries; collection charges, even for off-peak discounts and volume discounts, were never lower than accounting rates; inflation rates and exchange rates were relatively constant between countries [1].

These conditions are breaking down. Technological improvements and competition have led to lower costs and prices, but competition is not ubiquitous. This has led to an imbalance in the traffic flow between countries with lower prices and those with relatively higher prices, which in turn leads to settlement imbalances. The U.S. is not the only developed country to experience settlement deficit. Several other countries, including Sweden, Australia and Japan, have growing deficits on settlement payments, according to the ITU [1].

It is ironic that countries that are "efficient" in generating traffic or reducing their call charges are being penalized under the current accounting system. The "efficient" countries are rewarding the "inefficient" countries with settlement payments. This market distortion, however, is not the only problem. An increasing share of traffic is bypassing the accounting rate system via private line or the Internet. In addition, call-back operators and resellers are exploiting the fact that tariffs are not cost-based. The traffic imbalances are a symptom of another problem – the accounting rates are not cost-based.

The problem of settlement imbalances must be addressed from the developed and developing countries' sides. Countries with "efficient" systems are increasingly concerned about outgoing settlement payments. The U.S., with a settlement deficit which has grown to nearly five billion dollars (see Figure 1), is using all available fora – the ITU, the OECD, WTO and others – to push for a reform of the system.

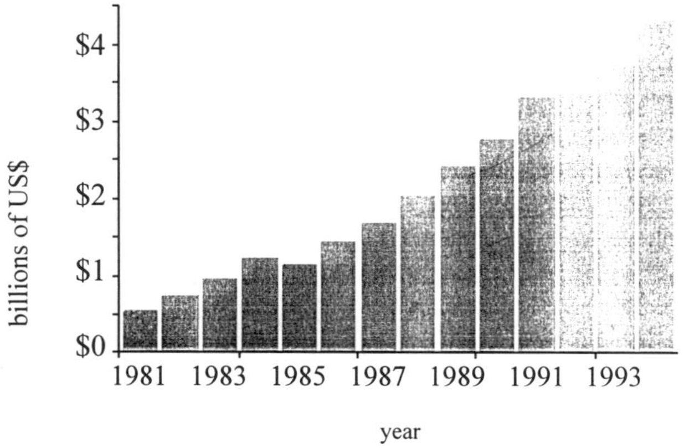

Figure 1. United States International Settlements Deficit
Source: FCC International Trends Report, August 22, 1996.

1.2 Accounting/Settlement Rates

The international settlement is the method by which telephone companies are compensated for the termination of traffic in their territory. The completion of an international telephone call will require the participation of two or more entities before the final connection is made. The entities involved often jointly own, operate, and maintain some of the equipment. In other instances the equipment is not jointly owned (i.e. AT&T pays an access charge to local carriers for originating and terminating calls). To handle the expenses attributed to joint ownership of facilities and equipment and termination of the calls the participants enter into settlement arrangements. These arrangements establish the terms and conditions for providing service between two companies or countries. A separate arrangement is negotiated for each service among each pair of countries. This negotiated rate is referred to as the "accounting rate." The settlement rate has historically been half of the accounting rate, although there have been proposals to change this division [3].

A phone call that originates in the United States is routed through facilities in the United States to a jointly owned circuit and transmitted to its destination in a foreign nation. The carriers of both countries incur expenses. The accounting rate represents a means of sharing the expenses.

The formula takes the difference in minutes between carriers and "settles" based on this difference. If more minutes have terminated in a country than originated in that country, then the telephone company is owed this difference times the settlement rate and *vice versa*. In practice, the companies settle on the difference in minutes between the countries on a quarterly interval. The net settlement payment represents the payment to the country that received more minutes of traffic than it originates.

1.3 Recent Activities

The present settlement process is under examination by all of the major telecommunications regulatory bodies. This section examines recommendations and policies developed to solve the settlement problem.

1.3.1 International Telecommunication Union

Accounting rate reform was listed as one of twelve concerns to be addressed by the ITU's 1995-99 strategic plan. The ITU specifically stated that the accounting rate process needs to be addressed on a multilateral basis. Accounting rate payments provide an important source of income to many developing countries, but they also tend to keep telephone charges to end users high, and can act as a disincentive to economic efficiency. "The challenge here is to reform the system while still maintaining the vital flow of investment funds to assist with network development" [2, p.3].

The ITU Standardization Sector, Study Group 3, has been given the task of creating a framework within which the international carriers can establish revenue-sharing mechanisms. The ITU has no role in setting or negotiating accounting rates but can make recommendations; however, ITU members have agreed to Recommendation (D.140) which recommends that accounting rates should move closer to costs in competitive markets [4, p. 2].

The ITU has suggested three alternatives: facilities-based interconnection agreements, progressive reduction of the accounting rates to the point at which they are so low that countries are willing to move to "senders-keeps-all", and a system of call termination charges [5].

More recently, the ITU organized an Informal Expert Group to discuss alternative methods of settlements. Regulatory bodies, PTTs, the World Bank, OECD and telecommunications companies were represented. It released its recommendations on April 11 of this year.

The main recommendations are: commitment by the World Bank and the ITU and others to assist those economies likely to be harmed by the reform process; an acceleration of the work of ITU-T Study Group 3 on reform of international settlement arrangements; commissioning a series of case studies on the impact of lower international settlements revenue; and an enhanced role for the ITU in data collection and dissemination of policy information on settlements.

Its position is that settlement rates should be reduced to a cost-based level and universal service should be maintained with these changes [4].

1.3.2 United States Government

The executive branch of the United States Government has pushed for reform of settlements: "... But privatization is not enough. Competition is needed as well. To promote competition and investment in global telecommunications, *we need to adopt cost-based collection and accounting rates*." [6] [emphasis added]

The Federal Communications Commission's (FCC) concern regarding the U.S. international settlement deficit resulted in policy reform as early as 1990. Its reform called for tougher negotiations by U.S. international carriers; "resale" and bypass were promoted; and a commitment was made to ask the ITU for assistance.

The FCC released a policy statement on "International Accounting Rate Reform" early last year [7]. In addition to restating earlier policies, it added several more. Competition is to be embraced and encouraged. Foreign monopolies must not use their market power to obtain discriminatory accounting rate concessions from competing U.S. carriers (i.e. "whipsawing"). Services that bypass the traditional settlement process (i.e. private line resale) were encouraged. The FCC again committed itself to work with the ITU and other international bodies. Additionally, progress reports and benchmark guidelines to measure progress towards the goal of cost-based accounting rates were established [7]. It noted that efforts to bring accounting rates closer to costs have been unsuccessful. The FCC suggested stronger action:

> If we do not see substantial progress toward accounting rate reform, we will take stronger measures to reduce accounting rates. U.S. industry and consumers can no longer be required to bear the burden of above-cost accounting rate payments to foreign carriers. The measures we contemplate may include: supporting U.S. carriers which refuse to continue to pay high accounting rates when existing agreements expire, to tying reductions in accounting rate to growth in net settlement payments. [7, p. 9]

1.4 Alternative Settlement Procedures

While numerous regulatory bodies, domestic and international, are proposing alternatives to the present settlement process in the U.S., only the FCC has the authority to enforce implementation. And, while the FCC does not itself negotiate, it does have authority over U.S. carriers. It is within the FCC's power to disapprove a settlement rate as negotiated by a U.S. carrier.

The FCC recently ordered all U.S. carriers to suspend settlement payments to Telintar, Argentina's long-distance and international telephone company, which it regards as using its market power to discriminate against U.S. carriers [7]. Similarly with Peru and Bolivia, the FCC issued orders not to accept the settlement rates offered because they were discriminatory and too high. The international carriers in both countries were informed via U.S. carriers that they must renegotiate an acceptable agreement, which they did.

The ITU and FCC have suggested several alternative settlement procedures.

1.4.1 Interconnect Fees

Interconnect fees are priced according to volume of traffic sent and received between operators. The advantage is that they are conceptually easy to understand, but they are open to abuse because they are often negotiated with a built-in bias to the operator which receives more calls than it originates. Interconnection agreements are too often negotiated on the basis of market power rather than on costs or needs. This form of revenue sharing was developed by the mobile sector and is tailored to its need [4].

It is conceptually similar to the settlement process described earlier.

1.4.2 Sender-Keeps-All

With sender-keeps-all, the originator of a call does not pay anything to the operator that terminates the call. The only cash payments are from the user to the provider – and from a particular access provider to the nearest high-capacity backbone [4].

The postal system practices sender-keeps-all for international mail. Similarly, "[t]he Internet used a sender-keeps-all methodology. The call itself might pass through several intermediate carriers with different bits of message traveling by different routes" [4]. This method, too, creates problems if traffic is asymmetrical between and among service providers. For example Telstra and KDD, respectively Swedish and Japanese international telecommunications providers, have complained to the FCC of the unfairness of this method since more traffic originates in the United States. If this continues, it could reduce the incentive to upgrade the network and systems of these providers.

1.4.3 Resale and Call-back

The FCC has been promoting "resale" and "call-back" services since 1990. The basic premise is that the resale/call-back will stimulate competition in a non-competitive market, forcing the foreign carrier to negotiate a better settlement rate.

Foreign governments' views on international call-back are not uniform, with legality questions being debated in international fora in recent years. Most notably, the 1994 ITU

Plenipotentiary Conference in Kyoto passed a resolution on alternative calling services which recommends that a member state having jurisdiction over a call-back provider whose operations infringe another member state's laws "inquire into the matter and take such actions as may be appropriate within the constraints of its national law" [8].

A foreign government must hold resale and call-back providers to be "legal" before these United States providers can do business in that particular country. The reasoning behind this is that call-back providers take business away from the foreign nation. This, however, is a paradoxical situation as we will show below.

The European Commission has noted "that without exception, the availability of call-back services exists within the territories of its member states. A study of call-back and other alternative calling procedures by the OECD noted that they are driven by price distortions in monopoly-based markets. The OECD study concluded that by introducing competition for international telephone service, these services "play a useful role in the marketplace and need to be encouraged" [9].

In 1995, the FCC concluded that resale and call-back services are in the public interest because they promote increased competition and create incentives for the reduction of foreign collection rates to the benefit of U.S. consumers and industry [10].

The FCC and the OECD, however, are not totally correct in their assessment that the existence of call-back providers will hasten the obtainment of better accounting rates. In fact, the existence of call-back providers results in increased call volume into the foreign country, hence generating additional settlement revenue. A case for the latter situation can be made from examining the South American countries, many of which have high call-back volume and are receiving high settlement payments.

A simple example illustrates the point. Let p_m equal the retail price or collection rate, c_m average incremental costs (and product-specific cost) and a_m the settlement rate in the monopoly country. If $p_m - a_m < a_m$ or $p_m < 2a_m$ and a call-back company "moves" a minute of traffic to another (lower priced) country from the monopoly country, the monopoly country makes a net gain of the difference of the inequality $(2a_m - p_m)$. It loses the revenue p_m but gains the revenue a_m and it does not owe a_m. Thus, there is a net gain. Moreover it does not have the billing and collection expense from its customers; it collects the settlement from a major carrier such as ATT, and any increase in traffic into the country due to the reduction in price will contribute an additional a_m times the increase in minutes (Δq_c) to revenues and produce a net economic profit of $(a_m - c_m)\Delta q_c$ for the monopolist country.

Thus, call-back can improve the monopolist country's revenue and profit. However, this should not be seen as a condemnation of call-back operations, but as an indictment of the accounting/settlements mechanism which, we demonstrate below, is inappropriate on other grounds. These countries' service providers should welcome call-back, not declare it illegal! Although call-back exacerbates the settlements outflow from the low-priced country, it will improve the country's balance of trade, since the call-back company will collect the revenue from its overseas customer in the monopoly country. Thus, even if call-back improves the telecommunications' settlement for the monopoly country, it deteriorates the balance of trade for the monopoly country.

1.4.4 Flexibility

Rather than adhering to a strict formula, the FCC will now consider a waiver of the settlement rates if conditions are sufficiently competitive. All of the above approaches are designed to move toward cost-based settlements.

1.4.5 Cost-based Pricing /Benchmarks

More recently, both the FCC [6] and ITU [4] have recommended cost-based pricing where a portion of the prices would be based on the cost of terminating the call in the country.

The proposed benchmark settlement rates are based on the three components of international traffic costs identified by the ITU: transmission facilities, international switching facilities, and national transport and termination. It suggests that the tariffed rate for long-distance traffic be used for the national component. The idea is that the highest long-distance rates in a country should be the ceiling price for the national termination portion of the settlement rate, since, in principle, it covers all the necessary costs to complete a call. The FCC promoted benchmark rates in 1991which are currently based on the high/low-density routes, developed/developing countries as a proxy for costs.

The FCC is considering other elements to determine benchmark rates such as market structure, technological change, and productivity improvements. However, determining the cost, as we know in the United States and Canada, can be a complex, contentious process, particularly where joint and common costs are present and economies of scale and scope are possible.

2. The Modelling Approach

This paper is the first of a planned series of research papers in which we will address the settlement issue. In this paper, we develop a stylized benchmark model. This model asks what is the optimal access charge in the international context. This benchmark translates into workable recommendations and casts doubt on some traditional policy proposals. In particular, it supports the policy of cost-based access/settlement charges; and shows that both the fifty-fifty split and bill-and-keep are inappropriate from the economic efficiency perspective.

The second model is a monopoly provider facing a country with a competitive market in international services. This is characteristic of the United States, the United Kingdom, and to a lesser extent Australia and Sweden facing monopoly carriers. The insight of this model is that the monopoly carrier charges more than the traditional monopoly price because of the settlement effect.

We review the various proposals that have been put forth in the context of these models.

2.1 Accounting Rates: An Intermediate Price

Accounting rates represent an intermediate price in the production of international telecommunications services. This simple observation suggests that much of the work applied to the interconnection issue would be applicable to international accounting rates. Major differences in structure exist: in the international case, both parties have bottleneck facilities; no overall regulatory authority exists; and the market structures vary among countries. Nevertheless, the insights offered by the interconnection literature are useful in the analysis of accounting rates.

2.2 Intermediate Pricing

Intermediate or interconnect pricing represents the price of the intermediate good or service needed by a firm to provide its service. In the telecommunications industry, the price is also known as the "access price" and would be the price charged by one service provider, usually the incumbent, for connection to its network in order for the other firm to complete the service for its end-user customers. For example, in the United States, it would represent the price that the long-distance carriers must pay the exchange carriers to complete a call on the public switched telecommunications network (PSTN). Another example would be the connection of a mobile provider to the PSTN.

The pricing becomes more complex when the company charging the interconnection price also competes with the company to which it supplies the intermediate service. The company charging for interconnection has an obvious incentive to overcharge the competing company – not only to enhance its own revenue, but to make the competing company's cost, and hence its price, higher.

In the international arena, the intermediate price is the settlement rate or one-half the accounting rate; see Section 1.4 Alternative Settlement Procedures, for the mechanics of the settlement process.

2.3 The Problem

International settlements have many features of the intermediate/interconnection problem that has been extensively addressed in the literature.[1] The differences are, first, that no omnipotent regulator exists in this arena, although the WTO agreements are approaching this function. Secondly, the intermediate service is "sold" by both parties. Finally, the demand for the service is not only dependent on the price in the home country, but has a strong cross-elasticity with the price in the receiving country [11,12].

2.4 Brief Survey of Literature

International settlements have only recently been a concern of economists. Demand modelling of international traffic has a longer history, but even this literature is not extensive.[2] Empirical research and modelling on settlements began with Johnson [14]. However, the issue has moved much faster than the modellers have. As discussed above, the ITU, WTO and FCC have addressed the issues extensively because of the imbalance in settlements and traffic.

---

[1] While the problem is a special case of interconnections, we will not review this literature – see the web page developed by Nicholas Economides at http://raven.stern.nyu.edu/networks/ papers.html for a listing of most of the relevant literature in this area. Also see James Alleman's site at http://www.Colorado.EDU/ engineering/alleman/ print_files/interconnection.pdf for a summary of the issues.

[2] See [12] pp. 264-266. and 338-341.

The research that has been done includes Johnson [14], who examined the collection rates (retail prices), costs, the imbalance in settlements, and analyzed the problem of the market structure. He notes the implications for policy and efficiency. Alleman et al. [15] developed a simulation model that examined alternative settlements and price policy impacts in a two-country model – the United States and the rest-of-the-world. They found that significant Pareto improvement could be achieved with declining accounting rates, declining collection rates or both for the two countries. Hakim and Lu [16] used a duopoly market structure to analyze settlements. They note that the fifty-fifty accounting rate is simply a rule of thumb, and show that the fifty-fifty share increases the collection rate (with high accounting rates). Hakim and Lu [17] extended their earlier model to heterogeneous outputs. Frieden [18] noted the lack of correlation of declining settlement rates and collection rates – while settlement rates fell, collection rates took longer to fall and did not fall by the same proportion. Yun et al. views settlements as a bilateral game of "pie splitting" and are not concerned with the inefficiencies of settlements [19]. Ergas and Paterson [20] attribute the traffic imbalance to the collection and accounting rates. However, Cheong and Mullins [21] noted that the problem with the settlement process is due to the regulatory structure – it does not allow entry or competition – rather than to the rates *per se*. This latter work is in the spirit of our effort. Virtually all agree that the settlement rate is significantly above cost of its provision.

In this paper we analyze the relationship between the collection and settlement rates and asymmetrical traffic. We differ in our analytical development in that we develop a first-best model to compare with suggested policy options. Secondly, we examine a simple two-country model with a monopoly provider facing a competitive market structure in contrast to the duopoly models generally employed.

2.5 Efficient Prices for Interconnection

Before reviewing these models, we address the issue of economic efficiency. Economic efficiency is maximized when prices are set at marginal cost.[3] However, this "first-best" pricing is not sustainable with a production function that shows increasing returns to scale and economies of scope, since the total cost for the firm will not be covered. Thus, the economists have turned to the notion of "second-best" pricing where the idea is to keep the firm "whole" – that is still covering all of the costs of the firm – while at the same time maximizing consumers' surplus. This can be approached as a Lagrange multiplier problem whose solution, in general, is to mark up the price of the outputs over marginal costs inversely proportional to the respective demand elasticities. Thus, the inelastic products are priced proportionally more than their product-specific marginal cost, in comparison to the pricing of elastic products. The intuition of this approach is that by "over-pricing" the more inelastic products, less consumers' surplus is lost. Formally, if demands are independent,

$$[p_i - (\partial c_i / \partial q_i)]/p_i = [\lambda/(1+\lambda)](1/\eta_i)$$

[3] See [22] or [23, pp. 68-69] for example.

for all i, $i = 1,\ldots n$, where p_i is price, $\partial c_i / \partial q_i$ is the product specific marginal cost, η_i is the demand elasticity of the ith output and λ is the shadow price of the budget constraint. The λ is adjusted until the equality holds.

It is on the basis of this formulation that the optimal price structure can be determined.

2.6 First-best Model

A model such as suggested by Laffont and Tirole can be applied to the international settlements issue [24]. First-best pricing would be marginal-cost pricing for both the intermediate and final products. This insight and result holds in this case.

We can gain insight into the nature of the problems by adopting a simple stylized model.[4] The benchmark first-best model assumes constant average incremental costs (marginal costs), no fixed costs, no distortions, no bypass, no entry costs and no market power on the part of the providers.

Define the level of one-way traffic on the different segments as q_i for i = 0, 1 and 2 using the model characterized by Figure 2. That is, q_1 is the traffic from country 1 to country 2 and *vice versa*, where

q_0: joint international service (two-way) traffic
q_i: international service one-way traffic, country i, i = 1, 2 and $Q = q_1 + q_2 \equiv q_0$

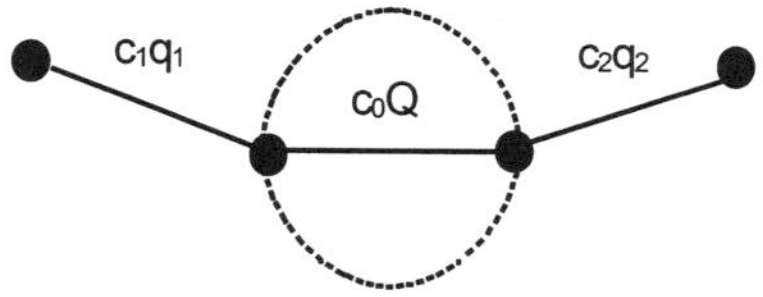

Figure 2. Stylized International Network

The costs are: c_0, c_1, c_2: average incremental cost (and product-specific (constant) marginal cost). Defining the prices as:

a_i: settlement price
p_i: international price
(collection rate), country i, i = 1, 2

then the first-best prices are equal to the marginal costs, namely,

$$p_1 = c_0 + c_1 + c_2$$
$$p_2 = c_0 + c_1 + c_2$$

This result is consistent with the classical solution, and gives a benchmark to be compared with the monopoly model developed below and with public policy recommendations.

[4] Following [24].

The first, obvious, point to note is that the first-best prices are the same. This is as it should be, since the costs do not change with the difference in the direction of traffic (aside from the billing, collection and certain other costs, which would be negligible compared to the prices) .

Second, the first-best settlement rates are then $a_1 = c_0 + c_2$ and $a_2 = c_0 + c_1$ which would be equal if and only if $c_1 = c_2$.

Third, the sender-keeps-all would be efficient only if the profit margins were equal to the cost of termination in the other county.

Finally, the efficient component pricing rule (EPRC) would provide efficient settlement rates in this case. Namely, the interconnection charge would equal the marginal plus the opportunity costs. However, this is true only because of the simple specification of the cost function and zero opportunity costs [25,26,27].

Three policy results are derived from this simple model:

- Serious divergences of rates between two countries are *prima facie* evidence that the prices are inefficient.
- The division of the accounting rate in half is inappropriate, unless, in the unusual case, $c_1 = c_2$. Similarly,
- The "Sender-keeps-all" or "Bill-and-keep" solution would be efficient only if the "margins" on the international calls were exactly equal to the cost of provision in the other country, again a highly unlikely situation.

2.7 Monopoly Carrier

In the United States, the United Kingdom and other countries that have opened their markets to competition, the international situation can be characterized as competitive providers facing a monopoly, since the international telecommunications market is not competitive in most countries. With this market structure, the monopolist has control over its prices and *the settlement rate*. By raising its price higher than its correspondent in the other country, traffic imbalances are increased in favor of the monopolist. Rather than treating this as a cost in the monopoly country, as some authors have, we propose to treat it as a price. So long as the asymmetry in international calling between the high- and the low-priced countries persists, the settlement rate appears to the monopoly carrier as a price. The formalization of this structure is for the monopoly firm to maximize its profit based on its cost, price and *the settlement rate*. The carriers in the competitive country are price-takers. Their prices are cost-based, which include their costs and the settlement cost set by the monopoly carrier. The optimization problem need only be formulated in terms of the monopoly. It must consider not only the price it charges its customers (the collection rate) but also the settlement (rate) price which will influence the demand in the other country, which in turn will determine the differential in the traffic.

Using the previous notation, ignoring the shared costs and using the subscripts m and c for monopoly and competitive markets respectively, the formalization of the profit-maximizing problem for the monopoly country is to maximize:

$$\pi = D(q_c, q_m)\, q_m + D(q_c, q_m)\,(q_c - q_m) - C(Q)$$

Without the second term (the settlement term) in the profit function and the competitive quantity argument, it appears to be the standard profit maximization. The problem is simplified since the settlement term is based on the differences in the quantities; maximize

$$\pi = D(q_c, q_m)\, q_c - C(Q)$$

The first-order conditions are:

$$\partial\pi/\partial q_c = [\partial D(q_c, q_m)/\partial q_c]\, q_c + D(q_c, q_m) - \partial C(Q)/\partial q_c = 0$$

$$\partial\pi/\partial q_m = [\partial D(q_c, q_m)/\partial q_m]\, q_c - \partial C(Q)/\partial q_m = 0$$

The first term is the marginal revenue of the competitive market that is set equal to marginal costs of the monopoly provider – the standard monopoly profit-maximizing price determination. Considering the process in steps, once the monopolist price is set, the monopolist finds a price that maximizes profits from the competitive country. Of course, the settlement price is strongly dependent on elasticity of demand in the competitive country's market. The "choking-off" of the demand in the monopoly country via its high price (collection charge) – higher than if it were only concerned with the in-country price – increases the differential in traffic between the competitive and monopoly country and hence, via the settlements process, its profits.

3. Summary

The use of call-back to correct or reduce accounting rates has limited or no value, as the first model demonstrated, although it will improve the balance of trade in the call-back originating country.

The modelling confirms the correctness of the cost-based pricing approach; moreover, it provides a simple rule for inefficient international prices – asymmetrical rates between countries are a sufficient condition for inefficiency. The splitting of the accounting rate is inappropriate on efficiency grounds. It also demonstrates that the sender-keeps-all approach is incorrect.

The second model illustrates the difficulty of bilateral negotiations with a monopoly carrier facing a competitive market structure, even if the whipsawing effect is eliminated. The monopoly will suffer large losses if it moves to cost-based rates. Moreover, when monopoly carriers face a competitive country, the settlement rates are set at profit-maximizing rates for the low-priced country and the price is higher in the high-price country than it would be without the power to set the settlement rate.

Future research needs to be undertaken to refine the models, to determine the gains and losses from cost-based settlements, and to estimate the magnitude of the losses to developing countries to aid in any transitional process.

References

[1] P. Tarjanne, How will the Accounting Rate System Need to be Modified in a Liberalized Market? In: Liberalization & Privatization of the European Telecommunications Sector, Preparing for 1998 & Beyond. Rome, 1996.

[2] International Telecommunication Union, How Will the Global Information Infrastructure be Priced? 1995.

[3] Organization of Economic Cooperation and Development, International Telecommunication Tariffs: Charging Practices and Procedures, *Information and Computer Communications Policy* **34**, 1994.

[4] International Telecommunications Union (ITU), Report of the Informal Expert Group on International Telecommunication Settlements. Geneva, Switzerland, 1997.

[5] International Telecommunication Union, World Telecommunications Development Report, Second Edition. October 1995.

[6] A. Gore, World Telecommunications Conference, Buenos Aires, March 21, 1994.

[7] Federal Communications Commission, Policy Statement on International Accounting Rate Reform, 96-37. January 31, 1996.

[8] International Telecommunications Union Plenipotentiary Conference (ITU), Report of the Informal Expert Group on International Telecommunication Settlements. Kyoto, Japan, 1994.

[9] Commission of the European Communities, C.o.t.E., Green Paper on the Liberalization of Telecommunications Infrastructure and Cable Television Networks. 1995.

[10] Federal Communications Commission, FCC (1995), in the Matter of Via USA, Ltd. Telegroup Inc, Docket 95-224, June 12, 1995; on recon., Rcd. 9540 (Order on Reconsideration) Comtelca/Intel Reply Comments, at 21. 1995.

[11] A.C. Larsen and D. Lehman, Asymmetrical Pricing and Arbitrage. Presented at Sixth International Conference on Forecasting and Analysis for Business Planning in the Information Age, Southwestern Telephone Company, St. Louis, 1988.

[12] A.C. Larsen, D.E. Lehman, and D.L. Weisman, A General Theory of Point-to-Point Long Distance Demand. In: (eds.) Alain de Fontenay, Mary H. Shurgard and David Sibley, Telecommunications Demand Modelling. North-Holland Publishing Co., Amsterdam, 1990.

[13] L. Taylor, Telecommunications Demand. Kluwer Academic Publisher, Dordrecht, 1994.

[14] L.L. Johnson, Competition, Pricing, and Regulatory Policy in the International Telephone Industry. Vol. R-3790, Santa Monica, CA: The Rand Corporation, 1989.

[15] J.H. Alleman, P.N. Rappoport, and K.B. Stanley, Alternative Settlement Procedures in International Telecommunications Service. In: Communications Policy in Europe. Kronberg, FRG, 1989.

[16] S.R. Hakim, and D. Lu, Monopolistic Settlement Agreements in International Telecommunications Agreements, *Information Economics and Policy* **5** (1993) 145-157.

[17] S.R. Hakim, and D. Lu. Duopolistic Pricing in International Telecommunications. In: Proceeding of Tenth International Telecommunications Society, Sydney, 1994.

[18] R. Frieden, International Toll Revenue: Tracking the Inequities and Inefficiency, *Telecommunications Policy* **17** (1993) 221-233.

[19] K-L. Yun, H-W. Choi and B-H. Ahn, The Accounting Revenue Division in International Telecommunications: Conflicts and Inefficiencies. Memo, undated.

[20] H. Ergas and P. Paterson, International Telecommunications Settlements Agreements, *Telecommunications Policy* **15** (1991).

[21] K.A. Cheong, and M. Mullins, International Telecommunications Service Imbalances, *Telecommunications Policy*, **15** (1991) 107-118.

[22] S.J. Brown, and David S. Sibley, The Theory of Public Utility Pricing. Cambridge University Press, Cambridge, Massachusetts, 1986.

[23] O. Shy, Industrial Organization: Theory and Applications. Cambridge, MA, MIT Press, 1995.

[24] J-J. Laffont, and J. Tirole, Creating Competition through Interconnection: Theory and Practice. In: Access and Interconnection Workshop. Wellington, New Zealand, 1995.

[25] W.J. Baumol, and J.G. Sidak, Toward Competition in Local Telephony. Washington, D.C.: The MIT Press, Cambridge & London, and The American Enterprise Institute for Public Policy Research, 1994.

[26] J. Alleman, Toward Competition in Local Telephony: A Book Review, *Information Economics and Policy* 7, No. 2, June (1995).

[27] R. Willig, The Theory of Network Access Pricing. Michigan Sate University Public Utility Papers: Issues in Public Utility Regulation. Lansing, MI, 1979.

Part V

The Internet: Demand and Policy

Telecommunications Transformation: Technology, Strategy and Policy
E. Bohlin and S.L. Levin (Eds.)
IOS Press, 1998

CHAPTER 14

The Demand for Internet and On-line Access[1]

Paul N. Rappoport
Department of Economics, Temple University, Philadelphia, PA 19122, USA

Lester D. Taylor
Department of Economics, University of Arizona, Tucson, AZ 85721, USA

Donald J. Kridel
Department of Economics, University of Missouri
St. Louis, 8001 Natural Bridge Rd., St. Louis, MO 63421, USA

William Serad
PNR and Associates, 101 Greenwood Ave., Jenkintown, PA 19046, USA

Abstract. This paper examines the demand for on-line access to the Internet using information from two proprietary surveys undertaken by PNR and Associates of Jenkintown, PA. Internet demand is strongly associated with age, education, and household income. Users seem to have a strong preference for flat-rate access to the Internet, and appear resistant to prices that are much higher than present levels. There is evidence as well that users will resist paying premium rates for higher-speed access.

1. Introduction

The increased popularity of the Internet has been widely tracked. A recent *Business Week* survey estimates that 40 million people in the US are Internet users.[2] An April 1997 FIND/SVP survey estimates that there are among 40-45 million adult users[3]. These numbers are up sharply from estimates obtained from surveys administered in November of 1996. For example, a November Louis Harris & Associates' survey estimated that there were 35 million adult users[4]. An IDC Research estimated that there were 31.4 million adult users in October 1996.[5]

---

[1] We are grateful to Don Michels of PNR and Associates for data organization and analysis.
[2] The starting place tracking surveys as they relate the Internet http://www.nua.ie. Nua maintains an extensive Internet survey database. The link is http://www.cyberatlas.com/demographics.html.
[3] The link is http://etrg.findsvg.com/internet/interest.html.
[4] *Op. cit.* cyberatlas.com.
[5] The link is http://www.idcresearch.com.

PNR and Associates' most recent estimate of households that use the Internet is 16%.[6] This is up from 14.8% in 1996[7]. These percentages are in line with the Harris Associates' estimates. There is ample evidence to support the increased popularity of the Internet and the resulting demand in Internet applications. Unfortunately, however, the demand for Internet access is very difficult to measure. To begin with, there is the question of what constitutes access. A global definition would include subscriptions to on-line service providers such as AOL and CompuServe, access to the World Wide Web via any national or local Internet service provider (ISP), access via schools and libraries, and access through accounts at work. Secondly, there is the question of how to obtain accurate measurements of access. And finally, there is the matter of devising an appropriate framework for modeling Internet demand.

According to *Internet Week* as reported by ZDNet AnchorDesk[8], as of May 29, 1997 the top 10 Internet service providers were:

(1).	American on-line	8 million subscribers
(2).	CompuServe	5.3 million subscribers
(3).	Microsoft Network	2.2 million subscribers
(4).	Prodigy	1 million subscribers
(5).	AT&T Worldnet	900,000 subscribers
(6).	Netcom	582,000 subscribers
(7).	EarthLink Network	310,000 subscribers
(8).	Concentric	227,000 subscribers
(9).	MCI	200,000 subscribers
(10).	Erol's Internet	200,000 subscribers.

These numbers are only estimates and do not take into account multiple users (or even subscriptions) per household, work vs. residential accounts, etc. The top 10 service providers are just the tip of the iceberg of companies providing Internet access, as altogether there are more than 3000 ISPs, most of them with under 10,000 customers. Obviously, this large number of ISPs makes tracking the number of Internet subscribers difficult, and with the Regional Bell Operating Companies positioning their Internet services as part of new composite service bundles, tracking will almost certainly become even more difficult in the future.

The purpose of this paper is to examine the demand for Internet access using information from two proprietary datasets of PNR and Associates of Jenkintown, PA. The first dataset is PNR's ReQuest III™ Survey that was fielded in May of 1996. The second dataset is part of a market share tracking study, Trends and Linkages in Communications (TLC™), initiated in January 1997 by Market Facts, Inc. and PNR.

In the ReQuest III survey, respondents were asked whether they subscribed to any (from a list) of consumer on-line Internet access services. They were then asked to provide the amount of their last on-line service bill. Additional questions were asked which focused on Internet services and interest in second lines, interest in and willingness-to-pay (specific

[6] The percentages are obtained from the PNR and Market Facts survey, *Trends and Linkages in Communications*, January 1997.
[7] This figure is derived from the ReQuest III survey of PNR and Associates.
[8] The link is http://www3.zdnet.com.

amounts) for unlimited Internet access, and interest in and willingness-to-pay for fast Internet access. In the TLC survey, respondents were asked if they had access to the Internet from their home.

Of the 45,000 households to which the ReQuest III survey was sent, approximately 31,000 responded. In particular, respondents in the survey were asked to provide their monthly expenditures for on-line services and to identify their on-line providers. In contrast with ReQuest™, TLC™ is a continuous survey of the communications market. Forty-eight thousand survey responses were used in the present analysis. Both surveys are weighted to reflect demographic national distributions.

As both surveys are large and nationally representative, they accordingly provide good bases for examining Internet demand. In the remainder of this paper, information from the two surveys will be used to provide both an overall picture of Internet demand and how this varies in relation to economic and socio-demographic characteristics. A good place to begin, however, is with a brief discussion of contrasts and similarities between Internet access and basic telephone service.

2. Analogy with Basic Telephone Service

An obvious analogy to pursue in analyzing the demand for Internet access is the demand for basic telephone service. While there are a number of similarities that support using the modeling framework that has been developed for telephone access as the framework for modeling the demand for Internet access, there are also some important differences. For both, there is (at least in principle) a distinction between access and usage. For usage to be possible, there has to be access, yet the demand for access depends upon the benefits from usage. In modeling the demand for telephone access, this distinction is taken into account by formulating a two-stage procedure in which usage is modeled (as function of income, price, and other relevant variables) conditional upon access and access is then modeled in a probit or logit framework in which the consumer surplus from usage is juxtaposed against the cost of access.[9] Also, for both, networks are involved, which makes for modeling complications because of the possible presence of externalities. To begin with, there is the so-called subscriber (or network) externality, which makes access to a large network more valuable than access to a small network. The existence of this externality can cause a network to grow endogenously, even though nothing may be happening to the 'objective' drivers of the system, such as income and prices. While this externality may be a factor in less developed countries in which telephone systems are primitive and penetration rates are low, it is almost certainly of only limited importance in mature economies, such as the U.S., Canada, Western Europe, and Japan. With the Internet, however, this is undoubtedly not the case, as much of the current torrid expansion of the Net is probably endogenously caused by expansion itself.

Additional externalities are associated with usage. With a telephone system, these are essentially of two types. The first, and the one most frequently discussed, is the so-called usage externality, in which a party calling another party affects the called party's welfare. One extreme of this is where the party called benefits from the call simply because the calling

[9] For a discussion of these models and their applications, see [4].

party is the one who does the paying. In some instances, implicit contracts may come into being, in which it is understood that if A calls B today, then the next time communication is needed B will call A.[10]

The second type of usage externality, and the one which is almost certainly of importance in Internet usage, is what Taylor (1994) has termed the 'dynamics of information exchange'. With this externality, the need for communication is created 'on the fly', in the sense that a call from A to B creates a further need for B to call A, or maybe for B to contact C, who in turn contacts D, etc. With the Internet, one of the forms this phenomenon can take is for a visit to one web site to create a need to visit several other web sites, which in turn creates the need to visit still more sites, etc.[11]

One of the important differences between the telephone network and the Internet relates to definition and measurement of usage. With the telephone network, usage is usually straightforwardly defined in terms of a connection from one telephone number to another and measured either in terms of the number of calls or the number of call minutes. With the Internet, usage is much more ambiguous. Is it to be defined in terms of 'hits' on a web site, connection to another computer or computer system, or just what? In addition, connection to the Internet is often continuous, but nothing is happening, as when a computer openly connects into an e-mail network. At this point, we have no answers to such questions involving usage, and will confine our attention in the remainder of this paper to Internet access, which fortunately is pretty straightforward to define and measure.

The demand for Internet access has received virtually no empirical attention, and the dynamics of the Internet market has only been partially studied. As will be seen in the next section, there is a strong relationship between Internet access and income, as well as strong associations between Internet users and socio-demographical characteristics such as age and education. A number of interesting and useful profiles of Internet users are also available.[12] However, we are still far from understanding the 'drivers' of this market. The quantification of factors such as income and price vis-à-vis the demand for on-line access is required for public policy analysis (competition, universal access to the Internet), pricing, and target marketing. How much of this growth is due to non-price factors? How much of the overall growth is price-sensitive (as opposed to being driven by income and the subscriber and usage externalities described above)? And, of course, what is the price elasticity for Internet access?

In studies of the demand for local telephone usage and access, the price variable used could be a flat-rate price with no per-call usage charges, or it could be flat-rate with a measured option, or it could be a measured rate calling plan. The price for access to the Internet initially involved a monthly charge with a fixed number of hours and a per-hour charge for hours that exceeded this fixed number. However, just as local telephone markets usually find consumers unwilling to select a measured-service price plan, Internet service providers have found that their customers overwhelmingly favor a monthly flat-rate plan. That, plus the fact the market for Internet services is highly competitive, has led to a simplification of prices facing the Internet consumer.

---

[10] For a detailed discussion of the implicit contracts which may be associated with the usage externality and their implications, see [1].

[11] The 'dynamics of information search' seems the appropriate descriptor for this phenomenon.

[12] See the GUV 6th WWW User Survey. The link is http://cc.gatech.edu/gvu/user-surveys.

Consider, for example, current on-line service provider prices. American On-line reversed its pricing policy in late 1996 and now offers an unlimited (hours) Internet plan for $19.95 per month. CompuServe prices their access for $9.95 for 5 hours and $24.95 for 20 hours. Microsoft has a number of options, including a charge of $19.95 for unlimited service and $6.95 for 5 hours of connection. Netcom charges $19.95 for unlimited monthly access. Prodigy also has two rates plans. They charge $19.95 for unlimited access and $9.95 for 5 hours of access. The cost per incremental hour varies from $1.95 (SpryNet) to $2.95 (CompuServe and Prodigy). Finally, most regional ISPs offer flat-rate pricing plans.

3. Characteristics Of Internet Users

In this section, we present a series of tables and figures showing distributions of Internet users in relation to a variety of economic and socio-demographical characteristics, beginning in Table 1 with the distribution of on-line charges obtained from the latest ReQuest III™ Survey. The distribution underscores the trend in the industry towards flat-rate pricing.

Table 1. On-line Services Distribution

Expenditure	Percent	Mode Value
<$5.00	6.2	$4.95
$6-$10	38.3	$9.95
$11-$15	24.8	$14.95
$16-$20	19.4	$19.95
$21-$25	7.1	$24.95
>$25	4.2	$29.95

In ReQuest III, respondents were asked first to indicate if they subscribed to an on-line Internet service and then to identify which particular service. Table 2 displays the distribution for respondents' choices.

Table 2. Choice of Internet Provider

Provider	Percent
AOL	50.2
CompuServe	19.5
Other (include ISP)	30.3

Table 3 displays penetration rates for multiple telephone lines, having a computer with a modem, having a cellular phone, and the presence of a fax machine at home.

Table 3. Penetration Rates for Selected Features

Variable	Percent
Second line	14.31%
Fax machine	13.0%
Computer	43.6%
Computer modem	25.7%
Cellular phone	27.1%

Figures 1 – 6 display Internet penetrations rates by selected demographic measures. Figure 1 confirms the strong relationship between Internet penetration and household income, while Figure 2 underscores the strong relationship between second-line penetration and Internet access, by age category.

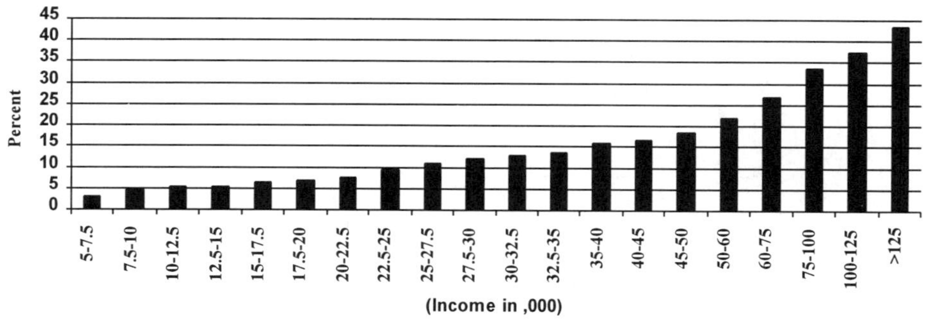

Figure 1. Internet Penetration Rate as a Function of Income

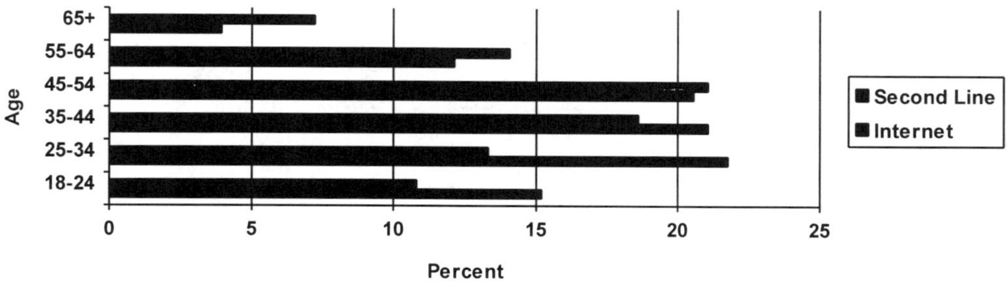

Figure 2. Internet and Second Line Penetration Rates by Age

Internet penetration is clearly skewed to younger households, whereas second-line penetration, interestingly enough, is more prevalent in the 35-54 age categories. Figure 3 gives Internet penetration by marital status, and shows that never-marrieds have the highest penetration rate. Figure 4 presents Internet penetration rates by occupation (in relation to an overall penetration rate of 15.8%). Three occupations clearly stand out, namely, professionals, executives and managers, and technical and related support functions.

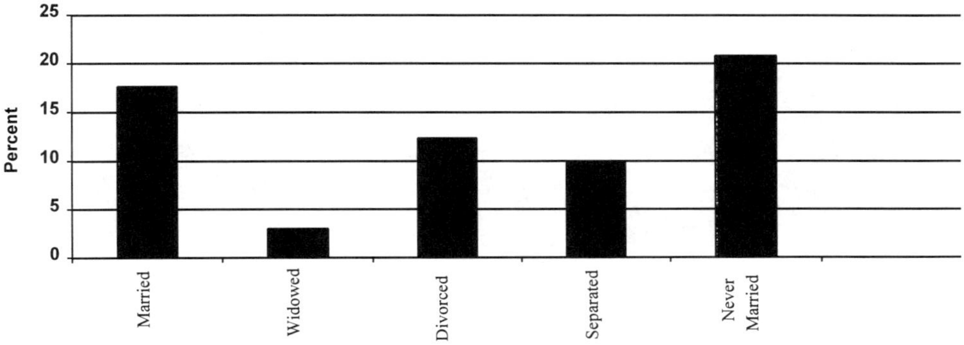

Figure 3. Internet Penetration Rates by Marital Status

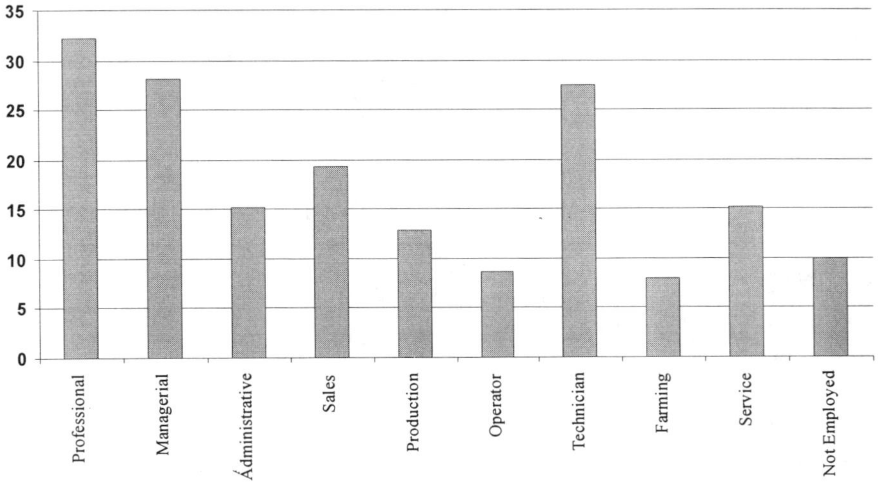

Figure 4. Internet Penetration Rates by Occupation

Figures 5 – 6 depict Internet penetration rates according to geographical area and population density. Figures 5 and 6 give penetration rates by Census divisions and population density. Not surprisingly, rural areas (non-MSAs) are seen to have a significantly lower Internet penetration rate than higher density areas.

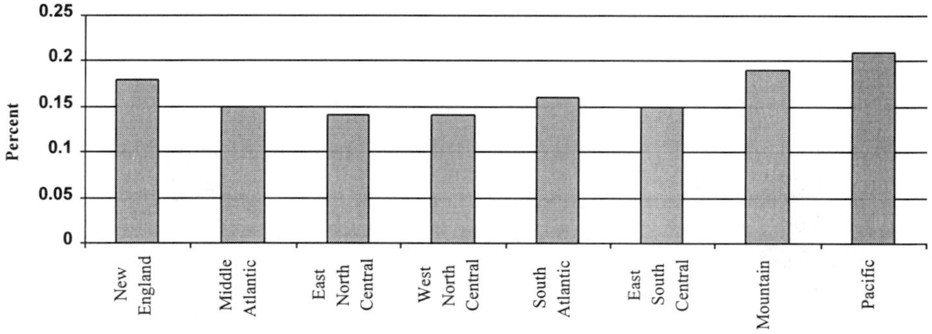

Figure 5. Internet Penetration Rates by Census Division

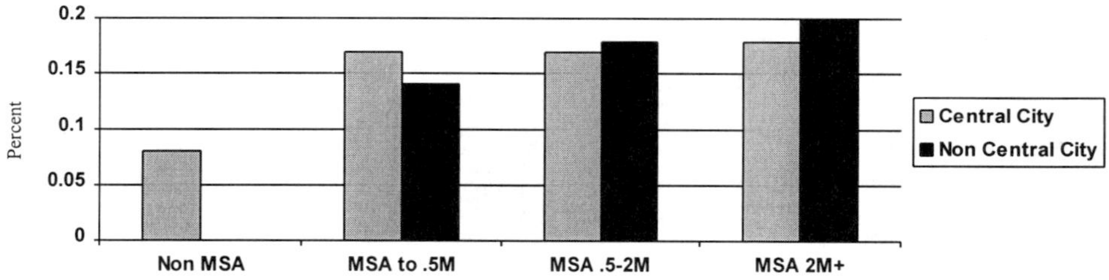

Figure 6. Internet Penetration Rates by Population Density

Tables 4 and 5 show some additional indicators associated with Internet demand. The data in these tables are based upon information obtained in the Market Share Tracking survey and are based upon January 1997 tabulations.

Table 4. The Importance of Second Lines and Computers In Explaining Internet Access

Unconditional Probability		**Conditional Probability**	
Second line	14.31	Internet Given Second Line	48.30
PC with Modem	17.82	Internet Given PC with Modem	67.1

Source: Market Share Tracker

Table 4 focuses on the relationships among Internet access, second lines, and computers with modems. Both conditional and unconditional probabilities are presented. If the conditional probability is greater than the unconditional probability, then the conditioning variable can be viewed as a significant 'causal' measure.[13] The probability of subscribing to the Internet, given a household computer with a modem, is 67.1%. This compares with the unconditional probability of subscribing to the Internet of 15.87%. The 'lift' associated with having a computer with a modem is accordingly 422% (67.1 / 15.87). Among other things, the implication of this is that, for target marketing purposes, the best prospects for Internet marketing are households with computers and modem and households with second lines.

Finally, Table 5 gives the average bills for both Internet subscribers and non-subscribers for local and long-distance telephone, cable, cellular, and on-line services.

Table 5. Communications Expenditures

Communications Expenditures	Internet Subscribers	Do Not Subscribe to the Internet	Overall
Average local phone bill	$38.24	$32.33	$33.04
Average long-distance bill	$39.69	$28.24	$29.71
Average cable bill	$35.57	$33.93	$34.15
Average cellular bill	$52.20	$44.06	$45.88
Average on-line services bill	$19.14		$19.14

Source: ReQuest III

4. Some Estimated Price Elasticities for Flat-rate Internet Access

In ReQuest III, the question was asked: "Would you be interested in getting an additional telephone line if you could have a regular phone line with unlimited Internet service for a flat monthly price of $xxx per month", where xxx was $20, $25 or $30. Each respondent was given one of the three prices. The responses that were obtained imply arc price elasticities for current Internet subscribers of -0.44 for prices between $20 and $25 and -1.0 for prices between $25 and $30. These estimates of 'access' elasticities suggest that prices greater than the current level of $9.95 or $19.95 would have dampening impact on Internet growth.

In ReQuest III the question was also asked: "Would you be interested in getting an additional telephone line if you could have a phone line with unlimited Internet access for a flat monthly rate of xxx which is 4 times faster than normal", where xxx was either $40, $50 or $60. Once again, each respondent was given only one price.

For current Internet subscribers, the estimated arc elasticities for faster Internet access implied by the responses are -0.51 for prices between $40 and $50 and -1.23 for prices be-

[13] This line of reasoning follows Suppes [3]. According to Suppes, if event A occurs before event B, then if $P(B|A) > P(B)$, A is a *prima facie* cause of B.

tween $50 and $60. The elasticity of -1.23 indicates that fast Internet access at a premium price is clearly not an attractive option to current Internet subscribers.

5. A Cluster Analysis of Internet Demand

In [2], a logistic model of Internet access is estimated using data from ReQuest III™ in which three of the most important predictors are price, income, and household size. In this section, we present multivariate cluster analysis to model the relationship between Internet access and these variables. A K-means analysis has been employed in defining six clusters as follows:

Cluster 1: Typical Empty Nesters

This cluster is composed of households where the age of the head of household is 54 and up; where income is between $35,000 and $70,000 and where the household size is 1 or 2.

Cluster 2: The Median Family

This cluster includes families (household size 3 and more) where the age of the head of household is 35-44 (middle age) and where household income is between $35,000 and $75,000.

Cluster 3: Better Off Families

This cluster is similar to Cluster 2 except that household income is greater than $75,000.

Cluster 4: Out to Pasture

This group consists of individuals who have retired. The age of individuals in this group is 65+; household income is less than $35,000. Household size is 1 or 2.

Cluster 5: Comfortable Empty Nesters

This cluster is similar to the Empty Nester cluster except for household income. Household income in this group is greater than $75,000.

Cluster 6: Family Startups

This cluster is defined by younger heads of household, lower incomes (< $35,000) and family sizes 3+.

Table 6 displays summary characteristics for the six clusters.

Table 6. Internet Cluster Profiles

Cluster	Internet Percent	Overall Percent	Index	Local Bill	LD Bill
1. Typical Empty Nesters	16.7%	17.3%	97	$31.77	$31.13
2. The Median Family	33.0%	19.3%	171	$37.77	$34.30
3. Better Off Families	18.6%	6.4%	291	$40.58	$41.65
4. Out to Pasture	10.5%	34.0%	31	$26.30	$16.68
5. Comfortable Empty Nesters	9.3%	4.9%	190	$39.53	$36.33
6. Family Startups	11.9%	18.1%	66	$30.04	$35.51

Source: ReQuest III

The cluster represented by The Median Family accounts for 33% of the Internet households. The ratio of column 2 to column 3 (multiplied by 100) in Table 6 provides an index of similarity between the distribution of Internet subscribers and the general distribution of the cluster. An index of 100 would imply the same distribution. An index greater than 100 would imply a greater percentage of Internet subscribers for this cluster relative to the percent of the cluster in the overall population.

Table 7 looks at custom/class feature penetration by cluster. Correlation of telephone features and services with Internet access provides local exchange companies with an inherent advantage in the marketing of Internet access services.[14]

Table 7. Feature Penetration Rates by Internet Cluster

Cluster	CW	CF	3W	CB	CID	VM
1. Typical Empty Nesters	86	97	98	95	94	100
2. The Median Family	115	84	78	79	88	76
3. Better Off Families	114	148	172	138	109	128
4. Out to Pasture	86	74	43	57	109	94
5. Comfortable Empty Nesters	88	130	122	97	91	132
6. Family Startups	101	83	123	185	123	98

Source: ReQuest III

The numbers in Table 7 are indexed values where the index is the penetration rate of that feature for a particular cluster group divided by the overall penetration rate for the feature. This ratio is then multiplied by 100. As expected, Better Off Families have a higher than average tendency to subscribe to telephone features. Clusters 2, 3 and 5 are all of interest for Internet targeting. The demographic-based clusters are also differentiated with respect to

[14] In Table 7, the codes are as follows: CW is call waiting, CF is call forwarding, 3W is three-way calling, CB is call block, CID is caller ID and VM is voice mail.

their average local- and long-distance bills. As expected, the size of these bills is correlated with the income and age components in the clusters.

6. Some Internet Usage Profiles

The Internet is a great source for statistics about itself. Figure 7 shows counts of the number of Internet hosts, which is seen to be increasing at a high exponential rate.[15] Figure 8 presents the distribution of local residential weekend calls by time-of-day against the number of weekend hits (proxy for calls) to "the number 1 adult web site". The two distributions differ, with most of the local calling distribution taking place between 8 AM and 8 PM. The adult usage is more uniform.[16] Figure 9 continues this comparison of local call distributions by summarizing the time-of-day patterns for the number one ranked shopping site, the number one ranked sports site and the number one ranked adult site for a typical weekday. Of interest are the time-of-day differences among the categories. The non-adult patterns follow more closely the local call pattern.

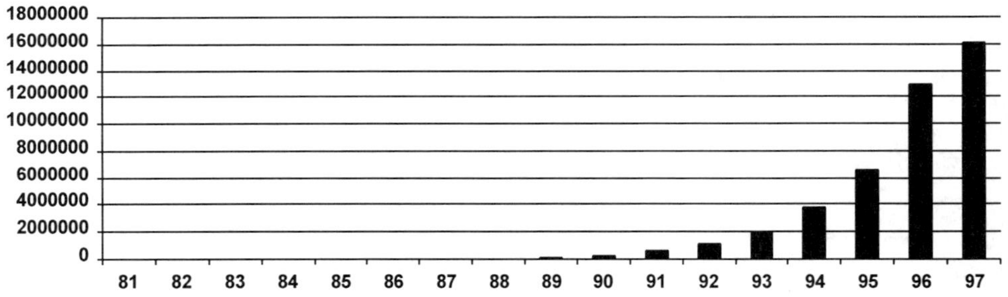

Figure 7. Growth in Internet Hosts

[15] Source: HTTP://nw.com/zone/host-count-history. A host is a computer that is directly linked to the Internet with a domain name with an IP address associated with it. There are a number of sources for web statistics. Actual usage statistics are provided WebSide Story (HTTP://wss5.websidestory.com.) This site represents a collection of sites listed by category. For each category, the numbers of sites and hits are tracked. Reports are available that summarize activity by time of day, day of week and so forth.

[16] The local calling distribution was computed by analyzing two hundred million flat-rate local calls.

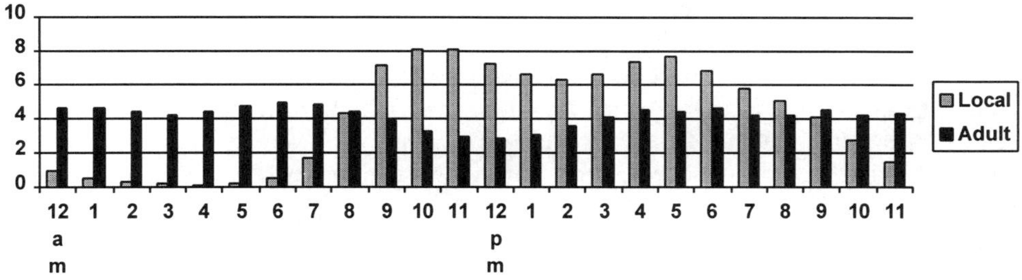

Figure 8. Distribution of Local Calls: Local Residential vs. Adult Services (Weekend)

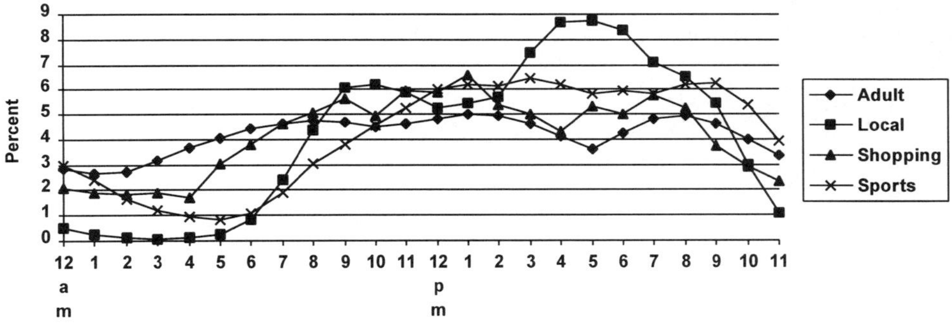

Figure 9. Distribution of WEB Usage vs. Local Calling Usage

7. Conclusions

In this paper, we have used data from two proprietary PNR surveys to analyze how Internet demand varies with economic and socio-demographical characteristics. Internet access is strongly associated with education and age, as well as with the level of household income. Internet usage profiles suggest that current Internet growth is likely to continue. The consequences of this growth on local networks will eventually be a concern, and at some point the Internet access will almost certainly have to be priced to reflect costs.

However, arc elasticities calculated from responses to questions regarding unlimited Internet access offered in conjunction with a second telephone line suggest that monthly access charges much in excess of those currently in force will face subscriber resistance. In line with households' strong preference for flat-rate telephone service, there is evidence that Internet users will resist efforts to charge for usage on a time-sensitive basis. Finally, there is evidence that current subscribers will be unwilling to pay much of a premium for increased access speed.

One of the challenges in modeling Internet demand will be to account for an expected rapid change in technology, including faster modems, ISDN service, cable modems, and so forth. For now, we appear to be in a 'field of dreams' phase: "If we build it, they will come." Currently, Internet users come to browse and chat, do research, and send and receive e-mail. When Internet commerce, which is now just in infancy, is firmly established, current thoughts on how to model Internet demand will almost certainly have to be reevaluated.

References

[1] A.C. Larson, D.E. Lehman and D.L. Weisman, A General Theory of Point-to-Point Long Distance Demand. In: Telecommunications Demand Modeling, ed. by A. de Fontenay, M.H.Shugard, and D.S. Sibley. North-Holland Publishing Co., Amsterdam, 1990.
[2] P.N. Rappoport, L.D.Taylor and D.J. Kridel, An Econometric Analysis of the Demand for Access to the Internet. PNR & Associates, 101 Greenwood Ave., Jenkintown, PA, 19046, July, 1997.
[3] P. Suppes, Probabilistic Theory of Causality. North Holland Publishing Co, 1970.
[4] L.D.Taylor, Telecommunications Demand in Theory and Practice. Kluwer Academic Publishers, 1994.

Telecommunications Transformation: Technology, Strategy and Policy
E. Bohlin and S.L. Levin (Eds.)
IOS Press, 1998

CHAPTER 15

The Internet, Externalities and Public Policy

Steven Globerman
Western Washington University, College of Business and Economics
Bellingham, Washington 98225-9170, USA

Abstract. The emergence and growth of the Internet has spawned numerous calls for extensive government regulation and subsidies. Such calls presume that "externalities" are more significant and ubiquitous in the case of the Internet than in the cases of "more conventional" communications media. This study evaluates potential externalities associated with the Internet that could support a case for government intervention. It concludes that the case for government intervention remains weak, and that further development of the Internet is best left largely to market forces.

1. Introduction

An increasing number of public policy issues are accumulating surrounding the Internet and the associated World Wide Web (WWW).[1] To date, identification and evaluation of these issues have been balkanized by the variety and specificity of concerns. For example, calls for government intervention have been linked to concerns about the Internet being used to distribute pornography and hate literature, to disseminate false or misleading information about commercial transactions and financial securities, to distribute copyrighted material illegally and to violate the privacy of individuals.

Concerns have also been expressed about the need to subsidize access to the Internet on the part of both potential users and potential suppliers of components and services. At the same time, governments have indicated concern about protecting domestic Internet content in the face of foreign content supply. For example, Canadian policy makers are concerned with ensuring that the growth of the Internet and its evolution into a full-fledged broadcasting medium does not undermine policies to promote Canadian content as they have been applied to more traditional broadcasting media [2].

[1] For convenience, we will simply refer to the Internet as a short-hand reference to all computer-based public communications networks including the backbone networks such as NSFNET. MacKie-Mason and Varian [11, p.28] define the Internet as a worldwide network of computer networks that use a common (TCP/IP) communications protocol.

While the scope of public policy concerns surrounding the Internet appear to be eclectic and somewhat unrelated, the contention of this paper is that most, if not all, reflect an implicit presumption that the Internet is characterized by significant externalities on both the "demand" and "supply" sides of the market. A related presumption is that government intervention of one sort or another is required in order to address the problems associated with the presence of substantial externalities.[2]

The broad objective of this study is to consolidate various public policy concerns surrounding the future development and growth of the Internet into a relatively circumscribed economic framework and to use that framework to identify and assess the important conceptual and empirical issues that policymakers should consider in deciding whether and how Internet activities should be regulated. The analysis raises strong cautions against government preemption of "free market" activity on the Internet.

2. Economic Attributes of the Internet

While the Internet obviously shares the infrastructure of other communications networks, most notably "conventional" telephone and cable networks, these other networks are potential substitutes for the Internet, to a greater or lesser extent, from the perspectives of final users. Perhaps most obvious, voice messages delivered over wired or wireless telephone networks are substitutes for electronic mail. Video images can be delivered through wired or wireless cable. Data files can be sent through high-speed private or virtual private digital telephone and cable networks. Documents can be sent as facsimile messages through the telephone network. And so forth. To the extent that other communications media are good substitutes for the Internet, the policy concerns outlined in the introduction are generic, and a focus on the Internet, *per se*, is too narrow if not misplaced entirely. Hence, it is relevant to identify what, if any, attributes distinguish the Internet from other media, particularly as they relate to public policy in the domain of communications.

2.1 Externalities: Demand Side of the Market

One obvious distinction between the Internet and conventional telephone and cable distribution networks is subscribership. While the share of the population subscribing to the Internet is growing rapidly, it remains relatively low compared to the share of the population subscribing to conventional telephone service and cable television. [3, p.248], for example, estimates for the United States that about 10 million residential telephone lines (out of 95 million households with telephone service) had Internet Service Provider (ISP) access at the end of 1995.[3] [11, p.29] put the number of people directly or indirectly connected to the Internet (as of October 1994) at around 14 million with as many again able to access the Internet via e-mail. While such estimates are rendered obsolete quickly by the continued

---

[2] Other possible sources of inefficiencies and inequities associated with an essentially unregulated Internet have been addressed elsewhere. See, for example, [5].

[3] Extrapolating recent trends would result in Internet usage levels in the year 2001 attaining those of telephones [3, p.247]. One can certainly be sceptical of simple extrapolations of early period experience with the Internet. Indeed, some observers argue that business users are becoming less enthusiastic about the potential for profitable widespread use of the Internet. See, for example, [13].

rapid growth of Internet membership, the point remains that, for the foreseeable future, access to and usage of the Internet will be below comparable access to and usage of basic telephone and cable networks.

A potentially important feature of emerging communications networks is the existence of network-related externalities. Communications network externalities are usually described as either access or usage externalities.[4] Access externalities are associated with the following phenomenon: when individuals join a network, they create value for others already on the network or who might join the network, as well as for themselves, because the former can now contact (or be contacted by) at least one more person. In essence, networks become more valuable with larger subscribership, all other things being constant, because more potential communications contacts become technically possible, and the increased value benefits others besides simply the last person to join the network. Usage externalities are associated with the phenomenon that when a "call" is made over the network, it generates benefits for both the caller and the called parties, but it is usually the caller who pays for the service. hence, the benefit of a typical call extends to those who do not necessarily pay for the call. All other things constant, larger networks will therefore generate larger usage externalities.

The relationship between externalities and network size may take complex forms. For example, networks may be of little or no value to subscribers unless they are of a minimum critical size [1]. This is because a network may be too small to generate sufficient "subscriber surplus" to make joining worthwhile by anyone. Conversely, beyond some critical threshold size of subscribership and anticipated usage, externalities may be sufficiently small in magnitude to be of relatively limited policy relevance. In this regard, [6, p.35] conclude that: "At penetration rates near 100%, an additional access line is unlikely to generate significantly large enough effects on the demands of those who themselves do not purchase access, to cause one of these individuals to now demand access, at unchanged income and price levels." Basically, a network may reach sufficient size that those on the network see little additional value from communicating with anyone not yet on the network.

Policy-related discussions of network externalities usually relate the magnitudes of the relevant externalities to subscriber penetration rates, i.e. the share of all possible subscribers already connected to the network. As noted above, higher existing penetration rates are taken to imply that potential subscribers to the network are less likely to make significant contributions to the value that existing network members already receive from the network. So, holding the overall size of the potential communications network constant, a higher penetration rate implies lower network externalities, at the margin. Analogously, growth of the size of the potential network implies a lower "effective" penetration rate given the existing number of subscribers. To the extent that the growing number of potential users have characteristics that put them in one or more "communities of interest" of those already on the network, externalities can be quite significant, at the margin, since new members are likely to be intensive communicators with those already on the network.

The general relevance of demand-side externalities to communications policy is that they constitute a potential source of "market failure". Specifically, if suppliers of communications services charge competitive prices to potential consumers, the latter will equate the

[4] For an extensive discussion of communications network externalities, see [16].

prices charged to the benefits they expect to receive from those services. If the latter do not include network externalities, competitive market prices will result in too little consumption of communications services from the standpoint of economic efficiency.

This point is simply made in Figure 1. The curve labelled *S* is the supply curve of an hypothetical communications service. Assuming that the service is competitively supplied, it will equal the long-run marginal cost to producers of the service.[5] The curve labelled *Dp* represents the demand curve for the communications service on the part of those who directly consume the service. Assuming that consumers are knowledgeable about the benefits to themselves of consuming the service, the *Dp* curve should equal the benefits that consumers expect to receive themselves from consuming the marginal unit of service. Put in other words, consumers would be willing to spend up to the amount shown on *Dp* for each additional unit of the relevant service. The curve labelled *Ds* represents the benefits that society as a whole realizes from each additional unit of the service that is consumed. The difference between *Ds* and *Dp* therefore represents the monetary value of the network externalities associated with different output rates of the service. As noted earlier, it is reasonable to think that network externalities are a decreasing function of penetration rates, all other things constant. This assumption is incorporated into Figure 1 by showing the *Ds* and *Dp* curves converging as output increases.[6]

The implication of the presence of network externalities is exhibited in Figure 1. Specifically, a competitive market would result in the production and consumption of *Q1* units of output. The socially efficient output rate is *Q2* units; i.e. where marginal costs and marginal *social* benefits are equated. Between output rates *Q2* and *Q1*, *Ds* exceeds *S* which (equivalently) means that the associated social benefits exceed the social costs of producing the extra output; however, consumers will not voluntarily pay higher average prices to coax the extra output from the marketplace, because the extra benefits that they expect to receive as individuals are less than the competitive prices they would have to pay beyond *Q1* units of output.

An implication is that governments may have to subsidize either producers and/or consumers of the service in question in order to elicit *Q2* units of output. Subsidies to consumers can take a variety of forms. One familiar to students of telecommunications pricing is to cross-subsidize the price of the service in question with net revenues from non-competitive services. Another less familiar way is to subsidize the "quality" of the service as perceived by the average user. In this context, quality can be thought of as a vector of attributes. In the case of the Internet, for example, quality might include the nature and variety of content, the speed of communications, reliability of access and so forth.

It might be noted that negative demand-side externalities might also exist, at least for some users. For example, some subscribers to the network may be actively bothered by messages that solicit business or intrude upon privacy unwantedly. In this case, if such annoying calls are a substantial portion of the traffic carried on the communications network, the *Ds* curve might actually lie to the left of the *Dp* curve. In this case, a competitive market would actually result in too much rather than too little output of the service being produced.

---

[5] Nothing is lost here by assuming that output of the service in question can be increased in continuous amounts rather than in discrete "lumps," even though the latter may be a more appropriate way to characterize the supply relationships for many communications services.

[6] Assuming that the size of the potential market is fixed, increased output implies higher subscription and/or usage rates for the service in question.

Intensive use of the network leading to delays and other manifestations of congestion can be thought of as a deterioration of the quality of communications. In terms of Figure 1, such congestion might lead to the *Ds* curve being to the left of *Dp* which, again, implies that a competitive market would result in too much output being produced.[7]

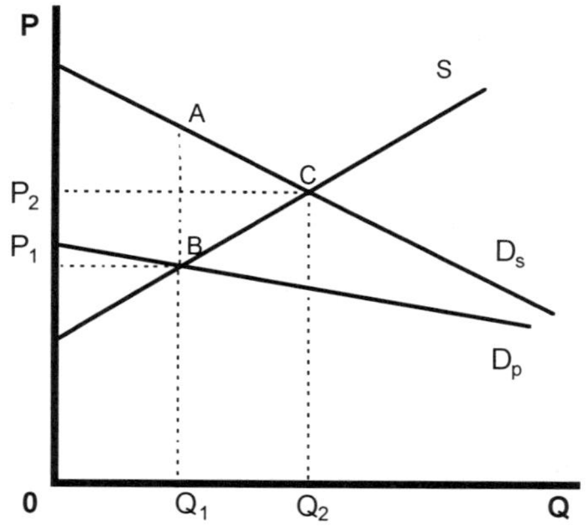

Figure 1. Implications of Externalities

The basic issue this study sets out to address is whether the Internet has distinctive characteristics that make subsidy and regulation especially relevant, at least compared to other substitutable modes of communication. Figure 1 hints at a possible answer. Namely, the gap between *Ds* and *Dp* is larger in a relatively new communications network such as the Internet compared to more mature networks. Subsidies directly address the issue of how to promote an increased subscription to and usage of a new network. Increased regulation might indirectly address the issue. Specifically, if concerns about encountering unpleasant content on the network, or about privacy being violated, or about the proliferation of misleading or fraudulent information on the network are important factors restricting the growth of the network, devoting government resources to mitigating these concerns is effectively providing an indirect subsidy to those potential subscribers who are particularly sensitive to these factors. That is, regulating content might improve the perceived quality of the network for these subscribers which is equivalent to lowering the effective cost.[8]

Conventional access and usage externalities are discussed in the network economics literature. In this context it is important to distinguish between real and pecuniary externalities. Such a distinction is not always easy to draw [9]. Basically, pecuniary externalities exist when parties not directly involved in a transaction suffer a change in income associated with that transaction. For example, the introduction and spread of personal computers con-

[7] As will be argued in a later section, congestion in communications networks is often the result of inefficient pricing which, in turn, reflects regulatory and other public policy restrictions on pricing flexibility.

[8] Indeed, it might be precisely equivalent to an indirect subsidy, since individuals could spend their own money monitoring content.

tributed to financial losses (and, in some cases, bankruptcy) suffered by makers of electronic typewriters and stand-alone word processors; however, there is little doubt that this substitution improved overall productivity and increased real income levels. That is, the loss of income for manufacturers of electronic typewriters was not associated with reduced economic efficiency.

Conversely, fisheries have been decimated, in some cases, by pollution from industrial activities. In this case, real output of a valued activity (fishing) is reduced, while the real output of polluting activities is increased. If fish are valued more (at the margin) by society than the output of the polluting industries, real income levels will be reduced by ignored pollution. In this case, the pollution externality has adverse impacts on overall efficiency. The loss in income of fishers and the associated gain in income of producers and consumers of the polluting products are associated with changes in real output, in this case a decrease in real output. So changes in income that are associated with lowered productivity and real output are associated with real externalities. The latter are public policy concerns, since they imply broad income losses for society, whereas pecuniary externalities may have concentrated adverse impacts on narrowly defined economic groups, but the impacts on overall economic efficiency may be either neutral or even positive.

A number of public policy concerns raised about the unregulated Internet can be seen as externalities, although it is not obvious whether they are real or pecuniary in nature. For example, individuals disseminating "pornography" on the Internet reduce the value of buying access to and using the Internet for those (potential many) individuals who would be deeply distressed by encountering such pornography, or who worry about their children encountering pornography. On the other hand, there is no doubt that some unknown number of individuals find value in the Internet precisely because of the readier availability of pornography on that medium. It is conceivable, although unlikely, that the overall access and usage of the Internet is unaffected by the presence of pornography on the Internet, so that the pornography externality has no net impact on Internet usage, or on the net social benefits derived from the network. In this case, it might be considered as largely a pecuniary externality; i.e. it increases the real incomes of some individuals and lowers the real incomes of others, while leaving the overall utilization of the network unaffected.

More likely, the bulk of potential Internet users are either disinterested or offended by pornography. It is likely, therefore, that a significant presence of pornography on the Internet will discourage access to and usage of the network, thereby resulting in reductions of realized net social benefits from the technology. In this case, we would consider the externality effects to be real or, equivalently, to have adverse real output effects in the economy. To be sure, technology is available which allows individuals to block access to undesirable Web sites; however, the technology is not "foolproof". Specifically, with new Web sites being created all the time and subject matter changing on existing sites, any Web screener will be imperfect. Moreover, individuals seeking technological protection from pornography (or hate messages) need to assume financial expenses in order to avoid Internet encounters for which they have an aversion.

Such significant expenses can themselves discourage both access to and usage of the Internet. Whether governments can reduce pornography or hate "pollution" at a lower cost than individuals can through voluntary acquisitions of technology or other screening procedures is an open question. Given that Internet traffic readily crosses international boundaries and that governments have different standards for defining socially offensive material, the long-run cost-effectiveness of "command and control" government censorship of the Internet is questionable, at best. In the absence of a cost-effective government response to content pollution on the Internet, a policy of no government action is preferable to content regulation, at least on economic grounds.

2.2 Externalities: Supply-side of the Market

The implications of significant externalities on the supply-side of the market are entirely analogous to the implications of the network externalities identified and discussed in the preceding section. Namely, if the private benefits to supplying services to the Internet are less than the benefits to all potential suppliers of services, i.e. those who currently are providing services and those who are not, an "unassisted" market may lead to too few services being provided. Put in other words, when applied to the Internet, there will be an underinvestment in the production of Internet services from society's standpoint.

2.2.1 Standards

At least two broad sources of supply-side externalities can be identified. One is related to the adoption of standards. To the extent that participants in a communications network harmonize their hardware and software standards, they mitigate risks associated with interoperability problems which would reduce the overall value of the network to those participants. A commonly agreed-upon set of standards enhances interoperability of the network and facilitates the participation of a larger number of interconnected users.

The effort and expenses that any one network participant puts into the standards-setting process can often be appropriated by other participants at relatively low cost. Thus, individual network participants will be tempted to "free-ride" on the standards-setting efforts and expenditures of other participants. Such actual (or even potential) free-riding could, in turn, discourage any significant efforts on the part of the collective of network participants to invest resources in setting effective standards. The potential for disruption or delay of the standards-setting process associated with free-riding leads some to call for government regulatory intervention and/or government subsidization of the standards-setting process.

The potential for free-riding is mitigated by several factors. One is the potential for voluntary cooperation when numbers of participants are relatively small. The Internet's early success has been linked to strong user involvement in the development process. This involvement, in turn, reflected the commitment of a relatively small number of non-commercial developers of the technology from its early stages. As the number of commercial participants on the Internet continues to grow, concern exists that the spirit of cooperation will wane.[9] However, another plausible scenario is that specific commercial interests will emerge as dominant participants and will be the *de facto* standards setters. In a number of markets, the conflict between closed and open architectures is being settled by the rise of "owned" architecture. In turn, other producers are developing technology needed to make interfaces between owned architectures work. With a relatively small number of dominant firms, cooperation in establishing standards may remain the rule rather than the exception.

[9] From 1985 to December 1994, the Internet grew from around 200 networks to well over 45,000 and from 1,000 hosts (end-user computers) to over four million [11, p.29]. Most of the technological decisions continue to be made by small committees of volunteers who set standards for interoperability.

The following examples of the emergence of dominant participant Internet technologies might be cited: (1) AT&T and Hewlett-Packard have agreed to integrate technologies to create electronic storefront systems that will allow merchants to host Web sites capable of secure electronic transactions. The partners will collaborate with other electronic commerce software developers to help merchants process orders and save customers. (2) Microsoft and Netscape Communications announced that they have agreed to support a technology that lets personal computer users manage what personal information gets disclosed to or withheld from a particular Web site.

2.2.2 Technology Spillovers

A second broad source of supply-side externalities is related to the imperfect appropriability of new information. The analogy here is to research and development (R&D) spillovers. In the context of the Internet, suppliers of new Internet services, such as new multimedia Web sites, convey information to other potential suppliers about what types of services might be in great future demand.[10] While the precise degree of success of an Internet site can presumably be kept confidential, reasonable inferences can be drawn from information such as waiting times to enter the site, the "elaborateness" of the site and so forth. While the precise details of the site cannot be duplicated, features of the site can be fairly quickly copied which, in theory, would dissipate some of the financial returns anticipated by the innovator. The appropriation of a significant share of anticipated entrepreneurial rents could significantly discourage innovation in the supply of Internet services.

In the R&D literature, the existence of spillover or innovation externalities is frequently used to justify government subsidies to R&D performance, as well as to strengthen intellectual property protection through, for example, broader patent protection or stricter enforcement of copyrights. In the context of the Internet, stronger copyright protection might be offered to content creators in order to encourage innovation in this dimension. Another approach is for government to make investments in new Internet content and capabilities and then make the results of these investments available for use by private sector firms. Does the Internet have a stronger claim for intellectual property protection or innovation subsidies than other economic activities? Similar to the discussion of demand-side externalities, the theoretical justification is that the relative immaturity of the Internet makes supply-side externalities more significant than they are likely to be in more mature communications activities.

In summary, the presence of economically significant externalities on both the demand- and supply-sides of the market for Internet services provides a potentially strong *a priori* case for a relatively activist public policy stance towards the Internet. In particular, network externalities and technological spillovers potentially justify government subsidies to potential users (on the one hand) and potential suppliers (on the other) to promote their presence in the marketplace. They also potentially justify greater (than normal) government efforts to monitor and regulate content on the Internet in order to discourage intellectual property violations, on the one hand, and "transaction costs" on the other hand related to the avoidance of offensive content material, potential invasions of privacy and so forth.

---

[10] Faulhaber [5] points out that the use of the Internet by specific consumers can help other consumers about the value of the Internet, thereby creating knowledge spillovers (or externalities) among consumers.

3. Evaluating the Externalities Arguments

The critical issue in any public policy evaluation of externalities is whether it is impracticable for participants to "internalize" fully the relevant costs and benefits of third-party effects in private bargaining.[11] Specifically, when it is relatively easy to identify those affected by external costs and benefits and when property rights can be created and traded at relatively low cost, externalities need not cause any loss of economic efficiency. Equivalently, scarce resources will continue to be used in their most efficient applications. For example, imagine that a group of UFO enthusiasts want to use the Internet to share information on UFO sitings more widely and more conveniently. The value of the network to potential UFO enthusiasts is presumably positively related (up to a point) to the number of other UFO enthusiasts on the network; however, large groups of such individuals may already be in contact with one another through magazines, mailing lists and so forth. This may facilitate their joining the Internet *en masse*, for example as groups or chapters of clubs. Even if they join as individuals, group pressure may ensure that significant numbers of individuals do not engage in "hold-out" tactics in order to be financially bribed to join the network. It may be easy for such hold-outs to be punished in other ways by members of their community of interest.

3.1 Assessing Demand-side Externalities

The fact that many organizational intranets are being constructed on top of the backbone (public) Internet network effectively means that many individuals join the Internet as part of a group in which access and usage externalities can be fairly readily identified and internalized among the members of the relevant communities of interest. Internalization of access and usage externalities can be accomplished as simply as the intranet manager, e.g. a corporation or trade association, making access to the Internet available to all eligible members as part of extending access to the corporation's or association's intranet.

In other cases, managers of an intranet can offer strong and direct financial incentives for individuals and groups to join the backbone network. For example, a company might be interested in transacting on-line with its suppliers. To encourage its suppliers to join the relevant communications network, the company could offer higher prices to those suppliers that receive and process the company's orders electronically. A blunter incentive would be provided by the company in question simply refusing to do business with any supplier that refused to join the intranet. In fact, such financial incentives are widely used by large companies to encourage small, vertically related companies to engage in electronic commerce, in effect using bribes and threats to internalize externalities.

Certainly, many individuals join the Internet on an individual basis or as part of a family unit; however, even in these cases, they rarely join without knowing specific individuals or families already on the network or thinking about joining. In these cases, network externalities can be effectively internalized by a shared sense of welfare among relevant communities of families and friends; e.g. I obtain direct satisfaction knowing that my brother is happy that he can stay in touch with me through e-mail.

---

[11] The seminal discussion of this point goes back to [4].

Finally, many individuals can be coaxed to join and use the network by monetary or other incentives extended by those who would like to communicate with them. For example, some enthusiastic basketball fans might be willing to join the network if it gave them electronic access to their favorite basketball players. While this might be a prohibitively costly set of transactions to organize on an individual fan-individual player basis, professional basketball players and team owners have an incentive to organize and offer an Internet Web site that can be accessed by electronically inclined basketball junkies. Indeed, the National Basketball Association (NBA) has set up a Web site with hyperlinks to Web sites of individual teams. Once at a site, the fan can leave a set of questions for a player (or his ghost writer) to answer, or leave some words of encouragement or advice and so forth. The NBA could charge for site visits. More likely, the players and team owners will more than recoup the cost of operating the site through the good will (and increased fan interest) created. As another example, Videotron and six consortium partners plan to offer a range of "infomedia" services delivered over cable. The new set-top box required by partners will be provided free-of-charge.[12]

The point to be emphasized here is that there are ordinarily strong incentives for potential and actual users of the Internet to internalize externalities and thereby benefit from the gains from networking. Moreover, economic and institutional arrangements make the costs of internalization relatively low in many cases. The fact that the Internet is a relatively new communications medium does not gainsay these assertions. Indeed, demand-side externalities may be of much less policy relevance than they were in the case of "plain-old-telephone service" (POTS). This is because the Internet is a much more flexible communications medium than POTS allowing both point-to-point and point-to-multipoint distribution, stored messaging and multi-media capabilities, among other things. An implication is that it is easier for communities of interest to organize themselves into readily identifiable sub-networks than was true in the case of POTS, and communications on the Internet have a much greater probability of being received and stimulating return communications than was the case in the early days of POTS.

The extraordinarily rapid growth of the Internet is itself suggestive evidence that access and usage externalities have not significantly constrained the utilization of this new communications medium and mitigates against the need for substantial new public sector subsidies to Internet users and suppliers; however, it should also be acknowledged that Internet access and usage have been directly and indirectly subsidized to this point in a variety of ways, and it is impossible to argue that these subsidies have had no significant impact on expanding and accelerating usage of the network.

In any case, even if transaction costs may have been a barrier to comprehensive internalization of demand-side externalities in the past, the growth of the Internet makes it more likely that externalities can be efficiently internalized by free-market transactions in the future. This is because the recognized commercial potential of the Internet enhances the incentives of entrepreneurs to find ways of internalizing the relevant externalities and realize economic rewards by doing so. An example is the widespread concern about protecting privacy when transacting on the Internet. In particular, consumers may be reluctant to disclose information about themselves to "Web-based" marketers for fear of confidential in-

[12] For a discussion of other examples, see [7]. Faulhaber [5] argues that the primary customer groups for Internet usage enjoy high penetration which makes the access externality argument for government intervention weak, at least for the current Internet customer groups.

formation being released to other parties. At the same time, by disclosing such information, consumers might be able to receive products and services more closely aligned to their individual tastes and preferences.[13] Hence, potential gains from trade may not be realized by concerns about invasions of privacy on the Internet. Unrealized gains from trade are an incentive to entrepreneurs, since would-be traders would gladly pay for solutions that allow them to trade more easily and cheaply. In this regard, a new service (TRUSTe) offered by the Electronic Frontier Foundation promises to hold information about consumers and reveal only what consumers authorize to companies who pay TRUSTe a fee for the service.[14]

3.2 Assessing Supply-side Externalities

The issue of externalities affecting suppliers to a market is certainly not unique to the Internet. For example, successful feature films spawn imitations, as do successful television series, books, magazines and other media products. What might be different about the Internet is that imitation can arguably occur more rapidly than in other communications media, thereby giving entrepreneurs less time to earn profits from early innovation.[15] In this case, the relevant issue is whether the innovator can take relatively low-cost actions to ensure that the returns appropriated will be sufficient to compensate for the costs and risks associated with innovating.

A potentially important possibility in this regard is the ability to link content on the Web site to the sale of specific product(s) or services that cannot be readily duplicated by rivals. In this case, imitation by rivals of Web site characteristics, *per se*, may have few adverse financial consequences for the innovator. As the Internet is used increasingly to demonstrate and promote products, Internet investments will be increasingly recouped in the form of revenues from goods and services that enjoy robust protection from conventional patents, trademarks and industrial secrecy. As technology advances, it may also be possible to mitigate problems of copying by "designing in" difficult-to-imitate features.

It might be argued that certain "critical" Internet network investments create such widespread opportunities for further commercial development that no single producer (or consortium of producers) could fully internalize the resulting commercial spinoffs. In this case, the associated costs and risks might discourage private sector innovation. In this case, it seems reasonable to look to the government to make the necessary investments.

Faulhaber [5, p. 283] argues that the historical record of government subsidy for high technology speaks more to the lobbying effectiveness of special interests than it does to the public interest served. The implication is that those calling for government subsidization of new communications hardware and software "infrastructure" should be obliged to identify the existence of specific learning spillovers and make a credible case that commercial firms would not find it profitable to make the relevant investments. He further argues that, for

[13] Technology that facilitates personal buying services is emerging from computer software developers. For some examples, see [17].

[14] See "Your Secrets Are Safe With Us," *Business Week*, June 23, 1997, p.48. To be sure, the Electronic Frontier Foundation may need to invest in certain credible commitments to assure potential customers that it will not act opportunistically and reveal information clandestinely.

[15] A possible illustration is the rapid proliferation of news-related Web sites established by the major broadcast networks. Yet notwithstanding this proliferation, the various companies involved all expect their Web sites to be profitable in the long run. See "All The News That's Fit to Browse," *Business Week*, June 16, 1997, p.133.

current services, hardware and operations, the production technology seems well-in-hand, no longer subject to learning-curve effects. While certain areas of operating a large and rapidly growing broadband network may still give rise to researchable issues for which government subsidies or grants may be appropriate, subsidizing the general operations of the Internet is difficult to justify on grounds of supply-side externalities.

3.3 Congestion and Pricing

Arguments for subsidizing Internet access and usage implicitly assume that there is no opportunity cost to putting more subscribers on the Internet. In fact, if Internet usage continues to grow with the same usage characteristics as at present, peak loading problems could arise with increased costs of congestion and delay for users already on the network. Indeed, there are already examples of access to popular Web sites being delayed or denied as a result of traffic congestion at that site. Internet experts are increasingly cautioning about the possibility of bandwidth scarcity, particularly in light of the flat-rate pricing system currently used by Internet providers which largely separates usage costs from prices charged, and the relative growth of "real-time" video and audio applications on the Internet which require data transfers of continuous bit streams for extended periods of time [14].

An obvious potential solution to scarcity problems is to make tie prices to the costs created by specific patterns of usage. For example, those who "camp out" on the Internet for long periods of time, especially during peak periods of usage, might be made to pay more than those who restrict their usage to short communications at off-peak periods. Those who demand relatively large portions of the available capacity should be made to pay more than those who use a sufficiently small portion of capacity as to create no significant access problems for others on the network. And so forth.

Internet pricing algorithms that are both economically efficient and technically feasible are just starting to be proposed.[16] It is beyond the scope of this study to offer a critical review of the various pricing and non-pricing algorithms that have been proposed to relieve current and prevent future congestion on the Internet.[17] Suffice to say that all allow for some role for government. A minimal government role in some pricing models is regulation to ensure that firms controlling bottleneck facilities do not engage in anticompetitive behaviour. What may be more to the point at this stage of the Internet's evolution is the likely difficulty that suppliers of backbone capacity will face in moving from flat rate pricing to some type of usage-based pricing. In Canada, for example, proposals by telephone companies to charge customers using large blocks of Internet time more than those using small blocks of time have met a firestorm of protest from Internet users, and the companies have backed down from their proposals. The long-experienced political difficulties in reducing the cross-subsidy to basic local telephone service caution against expecting efficient pricing algorithms to be implemented for the Internet. In the interim, the potential for increased congestion costs on the Internet counterbalance concerns about access and usage externalities leading to an inefficiently low rate of access and usage.

[16] See, for example, [10 and 8].

[17] A review and evaluation can be found in [14].

4. Distributional Concerns and Other Social Issues

Distributional considerations also figure directly and indirectly in calls for subsidizing and regulating the Internet. For example, there have been recommendations for public access to the Internet to be enhanced by making access more available through libraries, schools and other public places. Others call for increased funding of the infrastructure along the lines of the U.S. government's funding of the original NSFNET and the Canadian government's CANARIE program to upgrade the backbone of the Canadian Internet.

There are at least two aspects of distributional equity concerns surrounding the Internet. One is the conventional concern that poorer individuals and households may be significantly disadvantaged by being unable to afford the same access to the Internet as wealthier individuals and households. This view sees the Internet as a major potential education and productivity tool which, if not universally accessible, will lead to increased inequality of income over time as the economically better-off improve their labour market skills and incomes relative to the economically less well-off [15].

To the extent that the Internet leads to a widening of income disparities in society, it creates what we described earlier as pecuniary externalities. Even if these pecuniary externalities have minimal adverse impacts on real economic output and growth, large income inequalities are generally viewed as being socially undesirable in their own right. Moreover, widening income inequalities contribute to social tensions in society which may, in turn, promote anti-social behaviour such as crime which reduces real economic output and possibly real economic growth rates.

Other equity concerns have also been raised in support of government regulation of Internet content and public subsidies to suppliers of Internet infrastructure and services. One specific concern is that suppliers of telephone and broadcasting services over the Internet will be able to escape specific regulatory "burdens" that apply to conventional telephone and broadcast companies. For example, those currently using the Internet for broadcasting purposes have not been subjected to the same Canadian content requirements and related constraints that are applied to conventional broadcasters and cable companies. To be sure, at present there is very little activity on the Internet that would be defined as broadcasting under the intent of Canada's broadcasting legislation; however, many observers certainly expect Internet broadcasting to grow in importance over time, particularly as the information-carrying capacity of access connections into household residences improves.

The panoply of Canadian content regulations arguably burdens conventional broadcast distributors with higher costs and limitations on their use of imported programming. If Internet broadcasters were free of similar burdens, the resulting competition between new and conventional broadcasters could be considered unfair. Perhaps more important, it could contribute to the expansion of relatively inefficient producers at the expense of relatively efficient producers to the extent that the former enjoy regulatory privileges. This caution should not be interpreted as support for existing Canadian government policies in the broadcasting sector. On the contrary, it is a suggestion that whatever efficiency arguments might be made to expand existing government regulations to the Internet, it may be more efficient yet to eliminate the regulations in the sectors to which they currently apply.

A similar issue arises with respect to telephony. Namely, as voice telephony on the Internet emerges as a significant use of the network, conventional suppliers of telephone service will raise the issue of whether Internet suppliers of long-distance and "premium" service options are making a fair contribution to local access subsidies. If conventional telephone carriers must make such contributions, it would be unfair to exempt Internet providers from making them. It could also produce inefficiencies by making Internet carriage appear cheaper to consumers than it actually is because users can avoid an implicit tax by making calls on the Internet rather than on conventional networks. Acknowledging the merits of imposing symmetrical subsidy burdens on Internet service providers should, again, not be construed as support for the subsidy policy in place. The first-best policy is clearly to eliminate price subsidies in the telephone system, and (if necessary) extend direct subsidies to those whose telephone service might be disrupted by an inability to pay the resulting higher prices.

A concern about universal access to the Internet may increasingly provide intellectual support and comfort for those who want to maintain subsidized local telephone access. Specifically, it may come to replace the argument that poor residential subscribers will drop off the network if local rates are increased. Increasingly the concern will be that higher local rates will discourage low-income households from signing on for Internet access to the extent that access to the Internet is primarily taken through a second telephone line. It is beyond the scope of this brief paper to address any such rationalization of the perpetuation of the existing local access subsidy; however, it should certainly be noted that the local access subsidy has a variety of malignant properties as far as economic efficiency is concerned. One is that it discourages investment in new local distribution technologies by the conventional telephone companies, since they must divert net revenues earned on other services to subsidize local access. Another, as noted above, is that it creates potentially uneconomic asymmetries in incentives to invest in new local access technologies. Specifically, non-conventional access media, e.g. wireless, do not inherit the regulated price burden that conventional telephone companies carry with respect to pricing access, and it may be difficult (as a practical matter) to extract contributory payments from new local carriers. Investment capacity may therefore be diverted to new local distribution technologies, in part to bypass the toll to local access contribution requirement.

It is also beyond the scope of this paper to provide any detailed treatment of the issue of Canadian content on the Internet.[18] The concern about protecting its national culture is not unique to the Canadian government. Other national governments, especially in Asia, have expressed concerns about "unwanted" foreign culture entering their societies through the Internet. Perhaps the main observation one might make in this regard is that there may be very little that government regulation can do to protect and promote domestic cultural producers from foreign competition through the Internet. A more optimistic perspective is offered by the argument that the Internet enhances the ability of small country producers of entertainment programming relative to producers from large countries. This is because economies of scale and scope are arguably smaller for the Internet than for conventional broadcast distribution media such as television and magazines. Put in different terms, "narrowcasting" is an increasingly feasible business strategy using the Internet. In this case, perhaps the best strategy that governments can employ to promote national producers of media content is to allow competition on the Internet to flourish and thereby promote increasingly widespread use of the network.

[18] For a number of studies examining this issue, see *Policy Options*, October 1996.

5. Policy Conclusions

The main thesis of this paper is that the Internet offers no unique case for government subsidy or increased regulation.[19] This assertion might seem anomalous given the widespread perception that the government is largely responsible for financing the Internet. In fact, in the United States, the National Science Foundation (NSF) has been responsible for a relatively small portion of the financing of U.S. Internet infrastructure and services, although total government expenditures, including subsidies to university computing services, are multiples of the amount spent on the Internet by the NSF [12]. This does not gainsay the conclusion that the Internet has been primarily a privately financed endeavor, and that private financing will, if anything, play an even larger role in the future.

A specific argument of this paper is that the potential benefits of Internet usage and the supply of hardware and software in Internet applications are sufficiently large and sufficiently capable of being internalized that government initiatives to stimulate demand and/or supply are unnecessary. Nor is there any reason to believe that the market for Internet services is not fundamentally contestable in the long run. In this regard, perhaps the most important role for government is to promote competition in the supply of Internet services, particularly the hardware infrastructure in local distribution networks. In this regard, continuation of foreign ownership restrictions in telecommunications and broadcasting that many countries maintain is counterproductive. Moreover, the judicious use of competition law to ensure against anti-competitive abuses in the local distribution network by incumbent carriers is a potentially important instrument to ensure the continued efficient and vigorous growth of the Internet; however, competition policy should be neutral with respect to specific competitors. In particular, incumbent carriers should not be "handicapped" in their ability to compete against entrants and they should be allowed to charge efficient prices to both their own affiliates and their competitors. In the public policy debate surrounding the Internet, the role of competition policy is likely to take on increasing importance. It seems fair to say that telecommunications competition policy in the United States is in turmoil. Others might say it's in a shambles. Concerns about subsidizing and regulating the Internet are diversions from what is arguably the main policy concern likely to influence Internet developments, i.e. the implementation of an efficient and fair competition policy framework for local network distribution competition.

References

1. D. Allen, New Telecommunications Services: Network Externalities and Critical Mass, *Telecommunications Policy*, September (1988) 257-71.
2. D. Brassard, Information Superhighway. Minister of Supply and Services Canada, Ottawa, 1994.
3. R. Cawley, Internet Lies and Telephony. In: P. Enslow, P. Desrochers and I. Bonifacio, (eds.), 21st Century Communications Networks, IOS Press, Amsterdam, 1997, pp. 245-57.
4. R. Coase, The Problem of Social Cost, *The Journal of Law and Economics* **3** (1960) 1-44.
5. G. Faulhaber, Pricing Internet: The Efficient Subsidy. In: B. Kahin (ed.), Building Information Infrastructure, McGraw-Hill, New York, 1993, pp. 271-95.

[19] There remains substantial indirect regulation of the Internet. For example, many of the telecommunications circuits that make up the Internet are regulated in the United States, as well as in other countries.

6. M. Fuss and L. Waverman, Efficiency Principles for Telecommunications Pricing: Fairness for All. In: W. Stanbury (ed.) The Future of Telecommunications Policy in Canada, University of British Columbia, Vancouver, 1995, pp. 29-66.
7. S. Globerman, The Economics of the Information Superhighway. In: T. Courchene (ed.), Technology, Information and Public Policy, The John Deutsch Institute, Kingston, 1995, pp. 243-79.
8. A. Gupta, D. Stahl and A. Whinston, Priority Pricing of Integrated Service Networks. In: L. McKnight and J. Bailey (eds.), Internet Economics. MIT Press, Cambridge, 1997, pp. 3-24.
9. S. Liebowitz and S. Margolis, Are Network Externalities A New Source of Market Failure? University of Texas, Dallas, mimeo, undated.
10. J. Mackie-Mason and H. Varian, Pricing the Internet. In: B Kahin and J. Keller (eds.), Public Access to the Internet, MIT Press, Cambridge, 1995, pp. 269-314.
11. J. Mackie-Mason and H. Varian, Economic FAQs About the Internet. In: L. McKnight and J. Bailey (eds.), Internet Economics. MIT Press, Cambridge, 1997, pp. 27-62.
12. L. McKnight and J. Bailey, An Introduction to Internet Economics. In: S. McKnight and J. Bailey (eds.), Internet Economics. MIT Press, Cambridge, 1997, pp. 3-24.
13. G. MacDonald, The Internet is overrated, survey shows, *The Globe and Mail*, July 3 (1992) B9.
14. M. Sarkar, Internet Pricing: A Regulatory Imperative. In: S. McKnight and J. Bailey (eds.), Internet Economics. MIT Press, Cambridge, 1997, pp. 479-99.
15. D. Tapscott, How to Avoid Information Apartheid, *The Globe and Mail*, June 15 (1994) A10.
16. I. Vogelsang and B. Mitchell, Telecommunications Pricing: Theory and Practice. Cambridge University Press, Cambridge, 1991.
17. Wilderstrom, Service With a Click, *Business Week*, March 24 (1997) 19.

Telecommunications Transformation: Technology, Strategy and Policy
E. Bohlin and S.L. Levin (Eds.)
IOS Press, 1998

CHAPTER 16

'Lemons' in Cyberspace: A Call for Middlemen

Jing Lu*

Department of Economics, York University, Toronto, Canada M3J 1P3

Abstract. Contrary to the perception that the current information innovation in cyberspace is a massive process of disintermediation, this paper argues that there will continue to be a strong demand for intermediaries to perform market-coordinating activities. Due to the three prominent features of this anarchistic electronic world: low entry cost, identity insecurity, and the non-existence of a pricing mechanism, cyberspace has become a highly diversified and decentralized marketplace where frauds and 'lemons' abound. It is a huge market fair where an enormous amount of products is scattered around and where there are quality uncertainty, identity uncertainty and contract uncertainty. There is thus a natural demand for middlemen to solve this 'lemon' problem: seeking out reliable and valuable information, matching producers with interested consumers, and providing quality guarantees, identity guarantees and contract guarantees.

During late June and early July of 1996, the usually quiet LABOR economics discussion group via e-mail on the Internet suddenly went wild. Someone named "Bird Man" started out sending a 2,500-word get-rich-quick message to this list, urging subscribers to send money to some listed names in order to be sent money later on. The same chain letter duplicated four times on the list. Then angry subscribers countered by sending their angry comments along with the whole message to the whole group. As a result, every subscriber received this 15-screen-long message at least ten times. Finally, subscribers realized that they should direct their anger to the original sender of the message rather than to the whole group. And at last, the alleged sender replied, claiming that he did not send the message – someone else did it and faked his return address.[1]

---

*I would like to thank Janet Landa and Ryan McRonald for many helpful comments and suggestions.

[1]The whole incident can be retrieved from file archives at the Sam Houston State University which hosts this LABOR discussion group. Many similar incidents have been documented elsewhere, which involve both a flood of junk mail and the faking of identities. For example, Promo Enterprises, a "bulk e-mail company", was reported to have electronically mailed an ad to more than 170,000 people using another company's return address (*Business Week*, December 4, 1995, p.44). A self-styled "prankster" pretended to be the Premier of Ontario, Canada, and sent an outrageous note to the Ontario Usenet newsgroups, stating in part that: "I would ask those who engage in spending their time criticizing me to please assist those in need instead. That, I believe, is a more constructive use of your time." This fake note generated such a flood of e-mail to the Premier that he finally had to shut down his e-mail system (*The Globe and Mail*, October 27, 1995).

1. Introduction

"Welcome to Cyberspace", "The Year of the Internet", "The Web Changes Everything", "Death of Distance" – these are just a sample of many lengthy cover stories or special reports run by prestigious magazines like *Time*, *Newsweek*, *Business Week* and the *Economist*.[2] The long-heralded information age has finally come upon us. Much like the industrial revolution which liberated human beings from using slow and limited muscle power, this new information technology is liberating our brain power from slow and limited data processing. By creating an era of information wealth,[3] the current information technology in cyberspace has much reduced the cost of retrieving, storing and processing data. The Internet, which is an open-ended cross-connection of wide-range computer networks linking firms, banks, governments and consumers together, can much reduce the transaction cost of opportunity-tracking, deal-negotiating, data-transmitting, trip-making, etc. Agents can now reach each other by a mere click of the mouse rather than through many intermediaries – distances of time and space disappear. Thus, on the one hand there is a substantial reduction in physical transaction cost and on the other a creation of information wealth. No wonder the current innovation in computer technology is dubbed the "information revolution". The direct linkage between consumers and producers also has many people chanting "No more middleman", "Squeezing the middleman", "Making the middleman an endangered species", "Death of a middleman".[4]

But wait a second. What does a middleman have to do with the disappearance of distances in time and space in cyberspace? What is the role middlemen play in a market economy? According to the definition in Spulber [13], an *intermediary* is an economic agent who purchases from suppliers for resale to buyers or who helps buyers and sellers meet and transact. In other words, intermediaries are those market institutions that do not add value to products as do manufacturing, transportation and construction industries, even though they both employ capital and labor to carry out their activities. Rather, they coordinate transactions, seek out producers and consumers, and provide liquidity or availability of goods and services, all of which facilitate exchange on the market.[5]

[2]The term Internet simply means interlinked networks. It grew out of Arpanet, Bitnet, Usenet, and other networks, and is now a vast international network of networks. It enables computers of all kinds to share services and communicate directly, as if they were part of one giant, seamless, global computing machine. Internet is synonymous with cyberspace. The term cyberspace stems from William Gibson's science fiction *Neuromancer* [4] in which he portrays the computer-generated landscape as "cyberspace" where clusters of data reside. Other names like the Net, the Web, the World Wide Web, the information superhighway have also been used to denote the same Internet.

[3]There have been many discussions on the importance of information apart from being a consumption or entertainment commodity. Stigler ([14], p.183), for example, described the value of information as reducing the expected cost to the buyer of his purchases. Knight [6] foresaw the ability of information to serve to reduce uncertainty in the economy. He postulated that "the most thoroughgoing method of dealing with uncertainty is to secure better knowledge of and control over the future". Leijonhufvud [8] sees the importance of information through its coordinating role: "the central issue of macroeconomics is how, under what circumstances, and to what extent the economic system *coordinates* the activities of different agents. The job of coordination requires, above all, a great deal of *information*" (emphasis added by Howitt [5], p.33).

[4]Consecutively, the four title articles are from *Computerworld* July 17, 1995, p.55; *Futurist* May/June, 1995, p.15; *Business Week* June 6, 1994, p.114; and *Internet World* March 1997, p.38.

[5]In reality, the distinction between value-added industries and co-ordinating intermediaries may not be that clear-cut, in the sense that many manufacturers may carry out marketing activities such as advertising, selling,

Therefore, the reduction in physical distance between consumers and producers is not in the job description of middlemen; it belongs to the value-added services such as transportation to get commodities from one end to the other, and thus the direct linkage in cyberspace should amount to the shrinkage of these services, not the intermediaries. In fact, the current enchantment of the cyberspace is all about the reduction of many physical costs, like producing, transporting, storing, distributing, etc., which increases productivity and efficiency of capital and labor for all walks of life, consumers, producers, intermediaries, and governments alike. The cyberspace itself does not coordinate markets, does not match the right consumers with the right producers, does not signal price, quality and function of products; in sum, it does not perform the market-oriented activities described for middlemen and thus the establishment of its infrastructure does not have anything to say about the appearance or disappearance of middlemen in cyberspace. The middlemen would disappear only if the market is complete, frictionless, perfectly informed and fully coordinated, as postulated in the neoclassical Walrasian literature.

In this paper, we argue that there will continue to be a strong demand for middlemen to perform market-coordinating activities in cyberspace. This demand arises from the following three prominent features of this anarchistic electronic world: low entry cost, identity insecurity and the non-existence of a pricing mechanism. As a result, the cyberspace has become a highly diversified, abundant and decentralized marketplace where frauds and 'lemons' abound. It is a huge market fair where an enormous amount of products is scattered around and where there are quality uncertainty, identity uncertainty and contract uncertainty. There is thus a natural need for middlemen to solve this 'lemon' problem: seeking out reliable and valuable information, matching producers with interested consumers, and providing quality guarantees, identity guarantees and contract guarantees.

One supporting testimony for our argument for the importance of mediators in cyberspace is the LABOR incident narrated at the beginning of this paper. The same fraudulent scheme can occur in the conventional space, yet the high postage cost for the repeated messages to get to recipients significantly deters such behavior. In the cyber world, the small cost of duplicating and transmitting data can easily build up junk mail for recipients, tie up their computers, steal their time and distract their due attention to other messages. Similar problems have never occurred to other discussion groups like Post-Keynesian Thought or AFEE, precisely because they have moderators to filter out irrelevant messages and to maintain their quality. Another economics-related example is a real information broker, Professor William L. Goffe, at the Department of Economics, University of Southern Mississippi, Hattiesburg, U.S. Professor Goffe has made himself well-known in the economics profession by searching, sorting, indexing and rating useful information related to economics on the Internet and compiling it into an indispensable handbook, "Resources for Economists on the Internet".

The next section of this paper provides a detailed account of the three features in cyberspace: low entry cost, identity insecurity and no pricing mechanism. In fact, the conditions for Akerlof's [1] 'lemons' impasse are all met in cyberspace: goods of various qualities are priced uniformly, there exists asymmetric information between buyers and sellers on quality of goods, and buyers make decisions based on their anticipated quality. There is thus a natural call for some agents to specialize in the role of preventing market failure and

shipping, financing, etc., and that many intermediaries may perform value-added activities such as packaging, storing, delivering, etc.

enhancing market exchange. This section subsequently discusses many of the existing activities of middlemen in cyberspace. One important fact that illustrates a serious need for them is that the most popular Web sites are largely those that attempt to facilitate searches, filter information and bring order to the otherwise anarchic, scattered collection of information. It is precisely these activities that have prompted the expansion of the Web and its current exponential growth.

Section 4 sets up a mathematical model detailing the possible market failure in cyberspace arising from the overwhelming information with little attention and the asymmetric knowledge between buyers and sellers regarding quality of a product. It analyzes the role of middlemen and the extent of their activities in providing quality products, lessening producers' moral hazard and adverse selection, and reducing consumers' search and consumption cost, in such a way they prevent the 'lemon' impasse. The issue of identity uncertainty is not explicitly modeled but specifically discussed, and would only aggravate the 'lemon' problem.

Our last section summarizes this paper and concludes that as markets continue to become more diversified, fragmented and complicated while the value-added industries continue to be more efficient and squeezed, there will be a growing need for intermediaries to perform market-coordinating activities and an increasing share of the national income from the division of intermediaries.

2. Institutions of Cyber-middlemen

2.1. Theory

The current state of the Internet is anarchy, a borderless world with no ownership and sovereignty, no law and order.[6] It started out as an open place where people freely shared knowledge and exchanged information. It now accommodates more than 200,000 Web sites, 10,000 Usenet newsgroups and 20 to 30 million users in more than 160 countries worldwide,[7] thanks to recent innovations in Internet interface. The Internet thus has become a giant market fair where people meet, share and transact and where immense opportunities are to be explored. This is all the recent hoopla of the Internet is about.

The Internet, however, is also an immature economy. It is a decentralized marketplace where a huge amount of products remains to be organized and where there are problems of search uncertainty, quality uncertainty and identity uncertainty. It is like a playback of the frontier economy several hundred years ago, except that it is more fast-forwarding and condensed. The Internet is thus a test-field of economic theories on how markets function under these uncertainties and how intermediaries emerge to resolve these problems. Understanding these issues will help us understand many existing problems on the Net,

[6]The U.S. federal government is reluctant to step in to mandate any legal restraint for fear that regulation may stifle development and that personal security may conflict with national security (for details, see *Business Week* September 6, 1996, p.110, or *Internet Security – hearing before the subcommittee on science*, 1994, p.2). The Net policing issue is further complicated by the fact that any government's power of enforcement is restricted to its geopolitical boundaries. While the Internet recognizes no state or national borders, it puts most content beyond the reach of ordinary police jurisdiction.

[7]All facts in this section are read from *Wired*, a monthly magazine published in the U.S., unless otherwise specified. Detailed sources of citation are available from the author.

envision its future advancement and possibly develop a cyber economy that lives up to the expectation of "information revolution". To emphasize only the concrete and physical transaction cost, to claim the extinction of intermediation, is detrimental to our understanding of the working of a market economy and perhaps largely a result of the traditional economic literature in the Walrasian framework which ignores market friction and the role of intermediation (Rubinstein and Wolinsky [10]).

The Internet is a vast market for information. Information in cyberspace is made of electronic bits, unlike conventional commodities that are made of atoms. The bit-based products, such as electronic software applications and online magazine stories, are quite distinct from atom-based goods, such as tables and machines, in that they are intangible, easy to duplicate, remove, store and transmit. These properties certainly have their merits, but on the other hand they contribute to the following features of the current Internet: low entry cost, identity insecurity and the non-existence of a pricing mechanism.

The low entry barrier is manifested by the fact the it is technically simple and cheap to set up a Web site and transmit bulk messages,[8] and it can potentially reach more audiences than a printing plant. Anyone with a computer and a modem can be his own reporter, editor and publisher – spreading news and views to millions of readers around the world. As a result, this low entrance cost has prompted an explosion of information on the Net. A user can have access to 3,000 online newspapers daily, more than 200,000 Web sites containing 30 million pages and many discussion groups. Furthermore, new Web sites are still growing at the rate of one per minute, the volume of text at 20% a month. The very abundance and diversification of digital data exacerbates the most fundamental constraint on information commerce – the limits of human comprehension. Time spent searching for the right information may surpass the traditional cost of paying a visit or buying a brochure. As stated in Stigler ([14], p.171), if a given commodity has flooded large markets, then prices will not be competitive and the costs for searching would be extraordinarily high. As a result, transaction costs with cyberspace may increase rather than decrease.

Identity in cyberspace carries only electronic addresses, which however can be easily removed, faked and difficult to be traced back, unlike the conventional market where physical evidence cannot be as simply altered and where it is easier to define, detect or seek revenge of a crime or a fraud. As a famous *New Yorker* cartoon puts it, "On the Internet, no one knows you are a dog". As a result, there can be no stigma or responsibility attached to individual behavior, a person can play hit-and-run strategy without being recognized and punished, much faked information can co-exist with authentic data. All these undermine the public's confidence in the quality and reliability of the information provided over the Net. Internet commerce involving serious deals will simply fail before viable solutions emerge.

Furthermore, the pricing mechanism on the Net is still not well defined. On the one hand, the concept of "payment" has yet to be widely accepted. All information goods on the Net have been free and it has been the Net culture that "all information wants to be free". On the other, systems to support payments are still in their infant stage. It is difficult to define whether to charge consumption by bits or by time, and difficult to control for

[8]Setting up a Web site can cost as little as $5,000 – a pittance compared with the cost of a printing plant (*Time*, Spring issue 1995, p.60). A regular mailing to 1 million people would cost about $827,000, including postage and printing, but Internet mailing can reach the same number for the price of a local phone call plus a few hundred dollars for time spent online (*Business Week* September 9, 1996, p.110).

unauthorized viewing.[9] As a result, the nonexistence of a pricing scheme makes all products equally valued to a viewer in terms of his opportunity cost (e.g. lost leisure, online cost). People cannot pay for better information; the price premium relating to high-quality reputable products discussed in Shapiro [11] cannot be apprehended; promotion or advertisement cost will only aggravate the already existing budget deficit.[10] Even though publishers can turn to advertising for revenue – their sole source of income – it is barely enough to cover the cost.[11] There is thus low incentive for information providers to improve quality.

In sum, Akerlof's 'lemons' problem fits the current cyberspace well, in that goods of various qualities are priced uniformly, there exists asymmetric information between buyers and sellers on quality of goods, and buyers make consumption decisions based on their anticipated quality. As a result, one is likely to have the bad 'lemons' driving out the not-so-bad driving out the medium driving out the not-so-good driving out the good, in such a sequence of events that no market exists at all.

We therefore predict at least the following two roles for middlemen to perform in cyberspace. The first role is to streamline information so as to provide valuable information and reduce search cost. As Nobel laureate Herbert Simon ([12], p.40) puts it: "What information consumes is rather obvious: it consumes the attention of its recipients. Hence a wealth of information creates a poverty of attention, and a need to allocate that attention efficiently among the overabundance of information sources that might consume it." The technology for producing and distributing information is useless without some way to locate, filter, organize and summarize it. A new profession of "information managers" will have to emerge who combine the skills of computer scientists, librarians, publishers and database experts to help us discover and manage information (Varian [15]).

The second role for middlemen is to verify identity so as to provide reliable information. With anonymous or even pseudonymous suppliers of information, there is no cheap way for consumers to verify the reliability of information sources. There is thus a natural demand for some specialists to identify authenticity of information suppliers and provide trustworthy information products. Only after identity is secured will reputation or stigma be attached to demeanors and the repeated interaction between consumers and suppliers become possible. The repeated long-term interaction, as a consensus in the game-theoretical literature, makes possible forms of trust, cooperation and commitment that can be advantageous to some or all of the players (see, e.g., Fudenberg [3]). Therefore, in spite of the lawlessness in cyberspace, the secured identity will play a role of enhancing cooperation and improving efficiency. Even a formal legal system also requires well-defined identity to secure all the parties involved and to enforce its contract. In other words, identity security is fundamental to a market economy and is a prerequisite for the success of Internet commerce.

---

[9] In April, 1995, *USA Today* experimented with charging for access to its online newspaper. This was such a disaster, attracting only 1,000 subscribers out of its circulation of 2 million, that it had to make its Web site free in 4 months (see story in *Wired*, Feb. 1996, p.74).

[10] *USA Today*, for instance, must support more than 75 employees dedicated to running its Web site, plus a network of more than 225 freelance stringers filing special reports on hot online topics.

[11] "Anyone who tells you they are doing significant business on the Web is lying", declares Eileen Kent, vice president of *Playboy*'s New Media division, which is one of the few that has a strong demand for advertising links. *HotWired*, another popular Web site bragging about thousands of viewers per day, is also not profitable yet.

Even William Gibson, initiator of the term "cyberspace", wrote in his series of fiction about characters that interacted with their electronic "agents", called Johny Johny. These agents used artificial intelligence to second-guess their operator's preference and even adjust their own "personalities" accordingly.

There are many specific roles intermediaries play in a market economy (see Spulber [13] for a detailed discussion). In this paper, we focus on the role of middlemen to provide reliable and valuable information goods so as to reduce identity uncertainty, quality uncertainty and search cost. These roles of middlemen have been discussed in the literature. Landa [7], for example, attributes the success of the Ethnically-Homogeneous-Middleman-Group to their distinctive, homogeneous identity in coping with contract uncertainty. Rubinstein and Wolinsky [10] and Yavas [17] explain that middlemen reduce search cost and thus serve as time-saving institutions; Biglaiser [2] and Marvel and McCafferty [9] show that middlemen become experts in determining the quality of a good, thus reducing quality uncertainty. These roles are possibly more important in cyberspace because of the problems of identity uncertainty and overwhelming information with diversified quality.

2.2 Evidence

This section provides many real-world examples of the activities and popularity of the middlemen in cyberspace. With the natural call for intermediaries to reduce search uncertainty, quality uncertainty and identity uncertainty, entrepreneurs have already ventured into these venues and are competing fiercely for resolution. One important fact that demonstrates a serious need for cyber-middlemen is that the most popular sites on the Web are largely those that facilitate searching, filter information, and attempt to bring order to the otherwise anarchic, scattered collection of information.

2.2.1 Indexing and Searching

The Web search engines, like Yahoo!, Excite, Lyco, Webbrowser, AltaVista, Oracle, Context, Inktomi, etc., are among the most, if not *the most*, frequently visited Web sites on the Net; Yahoo!, for example, can attract almost 800,000 people a day. They all provide certain classification schemes, in an attempt to index all Web knowledge and make it act more like a well-organized library, so that consumers can look for their interests through indexes. The fact that they can attract customers makes them a lucrative business. Companies spend millions competing to create the best search engine and at the same time sell advertising space – their sole source of income on the Web.

There are also around 3,000 to 4,000 cyber-librarians in the U.S. specialized in searching for specific information for their clients. Brewster Kahle, for example, the founder of Wide Area Information Servers Inc., is one of the handful of people who have actually managed to get rich from information retrieval.

2.2.2 Filtering

Searching the whole Web even with the aid of search engines is likely to be time-consuming and costly, because people will end up with too many points to view. Some filtering becomes important, to provide an opinionated guide, a sort of "best of the Web", and not a complete catalog. This is precisely what the "cool site of the day" Web pages are, and why they have made for a popular business.

The cyberspace is like a conventional market where the distinction between value-added services and intermediary role may be blurred. American Online, Compuserve and Prodigy, the name three, are the companies that provide hard-wired services for consumers to get to the Net, and meanwhile they perform an information-filtering role. The newsstand on American Online, for example, now offers more than 35 prestigious newspapers and magazines for its customers – *Time, USA Today, The New York Times, The New Republic, Road & Track, Wired, National Geographic*, etc.

2.2.3 Matching

Gathering the right customers and channeling them to the right advertisers is a difficult process. If a company knows the specific interests of its customers, advertising can target the exact group, and matching can become more efficient. The *New Yorker* cartoon about how nobody knows you are a dog on the Net still holds true. That is, unless you tell others on the Net that you are a dog, along with what Zip code your doghouse is in, what kind of leash you wear, and what flavor Milk-Bone you prefer and why. When you disclose this, you become a very valuable dog indeed.

There have been attempts to build agent-based systems to help with the more repetitive, predictable tasks and behaviors. The system learns about its user's habits, interests, and behaviors with respect to that task. It can detect patterns and then offer to automate them on behalf of the user.

A successful example is the Web site of Individual Inc., the Burlington, MA, company that provides an online news service called NewsPage. Visitors fill out an electronic form in which they choose among 850 business news topics – from genetic research to network security. After receiving the profile of a certain user's interests, Individual's software scans a constant stream of 15,000 incoming stories from 600 daily and weekly publications, and provides a customized news digest on demand to each user. During the first two months of business, its Web site attracted 50,000 users who actually did pay $2.95 to $6.95 per month for the service. Subscription fees make up one-third of the site's revenue, the remainder coming from advertising. Individual took in around $250,000 in its first three months selling Web advertisement links.

A hard way to affect consumers and induce matching is by mass posting. Spamming is the now-familiar practice of deploying mass postings on Usenet and sending blanket mailings to Internet lists. Jeff Slaton, nicknamed "Spam King", has built himself a business out of spamming the Net. At $425 a shot, he floods the Net with the advertisement. He claims that he can do 15 mass postings a week, hitting up to 6 million potential consumers, adding income to more than $300,000 a year.

2.2.4 Pricing

Advertising income has been the major income for popular Web sites. The popular ones, such as *HotWired, Time, Playboy*, have all been charging between $30,000 and $100,000 for three-month placement of a postage-stamp- or business-card-sized button on their digital pages. People running these popular sites have been saying that demand for advertising links has been strong. *Time* reported that at least 35 major marketers booked for at least one quarter of 1995. *HotWired* sold ad placements to nearly 40 companies by the end of 1995. Demand for advertising has been so strong that *HotWired* was able to raise $7 million from a group of international investors in return for a 15 percent equity stake in the privately held Web venture. In spite of all these, the *HotWired* is not yet profitable.

Traditionally, an advertising price is based on how many people might see the ad. The television industry uses ratings to set ad prices, and the publishing industry uses audited circulation figures. Newspapers and magazines provide a smaller but more targeted audience. Advertising price in the Web is difficult to establish because few content-providers are able to tell advertisers much about the specific interests of their audience. Browsing the Web is pretty much an anonymous activity, and preferences are unable to be revealed through the buying pattern since almost all sites are free. Getting to learn the preferences of consumers is already a great leap toward marketing success.

Many publishers are starting to experiment with delivery information on the Internet on a compensated basis. Of course, CompuServe, America Online, and Prodigy are examples of services that provide access to information for a price. Building compensated information services into the Internet is an important aspect of making the Internet an effective business and information tool.

2.2.5 Verifying Identity

It has long been realized that identity security is essential to the proper function of transactions in the cyber-market. That is why a highly secure encryption code, the device intended to make contents readable only to the designators, was invented 20 years ago, even though it has not been approved for use by the U.S. government. Sooner or later, the U.S. government will ease its restriction on encryption codes (currently the limitation is 40-bit keys; a 129-bit key or longer will be safer) for the advancement of cyber-commerce. Competitors are already preparing for the explosion of the encryption market. Among the competitors of RSA, AT&T, Oracle, Novell and Microsoft, RSA Data Security currently dominates the market with 5 (out of 15) most coveted encryption algorithms and more than 200 licensees.

Thanks to encryption, digital documents can be trusted as much as paper ones. Digital signatures establish the authenticity of a message sender, allow an author to attach an unfakable name. Digital fingerprints tell whether a document has been altered. Electronic money, or e-money, can be used to facilitate cash-like transactions. There have been different names for e-money, smart-card, CyberCash, NetCheque, NetBill, etc., depending on various developers.

2.2.6 Guaranteeing Quality

Microsoft, IBM and a few others have been organizing a quality guarantee group, attempting to show with their stamps on the Web sites that they have checked background, approved of quality, and will warrant services. As Internet commerce keeps expanding, there will be many such Better Business Bureaus and Consumer Reports that establish consumers' confidence in products, build up brand names and reputation premiums, enforce co-operation and commitment; in all, create a dynamic and vigorous market system.

3. A Mathematical Model for Cyber-middlemen

This is a simple model which highlights the problem of getting valuable information and reducing search cost in cyberspace. With overwhelming information of diverse quality scattered around and with asymmetric knowledge between buyers and sellers regarding quality of products, there is no cheap way for consumers to locate a product of high quality. Competitive equilibrium between buyers and sellers without intermediation may well result in over-provision of low-quality goods and under-provision of high-quality ones – a 'lemons' impasse. There is thus a natural demand for some specialized agents to locate, filter, recommend and organize quality products.

The model subsequently discusses the endogenous determination of the extent of information management by middlemen, from the perspective of both buyers and sellers. By alleviating the asymmetric problem between buyers and sellers, middlemen not only locate valuable products for buyers and thus reduce their search cost and consumption loss, but also encourage provision of high-quality goods and reward premiums to high-quality sellers. In such a way, middlemen as an institution solve this transaction problem and enhance utilization of this virtual market.

The simple model here does not discuss explicitly the problem of the vast majority of irrelevant information in cyberspace which, though of interest to some customers, may inevitably waste others' attention. This issue, however, is naturally embedded in the model since there is only a small fraction of information that is valuable to a buyer, and it carries a dead weight loss (search and consumption) for the buyer not to be able to locate that information. Also, the identity uncertainty problem in cyberspace is not modeled here. Identity in this model is assumed to be fixed and secure so that good products will be remembered and revisited, while lousy products will be ignored in the future. However, identity problems would only exacerbate the already existing quality problem. When identity is insecure, low-quality products would always try to fake the identity of the high-quality ones, extracting the possibly existing reputation premium. In such a way, the quality-uncertainty problem becomes more severe, which further undermines buyers' confidence in finding a valuable product. Naturally, there will be a demand for intermediaries first of all to verify authenticity and purge unreliable information, before guaranteeing quality.

3.1 Market

Suppose there is only one type of information commodity in cyberspace (discussion groups or electronic magazines, for example), provided by many sellers. There are two qualities of that commodity: $\underline{q}$ and $\overline{q}$, with the former lower than the latter. The respective costs of producing $\underline{q}$ and $\overline{q}$ are $c(\underline{q})$ and $c(\overline{q})$; certainly $0 < c(\underline{q}) < c(\overline{q}) < \infty$.

There are in total M potential sellers in the market; among them a fraction δ can produce either high-quality or low-quality goods, and the fraction $(1-\delta)$ can produce only low-quality goods. The decision of high or low quality has to be made in the beginning of the period and the seller maintains that equilibrium quality forever. Let θ denote the portion of potential high-quality sellers who do produce high-quality goods. Thus the average quality of that commodity in the market is $q^* = (\theta\delta)\overline{q} + [(1-\theta)\delta + (1-\delta)]\underline{q}$; obviously, $\underline{q} \le q^* \le \delta\overline{q} + (1-\delta)\underline{q}$.

Buyers are homogeneous, with their size normalized to 1. They pay a startup cost C to enter the market; C can be the cost for a computer, a modem, or a regular hookup charge. They sample and consume only one commodity at a time after entering. They do not observe the quality of any specific good beforehand but will do so after they encounter one. They then decide their consumption level: $\overline{n}$ for $\overline{q}$, and $\underline{n}$ for $\underline{q}$. If a buyer locates a high-quality product, she will stick to that forever. If not, she will switch to another seller in the next period and keep switching until she finds a high-quality one. M is assumed to be large enough that the chance of her meeting the same low-quality seller is small and negligible.

All θ, q^*, $\underline{n}$, $\overline{n}$ are endogenously determined. Except for the uncertain quality facing buyers, there are no other uncertainties in the model.

3.2 Buyers and Middlemen

Even though there is no price charged for consuming informational goods on the Net and it has been a culture there that all information is free, consumers still incur a significant cost in the form of online expenses and the lost opportunity for other activity. Bills for Internet activity can build up very quickly. Thus, if the Internet product is of low value, consumers will not pay much attention to it. In the extreme case where surfing the Net yields so little return that it does not even pay for the opportunity cost, buyers will simply stay away from it and rather adhere to conventional markets.

To see this, let U^N denote the value function of a representative buyer who has not found a high-quality product. Let $u(\underline{n},\underline{\ell},\underline{x}|\underline{q})$ denote her one-time utility for consuming level $\underline{n}$ of cyber information, number $\underline{x}$ of other goods, amount $\underline{l}$ of leisure, based on the quality level $\underline{q}$ of cyber commodity. In such a way, the cyber information here is modeled only as a consumption good. Then

$$U^N = u(\underline{n}, \underline{l}, \underline{x}\,|\,\underline{q}) + \beta\theta\delta U^M + \beta(1-\theta\delta)U^N, \tag{1}$$

where β is the discount factor, U^M is the value function for a mature buyer who has found a valuable product, and $\theta\delta$ is the probability that the new buyer will discover a good product next period so as to become a mature buyer. Evidently, the value function for a mature buyer, U^M, is:

$$U^M = u(\overline{n},\overline{\ell},\overline{x}\,|\,\overline{q}) + \beta u(\overline{n},\overline{\ell},\overline{x}\,|\,\overline{q}) + \beta^2 u(\overline{n},\overline{\ell},\overline{x}\,|\,\overline{q}) + \cdots = \frac{u(\overline{n},\overline{\ell},\overline{x}\,|\,\overline{q})}{1-\beta}. \tag{2}$$

In its general form, the one-shot utility function $u(n,\ell,x|q)$ with quality level q is assumed to be concave, twice-differentiable with positive first derivative and negative second derivative of all its three arguments, n, ℓ, and x. For simplicity and tractability, we assume the utility function to be additively separable in its three arguments and take the functional form $u(n,\ell,x) = q\ell n(1+n) + \ell n(1+\ell) + \ell n(1+x)$, where $n \ge 0, \ell \ge 0, x \ge 0$. The higher the quality level q, the higher the satisfaction level u, *ceteris paribus*. The constraint for a buyer's consumption is $rn + px = w(1 - \ell - n)$, where p is price for x, w is wage, r is unit cost of connection for consuming n. Total leisure hours are normalized to 1. This budget constraint models the cost of consuming cyber goods in two forms: online cost rn and the lost leisure opportunity wn.

Proposition 1: *Buyers' consumption n of cyber commodity depends positively on q and p, negatively on r, and ambiguously on w. Further, n = 0 when q is sufficiently low.*

Proof: Both equations (1) and (2) are simply a one-period maximization problem subject to the budget constraint, that is,

$$\begin{aligned} &Max\ \ q\ell n(1+n) + \ell n(1+\ell) + \ell n(1+x),\ n \geq 0, \ell \geq 0, x \geq 0 \\ &S.t.\ \ rn + px = w(1-\ell-n). \end{aligned}$$

Standard procedures solving this optimization problem yield

$$n + \frac{2(1+n)}{q} = \frac{p+2w}{w+r},$$

with positive derivatives of n on q and p, negative on r, and ambiguous on w depending on the relative strengths of income effect and substitution effect. When $q \leq 2(w+\mathrm{r})/(2w+p)$ is a sufficiently low value, then $n \leq 0$ and the buyer's consumption takes the lowest possible value: $n = 0$

Q.E.D.

The higher the quality of cyber goods, the higher the utility buyers draw from consumption, and the more attention they will thus pay to them. When the price p of other commodities x decreases, the buyer's attention to cyber goods will decrease and consumption will shift more toward x. The consumption also depends relatively on the cost of connecting to a cyber commodity, r, which plays the same role as the price of cyber goods. Finally, the effect of wage w plays an ambiguous role due to income and substitution effects: a rise in w increases the buyer's income and thus should increase her consumption of all goods, including n; meanwhile, it also increases the opportunity cost of consuming n which takes up leisure time, and thus could decrease n. When the quality of cyber goods is sufficiently low, though still positive, buyers will not pay attention due to positive opportunity cost.

Proposition 2: *Market failure occurs when both $\theta\delta$ and $\underline{q}$ are sufficiently low.*

Proof: Market failure occurs when one side of the market ceases to enter the market. Let U^O denote the life-long utility of a buyer who does not enter into the cyber market and consumes only ℓ and x. Thus, $U^O = u(\ell, x)/(1-\beta)$. Let U^I denote a buyer's expected return on entering the market,

$$U^I = (\theta\delta)U^M + (1-\theta\delta)U^N - C$$
$$= \frac{\beta(\theta\delta)(1-\theta\delta)U^M}{1-\beta(1-\theta\beta)} + \frac{(1-\theta\delta)\underline{u}}{1-\beta(1-\theta\beta)} - C,$$

by making use of equations (1) and (2), where $\underline{u} = u(\underline{n}, \underline{\ell}, \underline{x} \mid \underline{q})$.

The limit of U^I when $\theta\delta$ and $\underline{q}$ take sufficiently low values is

$$\lim_{\underline{q}\to\frac{2(w+r)}{2w+p},\,\theta\delta\to 0} U^I = \lim_{\underline{q}\to 0} \frac{\underline{u}}{1-\beta} - C$$
$$= U^O - C < U^O$$

Therefore, when $\theta\delta$ and $\underline{q}$ are sufficiently low, it is not worthwhile to enter the market and incur a cost of C.

Q.E.D.

Proposition 2 shows that when the overwhelming majority of information-providers sell low-quality information (i.e. $\theta\delta \to 0$) which is of too little value to consumers (i.e. $\underline{q} \leq \frac{2(w+r)}{2w+p}$), consumers will simply shun the cyber market and resort to conventional markets.

The positive startup cost C prohibits buyers from entering into the market, resulting in a complete market failure.

Proposition 3: *Buyers' demand for middlemen to filter information quality depends positively on β and $u(\bar{n}, \bar{\ell}, \bar{x} | \bar{q}) - u(\underline{n}, \underline{\ell}, \underline{x} | q)$ negatively on $\theta\delta$.*

Proof: The welfare loss for a new buyer from equations (1) and (2) is:

$$\Delta^U = U^M - U^N$$
$$= \frac{1}{1-\beta(1-\theta\delta)}\left[u(\bar{n}, \bar{\ell}, \bar{x}|\bar{q}) - u(\underline{n}, \underline{\ell}, \underline{x}|\underline{q})\right] \geq 0,$$

because u is an increasing function of q and hence $\Delta^u = u(\bar{n}, \bar{\ell}, \bar{x}|\bar{q}) - u(\underline{n}, \underline{\ell}, \underline{x}|q) \geq 0$. Buyers would be willing to pay Δ^U to be informed of the quality of cyber goods. Their willingness to pay, Δ^U, is straightforwardly an increasing function of β and Δ^u, and a decreasing function of $\theta\delta$.

Q.E.D.

Thus, as long as there exists a search cost for the uninformed buyers to become informed, there will be a demand for middlemen to take up this informing role, exploit this rent opportunity and increase new buyers' welfare. This demand increases when not finding a good product becomes more of a problem in the future (i.e. β increases), when the discrepancy in utility due to the difference between high and low quality grows (i.e. Δ^u

increases), or when it becomes more difficult for the buyer to find the right product on her own (i.e. $\theta\delta$ decreases). In cyberspace, it is likely that both Δ^u increases and $\theta\delta$ decreases. Because of the low entry cost, the cyber world could be flooded with junk information so that the distribution of product quality becomes heavily centered in the lower tail. As a result, the difference between the high- and low-quality products increases and the probability of finding the high-quality product decreases. Thus, in the cyber world, there may even be a stronger demand for middlemen to act as a screening device than in the conventional market.

3.3 Sellers and Middlemen

A seller maximizes his stream of consumers' attention net of production cost. A high-quality producer will expect to have $\frac{\bar{n}}{M}$ amount of consumers' attention in the first period since the total number of consumers, which is normalized to 1, is distributed randomly across all suppliers M. This amount $\frac{\bar{n}}{M}$ will continue in the next period due to the high quality. There will also be an extra attention of $(1-\theta\delta)\frac{\bar{n}}{M}$ in the next period to the high-quality provider, due to the low-quality buyers in this period, amounting to $(1-\theta\delta)$, randomly switching sellers in the next period. They will stay and consume at $\bar{n}$ if they find a high-quality product, and keep switching if not. Therefore, a high-quality seller receives $\frac{\bar{n}}{M}+(1-\theta\delta)\frac{\bar{n}}{M}$ attention in the second period, $\frac{\bar{n}}{M}+(1-\theta\delta)\frac{\bar{n}}{M}+(1-\theta\delta)^2\frac{\bar{n}}{M}$ in the third period, and so on. His discounted net future utility, V^H, is:

$$
\begin{aligned}
V^H &= \left[\frac{\bar{n}}{M}-c(\bar{q})\right]+\beta\left[\frac{\bar{n}}{M}+(1-\theta\delta)\frac{\bar{n}}{M}-c(\bar{q})\right] \\
&\quad +\beta^2\left[\frac{\bar{n}}{M}+(1-\theta\delta)\frac{\bar{n}}{M}+(1-\theta\delta)^2\frac{\bar{n}}{M}-c(\bar{q})\right]+\dots \\
&= \frac{\bar{n}}{M(1-\beta)(1-\beta(1-\theta\delta))}-\frac{c(\bar{q})}{1-\beta}.
\end{aligned} \tag{3}
$$

A low-quality seller, on the other hand, has no customer loyalty. He only relies on consumers bouncing randomly to him. His discounted stream of utility, V^L, is:

$$
\begin{aligned}
V^L &= \left[\frac{\underline{n}}{M}-c(\underline{q})\right]+\beta\left[(1-\theta\delta)\frac{\underline{n}}{M}-c(\underline{q})\right]+\beta^2\left[(1-\theta\delta)^2\frac{\underline{n}}{M}-c(\underline{q})\right]+\dots \\
&= \frac{\underline{n}}{M(1-\beta(1-\theta\delta))}-\frac{c(\underline{q})}{1-\beta}.
\end{aligned} \tag{4}
$$

Proposition 4: *In equilibrium, the fraction of high-quality producers among the potential high-quality providers,* θ, *depends positively on* β *and* $\bar{n}$, *negatively on* $M, c(\bar{q}) - c(\underline{q})$, δ *and* $\underline{n}$. *It is stable and uniquely determined by*

$$\theta \begin{cases} = 1 & \text{when } V^H > V^L \\ \in [0,1] & \text{when } V^H = V^L \\ = 0 & \text{when } V^H < V^L. \end{cases}$$

Proof: Combining equations (3) and (4) yields

$$\theta \underset{>}{\overset{<}{=}} \frac{1}{\delta}\left(1 - \frac{1}{\beta} + \frac{\bar{n} - (1-\beta)\underline{n}}{\beta M \Delta^c}\right) \text{ when } V^H \underset{<}{\overset{>}{=}} V^L, \quad (5)$$

where $\Delta^c = c(\bar{q}) - c(\underline{q})$. It is straightforward that (5) implies $\partial\theta/\partial\bar{n} > 0, \partial\theta/\partial M < 0, \partial\theta/\partial\Delta^c < 0, \partial\theta/\partial\delta < 0, \partial\theta/\partial\underline{n} < 0$. An important fact used to show that $\partial\theta/\partial\beta > 0$ is that $\underline{n} - Mc(\underline{q}) > n - Mc(q)$, which means that the low-quality sellers can at least gain more in the first period and then gradually lose their ground in the future. Otherwise, the potential high-quality sellers would always prefer to provide high-quality goods and θ would always be equal to 1 – an extreme case.

The stability of θ depends on whether the system diverges further when deviating from the equilibrium position. From equations (3) and (4), we have $\partial(V^H - V^L)/\partial\theta < 0$, implying that a hypothetical increase in θ lowers the value of V^H relative to V^L, encourages more provision of low-quality goods, and lowers the fraction of high-quality providers θ, hence a stable equilibrium of θ.

For the determination of θ: when $V^H > V^L$, all potential high-quality providers must find it optimal to provide high-quality goods; thus $\theta = 1$ and $\frac{1}{\delta}\left(1 - \frac{1}{\beta} + \frac{\bar{n} - (1-\beta)\underline{n}}{\beta M \Delta^c}\right) > 1$. Likewise, when $V^H < V^L$, $\theta = 0$ and $\frac{1}{\delta}(1 - \frac{1}{\beta} + \frac{\bar{n} - (1-\beta)\underline{n}}{\beta M \Delta^c}) < 0$. Only when $V^H = V^L$ can θ be a fractional value which equals the right hand side of equation (5).

Q.E.D.

Therefore, θ is endogenously determined, which equilibrates value functions for the high- and low-quality producers. θ depends positively on β and $\bar{n}$ because, when the future becomes more important and when readers pay more attention to valuable goods, high-quality producers will value and draw more satisfaction from producing high quality, hence more high-quality producers. By the same token, θ is inversely related to $\underline{n}$. θ depends negatively on δ, the fraction of potential high-quality producers, because a low δ implies a lot of low-quality sellers, thus a high marginal gain from attracting consumers to high-quality goods, and thus a great incentive to provide good products.

More interesting cases are the negative dependences of θ on M and Δ^c. For the former, the vast number of information providers, M, reduces their share of consumers' attention, giving them disincentive to improve quality. As shown in equations (3) and (4), as M keeps increasing, the difference in total attention to different quality diminishes and that in cost becomes more prominent. The value of producing low quality will finally dominate that of high quality, the right side of equation (5) becomes negative, θ will approach zero and all producers will provide only low-quality goods. Therefore, when consumers' attention is limited, the mere abundance of informational goods from all over the world creates a lack of demand for each of them and a lack of motivation for them to improve quality. As a result, the market may fail for the high-quality products.

For the latter case, the increase in the cost differential between the high- and low-quality goods, Δ^c, gives producers rather a disincentive to provide high-quality goods. Valuable and responsible information is often much more costly to produce than the instinctive chitchat. In the conventional market where there are not only information production costs but also many printing, publishing and distributing costs, the cost of producing both high- and low-quality products is huge, and only the best products that sustain consumers' demand can survive. Thus this high cost acts as a natural barrier to guarantee quality of goods. In the cyber world, however, cost has mostly attenuated to production cost. This huge reduction in cost prompts a mushrooming of all varieties of information, many being those that would not have existed in the conventional markets, competing for people's attention. Furthermore, the production cost of good products relative to junk information now becomes more pronounced. This increase in Δ^c again propels provision of low-quality goods. In the end, valuable products may be rare in the massive pool of cyber goods and the market for good products may again fail to exist.

Proposition 5: *Only high-quality sellers would ever demand the existence of middlemen to filter information quality. Their demand depends positively on* $\bar{n}$ *and* θ*, negatively on* M*. Their demand further depends positively on* β *and negatively on* δ *when* $\delta \to 0$.

Proof: With the assistance of middlemen to filter information and guarantee quality, low-quality products will receive no attention but have a positive production cost, and they will cease to come to the market.

All the potential high-quality sellers will produce only high-quality goods. Now the value function for a producer, V, becomes:

$$V = \left[\frac{\bar{n}}{\delta M} - c(\bar{q})\right] + \beta\left[\frac{\bar{n}}{\delta M} - c(\bar{q})\right] + \beta^2\left[\frac{\bar{n}}{\delta M} - c(\bar{q})\right] + \ldots$$
$$= \frac{\bar{n}}{\delta M(1-\beta)} - \frac{c(\bar{q})}{1-\beta},$$

with the total amount of attention $\bar{n}$ now divided evenly among the high-quality producers δM. The difference between V and V^H, denoted by Δ^V, determines the magnitude of sellers' demand for middlemen,

$$\Delta^V = \frac{\bar{n}}{\delta M(1-\beta)} - \frac{\bar{n}}{M(1-\beta)(1-\beta(1-\theta\delta))}$$
$$= \frac{\bar{n}}{M(1-\beta)}\left[\frac{1}{\delta} - \frac{1}{1-\beta(1-\theta\delta)}\right]. \qquad (6)$$

The sign of Δ^V is ambiguous depending on the sign of $\left[1-\delta+\beta(1-\theta\delta)\right]$. This is because while middlemen attribute all attention to high-quality goods, they also increase the number of high-quality producers from $(\theta\delta)M$ to δM, and thus may decrease the proportional attention to high-quality sellers, especially in the long run. Nevertheless, as $\delta \to 0$, i.e., when an overwhelming majority of cyber information is of low quality that has been filtered out by middlemen, Δ^V is unambiguously positive and there is a positive demand for middlemen's filtering job.

It is straightforward from equation (6) that

$$\frac{\partial\Delta^V}{\partial\bar{n}} > 0,\quad \frac{\partial\Delta^V}{\partial M} < 0,\quad \frac{\partial\Delta^V}{\partial\theta} > 0,\quad \frac{\partial\Delta^V}{\partial\delta}\Big|_{\delta\to 0} < 0,\quad \frac{\partial\Delta^V}{\partial\beta}\Big|_{\delta\to 0} > 0,$$

though the signs of both $\partial\Delta^V/\partial\delta$ and $\partial\Delta^V/\partial\beta$ are ambiguous.

Q.E.D.

Only the high-quality sellers will demand the existence of middlemen in cyberspace to allocate consumers' due attention to high-quality products. This demand decreases as $\bar{n}$ decreases or M increases; in both cases the consumer's consumption of high-quality goods distributed to each seller decreases. When the original fraction of high-quality producers θ was already high, there will not be as much loss of consumers' attention to high-quality producers after the introduction of middlemen, and thus the demand for them will be high. Although sellers' demand for intermediaries is ambiguous on δ and β depending on the interplay of parameter values, it is nevertheless negative on δ and positive on β when the size of low-quality producers was large, because then many customers can be introduced to high-quality goods and future consumption becomes more important.

In any event, when the market is mixed with good and poor products and has many uninformed buyers, it would pay for a high-quality producer to have his quality announced so that he can instantly attract those customers, having at least a one-shot windfall. Therefore, high-quality sellers have an incentive to pay middlemen or some central authorities to proclaim their high quality, even if the final welfare gain when all other sellers do the same may not be so decisive.

4. Summary and Conclusions

The Internet is a mature technology, but an immature economy. It has created immense opportunities, yet also given rise to new problems and challenges. The Internet is thus a living proof of how an anarchic, wild world can or cannot transform itself into a prosperous marketplace. It is also a testfield of economic theories on the resolution of these challenges. Understanding economics in cyberspace is important. It will help us understand the evolution of a market economy, recognize many of the existing problems in cyberspace, envision its future advancement and possibly develop a cyber-economy that lives up to the expectation of "information revolution".

This paper offers an attempt to address the role of intermediaries in cyberspace. As the lawless cyberspace is highly diversified, abundant and decentralized where frauds and 'lemons' abound, there will be a natural demand for middlemen to perform market-coordinating activities. Without these middlemen, cyberspace will remain primitive and underdeveloped, and market failure may well result. All along, this paper discusses many living examples of the dynamic and active role of middlemen in current cyberspace.

The Internet is a technology that reduces distance in time and space. It thus reduces the need for many value-added services such as transportation, storage and distribution. It meanwhile connects the whole globe into one sphere, creating such an enormous market terrain that it is necessary for intermediaries to play a coordinating role. As production sectors continue to become more efficient and squeezed and markets more diversified and complex – as the trend has been and will be – there will be a growing need for intermediaries to reduce transaction costs, and an increasing share of the national GDP from the division of intermediaries.

References

[1] G. Akerlof, The Market for 'Lemons': Quality Uncertainty and the Market Mechanism, *Quarterly Journal of Economics* 84 (1970) 488-500.

[2] G. Biglaiser and J. Friedman, Middlemen as Guarantors of Quality, *International Journal of Industrial Organization* **12** (1994) 509-531.

[3] D. Fudenberg, Explaining Cooperation and Commitment in Repeated Games. In: J. Laffont (ed.), Advances in Economic Theory, Sixth World Congress, **1** (1992) 89-131.

[4] W. Gibson, Neuromancer. New York, Ace Books, 1984.

[5] P. Howitt, The Keynesian Recovery. Philip Allan, 1990.

[6] F. Knight, Risk, Uncertainty and Profit. New York, A.M. Kelley, 1964. (First published in 1921.)

[7] J. Landa, Trust, Ethnicity, and Identity: Beyond the New Institutional Economics of Ethnic Trading Networks, Contract Law, and Gift-Exchange. The University of Michigan Press, 1994.

[8] A. Leijonhufvud, Information and Coordination. New York, Oxford University Press, 1981.

[9] H. Marvel and S. McCafferty, Resale Price Maintenance and Quality Certification, *RAND Journal of Economics* **15** (1984) 346-359.

[10] A. Rubinstein and A. Wolinsky, Middlemen, *Quarterly Journal of Economics* 102 (1987) 581-593.

[11] C. Shapiro, Premiums for High Quality Products as a Return to Reputation, *Quarterly Journal of Economics* **98** (1983) 659-679.

[12] H. Simon, Designing Organizations for an Information-Rich World. In: M. Greenberger (ed.), Computers, Communications, and the Public Interest. John Hopkins Press, 1971.

[13] D. Spulber, Market Microstructure and Intermediation, *Journal of Economic Perspectives* **10** (1996) 135-152.

[14] G. Stigler, The Economics of Information, *Journal of Political Economy* (1961) 213-225.

[15] H. Varian, The Information Economy, *Educom Review*, January/February (1996) 44-46.
[16] J. Wallis and D. North, Measuring the Transaction Sector in the American Economy, 1870-1970. In: S. Engerman and R. Gallman (eds.), Long-Term Factors in American Economic Growth. Chicago. University of Chicago Press for the National Bureau of Economic Research, 1986.
[17] A. Yavas, Marketmakers Versus Matchmakers, *Journal of Financial Intermediation* **2** (1992) 33-58.

Telecommunications Transformation: Technology, Strategy and Policy
E. Bohlin and S.L. Levin (Eds.)
IOS Press, 1998

CHAPTER 17

Copyright Law and Innovation Policy: Challenge of the Internet

Jisuk Woo

Korea Information Society Development Institute,
1-33 Juam-Dong, Kwachun, Kyunggi-do, 427-070, Korea

Abstract. This paper examines the application of copyright law to the Internet. The merging of computer and communications technologies, and the resulting network of the Internet, have presented fundamental challenges to the existing legal framework of copyright law. This challenge leads us to ask essential questions about the concepts of innovation, value, and progress that have been lacking in the debates regarding intellectual property. The primary theme of the paper is that the copyright regime in the current state may no longer be a coherent mechanism to advance innovation in the face of changes in the production and transmission of information, and that the very concept of "innovation" might change to include more "communication" of information as well as more production of information.

1. Introduction

This paper examines the issues with respect to the application of copyright law to the Internet. While copyright law has been understood to promote innovation by ensuring rewards to works of authorship, particular characteristics of the new Internet environment might undermine the validity of the existing theory of innovation.

The move from analog to digital technology enables the convergence of different forms of content such as video, sound, image, text, data, and computer code in a common system. Computer and communications technologies have been merging, as computers interact more directly with one another and telecommunications carriers provide more data processing services. A "network of networks"[1] often describes the nature of the Internet, which links together universities, research organizations, and governmental agencies. The Internet helps disparate technologies and historically distinct industries to be converged.[2] The structural convergence among the domains of communications, information, and computing

[1] Information Infrastructure Task Force, Intellectual Property and the National Information Infrastructure: The Report of the Working Group on Intellectual Property Rights (Sep. 1995) [hereinafter White Paper].

[2] For review of the relationship between digitization and the information economy, see Don Tapscott, The Digital Economy (1996); Brian Kahin, The Internet Business and Policy Landscape, *Ann. Rev. of Inst. for Info. Stud. 47* (1997). Kahin considers "the Internet" as consisting of the Internet protocol (the connectivity offered by Internet service providers), the Web, and an array of related services and applications. *Id.* at 47.

also occurs, demonstrated by remote data processing, database publishing, interactive multimedia, broadcasting, videotex, and electronic publishing.[3]

Consequently, the Internet changes the economics of content, and posits challenges for owners, creators, sellers, and users of intellectual property.[4] The increasing use of the Internet brings about significant shifts in the relationship between information and users, and between authors and users, subsequently changing the value of information products and the nature of human communication.[5] Therefore, fundamental challenges are posed to the copyright system and the rationales behind the system. The current debates center around the questions regarding proper ways to compensate for works of authorship and to promote innovation.

This paper attempts to provide a useful perspective on thinking about the issues related to the controversy surrounding the Internet and copyright law. The primary theme of the study is that copyright regime in the current state may no longer be a coherent mechanism to advance innovation in the face of the changes in the production and transmission of information, and the very concept of "innovation" might change to include more "communication" of information as well as more production of information.

Part 2 analyzes theories and assumptions behind the alleged relationship between copyright law and innovation. Part 3 discusses changes in the nature of information and communication on the Internet that are particularly relevant to copyright law. Part 4 analyzes some of the difficulties in applying the copyright doctrines and concepts to Internet contents. Finally discussed is the relationship among the changes in the technology, subsequent changes in information practices and businesses, and their effect on copyright law.

2. Incentive Structure of Innovation in Copyright Law

The relationship between copyright law and innovation seems to stem from Article I of the U.S. Constitution, which provides that:

> Congress shall have Power … To promote the Progress of Science and useful Arts, by securing for limited Times to Authors and Inventors the exclusive Right to their respective Writings and Discoveries.[6]

Courts and commentators tend to agree on the general principle that the remuneration to the authors is not the ultimate goal of the Constitution, but is a tool to establish an incentive for authors to create. According to this principle, incentives to the authors should be provided to the extent to which the production of future works of authorship or of other kinds of products is not stifled. But there has been controversy regarding not only the appropriate amount of remuneration to ensure the incentive, but also the appropriate amount of incentives that one should try to establish. Although it seems to be generally believed that the more incentives there are, the better it is for innovation, whether the current protection of an

---

[3] Kahin, *supra* note 2.
[4] Esther Dyson, Intellectual Value, WIRED 3.07.
[5] Aryeh S. Friedman, Law and the Innovative Process: Preliminary Reflections, 86 *Colum. Bus. L. Rev.* 1, 3 (1986). See also Ethan Katch, The Electronic Media and the Transformation of Law (1989).
[6] U.S. Const. art. I, section 8.

author's rights could defeat, rather than help, the future production of works of authorship has been a source of controversy. Added to the current debates are views that other social values should be considered, such as the public's access to the existing works of authorship, and participatory democracy that is argued to be achieved by free dissemination of information. Differing views on the goals of copyright law and some studies that support them will now be discussed.

There are some normative arguments about why authors should be rewarded. These arguments themselves may not be directly related to the issue of innovation, but influence the author's incentive to produce work in an indirect way. The normative debate centers around the issue of whether an author should receive, and consumers should have to pay, the cost of producing a creative work, the effort of creating the work, or the value of that work. In making a claim for a broad scope for copyright protection, commentators typically assert that an author should be entitled to the value of the work of his own creation.[7] But as only a few workers are rewarded by the total value of their works, there is nothing inherently immoral when an author does not receive the total "value" of his work to society.[8]

In addition, the value of any new work of authorship is shared between those who have contributed to it. [9] An attempt to recognize other individuals whose efforts are bases of the market value of a work of authorship, such as those who have produced pre-existing public domain material or a pre-existing copyrighted work, will complicate the value-based approach.[10] In terms of value, copyright law focuses on protecting "original" and "creative" works rather than mere "labor and effort."[11] Creativity is considered different from the author's effort and labor, because a work may not be creative even after a large amount of effort and labor has been spent.[12]

A relevant issue that may justify providing rewards to the authors concerns a situation in which copying or modifying a work of authorship injures the author's pride, as well as his purse. Although American copyright law does not protect these personal interests of the author, there have been some arguments for acknowledging these moral rights to maintain the integrity of the work. The significance of this issue lies in the fact that it may not only threaten the author's moral rights, whether legally acknowledged or not, but also may stifle his incentive to produce more works in the future. Discontinuing the production of any

---

[7] Glynn S. Lunney, Jr., Reexamining Copyright's Incentives-Access Paradigm, 49 *Vand. L. R.* 483, 573 (1996).

[8] Stephen Breyer, The Uneasy Case For Copyright: A Study of Copyright In Books, Photocopies, and Computer Programs, 84 *Harv. L. R.* 281, 286 (1970).

[9] Lunney, *supra* note 7, at 573. See also Peter Jaszi, On the Author Effect: Contemporary Copyright and Collective Creativity, in The Construction of Authorship (Martha Woodmansee & Peter Jaszi, eds., 1994).

[10] Another argument is that if the work is of great social value, then the contribution should be acknowledged. However, Stephen Breyer argues that acknowledging the contribution by copyright law may do more harm than good especially if the work is a great one, because of its tendency to generate higher prices. He suggests that a better way to acknowledge great works of authorship is by taxes, or grants and prizes. Breyer, *supra* note 8, at 286-7.

[11] Courts have rejected the relevance of labor and effort often by commenting that the "sweat of the brow" does not make the resulting work copyrightable. See Jane C. Ginsburg, Creation and Commercial Value: Copyright Protection of Works of Information, 90 *Colum. L. R.* 1805 (1990).

[12] For example, applying the work-made-for-hire doctrine to computer software copyright cases has posed a challenge to the courts because of the difficulty in deciding whether giving instructions and actually writing the program makes the person the author of a computer program. See Clifford Scott Aymes v. Jonathan J. Bonelli, 21 U.S.P.Q. 2d (BNA) 1716 (D.N.Y. 1991).

more works because of the fear that his work might be "wronged" is theoretically possible. However, it would be reasonable to expect that only relatively few, extreme cases would result in the discontinuation of future materials, given the usually strong need of the authors to produce and disseminate their works whether to make a living or to have their ideas noticed.

To those who consider that the ultimate goal of author remuneration is the advancement and dissemination of culture and knowledge by stimulating the authors to produce work products, the critical question is what would happen if copyright protection were abolished. But surprisingly few empirical studies have been conducted to answer this question. Stephen Breyer asked whether in book publishing the author would not produce the work if he knew he would not be paid more for the book than the marketplace price.[13] He calls this a "persuasion" cost, the amount of money needed to persuade a man to write a book.[14] He argues that the author's moral and economic claim to be paid more than his persuasion cost is not any stronger than the claim of others also responsible for producing his book: the publisher, the printer, the bookseller, and those responsible for the literature of the past that inspired him. Breyer concludes that his cost-benefit analysis suggests that the case for copyright in books as a whole is weak. According to Breyer, to abolish protection would not cause a very large or a very harmful decline in most kinds of book production, but would benefit some readers by yielding lower prices. But the applicability of Breyer's analysis is limited because it was focused on the publisher rather than the author, and was conducted in the context of the book publishing industry. The mechanisms by which potential authors are encouraged or discouraged to have incentives to produce work do not seem to be clearly known, although this has been the center of debates in intellectual property law.

The debates do not end with the incentive structure itself. Even if broadening the scope of copyright is found to increase the incentive to produce more works, some scholars speculate that it may limit access to such works. The examination of incentives and access has been what commentators and courts have relied on in defining the appropriate scope of copyright. The limitation of access may occur by increasing the price of the works and by limiting the material that others can use to create additional works. But there is an argument that incentives and access may not be as contradictory as is suggested in an either-or analysis.

Glynn S. Lunney considers the incentive-access paradigm fundamentally flawed, because incentive and access will always oppose each other with equal force.[15] Because a work's desirability will indicate both the need to ensure the work's creation and the need to secure its widespread distribution, providing copyright protection when it does not threaten a compelling need for access results in more production of less desirable works and less production of more desirable works. But in fact, the more necessary the work, the more compelling the case for access, and the more limited the protection copyright should provide.

These arguments regarding access tend to focus on the possibility that the future production of works of authorship might be hindered. But there can be another concern that is related to enhancing the public's access to the existing materials, whether individuals of the public produce new works of authorship or consume the works for other purposes. How the

[13] Breyer, *supra* note 8, at 286-87.
[14] *Id.* at 285.
[15] Lunney, *supra* note 7.

public's access to the materials might be related to the issue of innovation will be discussed in the last part of this paper.

Pointing out the problem with the incentives-access paradigm, Lunney instead calls our attention to the notion of allocative efficiency, because in a world of finite resources, more works of authorship must mean less of something else. Therefore, from an allocative efficiency perspective, copyright provides the proper degree of protection when it ensures that individuals will produce works of authorship if, and only if, such production would represent the most highly valued use of their resources. Lunney considers that the basic purpose of a property system is to ensure that resources are allocated to their highest valued use.[16] But assigning social value is not an easy task.[17] Moreover, the allocative efficiency perspective focuses on the production side of the problem, thus avoiding the issue of the dissemination of the products. This perspective seems to be based on the assumption that as long as works of authorship of greater social value instead of lesser value are produced, the public will somehow benefit from it. But great works of authorship that are produced but not available will not generate high-valued use.

In conclusion, while copyright law is generally understood to provide remuneration to the authors to provide incentives to produce work, the incentive structure of works of authorship is not clearly defined. The issues regarding copyright law and innovation will now be examined against the challenges to copyright law posed by technological developments and subsequent changes.

3. Shifts in Intellectual Property on the Internet

This section discusses the nature and value of intellectual property on the Internet. Special attention is given to the ways in which the digital technology and computer networks change the processes of information transfer, which in turn change the economics of the production and the dissemination of information.

3.1 Ease of Copying and Transmission

A number of shifts in the nature of technology are driving the convergence of the computing, communications, and content technologies and industries.[18] The most important change is the digitization of the media and networks. The digitization of information makes it easier to make multiple copies of a work. Moreover, the copies can be made in a way that is indistinguishable from an authorized original. Also, the transmission of information becomes easier, faster, and cheaper, and thus works in digital form can be used by multiple users. These characteristics will give incentives for the copyright holders in such works to attempt to restrict access and use, and to derive revenue more from uses than from the sales of copies.[19]

---

[16] *Id.* at 579.

[17] Lunney distinguishes between information works and entertainment works (works versus non-works). *Id.*

[18] See Tapscott, *supra* note 2.

[19] Pamela Samuelson, Digital Media and the Changing Face of Intellectual Property Law, 16 *Rutgers Computer & Tech. L. J.* 323, 328 (1990).

3.2 Amorphosis and Metamorphosis of Information

Digitization means that all information, including text, audio, and video content, can be used in any order and be rearranged at will. Digital content can be transformed for use in another medium – for example, from text to voice or vice versa.[20] Works in digital form can be easily added, deleted, edited, or otherwise modified and manipulated. On the Internet, derivative work products with a large array of differing degrees of added creativity will be produced and transmitted. Digital information, unconstrained by packaging, becomes a continuing process.[21]

In addition, when information is in electronic form, it becomes something much less identifiable with any particular medium.[22] The physical forms that have been used to contain digital information such as floppy disks and CD-ROMS also disappear as more and more computers are connected to the Internet. Before, information and the physical form in which the information is fixed were not considered separate, and pricing and selling the information product was based on the physical form. Now that information is more and more detached from the physical place, creators and sellers of information should find a new way to unitize the information.

3.3 Hypertext

Text in electronic form makes possible many functionalities that were previously too time-consuming and tedious. One example is hypertext. Hypertext is an electronic version of text that generally consists of a set of "components" of text or graphics and a set of "links" connecting different parts of the text to each other.[23] The hypertext function on the Internet permits one to move from any spot in a text to any subject of one's choosing simply by pressing a key or button.[24] To make a successful conversion of the printed text to hypertext, the converter will need to create surrogates for the kinds of usability aids that readers are used to from books, but which tend to be lost when a work is converted from printed to electronic form.[25] For example, skills involved in developing the hypertext of an encyclopedia include providing detailed textual information, maps, pictures, or other kinds of information as hypertext components, creating "links" between hypertext components, creating bookmarks that allow one to retrace one's steps, most of which require the hypertext developer's intellectual judgment. This "skill" may be considered to involve some kind of additional "authoring".[26]

The combined characteristics of the ease of transmission, modification, and hypertext function also contribute to a change in the models of communication on the Internet, from mass communication to person-to-person communication, from linear to multi-directional

---

[20] Robert H. Anderson, Tora K. Bikson, Sally Ann Law, Bridger M. Mitchell, Universal Access to E-mail: Feasibility and Societal Implications 96 (RAND).
[21] John Perry Barlow, The Economy of Ideas, WIRED 2.03, at 9.
[22] Katch, *supra* note 5, at 31.
[23] See, e.g., Robert J. Glushko, Hypertext and Multimedia Design: An Engineering Approach (1992).
[24] Katch, *supra* note 5, at 18.
[25] See Pamela Samuelson, Some New Kinds of Authorship Made Possible By Computers and Some Intellectual Property Questions They Raise, 53 *U. Pitts. L. R.* 684 (1992).
[26] *Id.*, at 696.

communication.[27] As a result, the traditional distinction between senders and recipients of information, and between authors and users of information, becomes blurred.

3.4 Network Externalities and Economies of Scale

One of the important economic characteristics of the Internet is network externalities. Network externalities arise because an individual's decision to join a network is typically guided only by the benefit to that individual of communicating with others on the network.[28] This private benefit differs from the social benefit, which includes the benefits that others receive from being able to communicate with the individual in question.[29] But the more Internet users there are, the more valuable the Internet becomes to each user because of the increasing returns.[30] Also, the strong economies of scale might lead a company to attempt to focus on building up a customer base for more elaborate or complementary products. Distributors and creators of content begin to realize that the value of their work doesn't generally get recognized without broad distribution. Thus open interconnection and interoperability become important in the Internet.

3.5 Service and Relationship

The characteristics of digital technology and computer networks change the value of information that is based on its meaningfulness.[31] As the production and dissemination of information rapidly expand, points of view and authority can provide users with useful guidance. Information on the Internet might be infinite and unbounded, but is limited by the amount of human attention.[32] People's presence, time, and attention become valuable on the Internet. In the case of business information, value results from interpretation and analysis by

---

[27] Courts have also acknowledged this change. For example, the District Court of Pennsylvania, in its holding that the provisions of the Communications Decency Act of 1996 that restricted certain communications over computer networks violated the First Amendment, provided a list of different ways of communication on the Internet as follows:

> Once one has access to the Internet, there are a wide variety of different methods of communication and information exchange over the network. These many methods of communication and information retrieval are constantly evolving and are therefore difficult to categorize concisely. The most common methods of communication on the Internet (as well as within the major on-line services) can be roughly grouped into six categories:
> 1. one-to-one messaging (such as "e-mail"),
> 2. one-to-many messaging (such as "listserv"),
> 3. distributed message databases (such as "USENET newsgroups"),
> 4. real time communication (such as "Internet Relay Chat"),
> 5. real time remote computer utilization (such as "telnet"), and
> 6. remote information retrieval (such as "ftp," "gopher," and the "World Wide Web").

American Civil Liberties Union, et al. v. Janet Reno et al., 929 F. Supp. 824, 834 (D.P.A. 1996).

[28] Anderson et al., *supra* note 20.

[29] *Id.*

[30] See Kahin, *supra* note 2.

[31] Barlow, *supra* note 21.

[32] Esther Dyson, Intellectual Property On the Net, Release 1.0., at 13.

people, which depends on the specific efforts and skill of the people interpreting it.[33] Also, attending a concert and listening to a song on a tape may be quite different experiences.[34] The value of information is thus generated by the activity and presence of a person rather than by the ownership of the content.[35] The fact that real-time performance and services generate value presents an interesting shift in what the information economics will be based on: relationship rather than possession. Building a relationship with a customer becomes more important rather than controlling the copies of the content.[36]

In the relationship of sender and receiver of information, and the depth of their interactivity, the value of what is sent depends on the extent to which each individual receiver has the receptors such as shared terminology, attention, interest, language, paradigm, that are necessary to render what is received meaningful. [37] An important possibility is presented here: receiving information can often be as creative as generating it. Having access to the information, and the ability to choose, process, and edit the information become more important. Therefore, the question of where the creativity lies, which is a critical one in copyright law, becomes more complicated, and the distinction between creators of information and users of information becomes fuzzy.

4. Application of Copyright Law to Internet

Debates regarding the application of copyright law to the Internet center around finding ways to ensure the exclusive rights of copyright holders that are provided by copyright doctrines in this changing environment. But the difficulties in applying the copyright law to the Internet compel us to ask what it is that the copyright law is designed to achieve. In order to answer this question, rationales and concepts of copyright law should also be carefully examined. The doctrines and concepts that are frequently discussed include copying, fixation, idea versus expression distinction, authorship, fair use, and access.

---

[33] *Id.* at 15. Esther Dyson points out that although principles are easy to come up with, matching them to specific instances is the intellectual exercise that adds true value. *Id.* Barlow similarly points out the "question and answer" relationships between authorities and those who seek their expertise; *supra* note 21, at 15.

[34] Dyson, *supra* note 32, at 11-12.

[35] Barlow states that "the value of that relationship will reside in the quality of performance, the uniqueness of your point of view, the validity of your expertise, its relevance to your market, and ... the ability of that market to access your creative services swiftly, conveniently, and interactively." *Supra* note 21, at 15. Dyson agrees by saying that "new kinds of content-based value that can be created on the net would include "services, the selection of content, the presence of other people, and assurance of authenticity..." *Supra* note 32.

[36] Dyson explains that a subscription would be a good way to solve the problem generated by network externalities and a need to distribute broadly. *Supra* note 32, at 3. Also see Katch, *supra* note 5, at 225.

[37] *Id.* Kahin also acknowledges that the ultimate value of the Internet or products and services on the Internet depends on the user's software, her ability to use the software, and her skills at navigating the Internet. *Supra* note 2, at 66-67.

4.1 Copying

Copyright law prohibits making unauthorized copies of copyrighted materials.[38] Then what constitutes a "copy"? On the Internet, a user can access digital information only through a reproduction of the information in RAM[39] or on a hard drive, floppy disk, or magnetic tape. It is impossible to read, view, print, upload, download, transfer, or otherwise access digital expression without making at least one copy of it.

There has been some controversy about the meaning of "copies" in a digitized environment. According to the language of the statute, the fixation of a work occurs "when its embodiment in a copy ... is sufficiently permanent or stable to permit it to be perceived, reproduced, or otherwise communicated for a period of more than transitory duration."[40] Further, the legislative history of the 1976 Copyright Act explains that "The definition of 'fixation' would exclude from the concept purely evanescent or transient reproduction such as those projected briefly on a screen, shown electronically on television or other cathode ray tube or captured momentarily in memory of a computer."[41]

But the Ninth Circuit analyzed the issue of "copies" differently in its controversial decision of *MAI Systems Corp. v. Peak Computer Inc.*[42] The question was whether running an operating system program in the course of repairing computer hardware creates an unauthorized "copy". The court held that the loading of software into a computer's RAM constituted a "copy" under section 101, since it can be perceived for more than a transitory duration. The Clinton Administration's White Paper[43] has also been criticized because of its overly expansive view of the creation of a "copy".[44] The White Paper's claim suggests that any loading into a computer's memory of a copyrighted work is a sufficient fixation that may lead to a violation of the copyright holder's rights. If that is the case, any viewing or transmitting of information on the Internet would be a violation. When information is sent from a sender to a recipient across a network, it would be a violation that occurs only because the information passes through multiple intervening points on the Internet. In addition, one "copying" that occurs in retrieving any information in the Internet would violate the copyright holder's exclusive right to reproduce. It has been suggested that a fee might be charged for each "transmission" that is made, but then users face the problem of having to decide which information to view even before they know the content of the information. Furthermore, copyright law has traditionally not provided copyright owners with exclusive rights to control uses made of "copies" of their works.

---

[38] 17 U.S.C. section 106 provides that the owner of copyright has the exclusive rights to reproduce the copyrighted work in copies or phonorecords.

[39] RAM refers to random access memory that is the computer's active memory. RAM is erased when a computer is turned off.

[40] 17 U.S.C. section 101.

[41] H.R. Rep. No. 1476, 94th Cong., 2d Sess. 51, 54 (1976).

[42] 991 F.2d 511 (9th Cir. 1993).

[43] White Paper, *supra* note 1.

[44] For example, see James V. Mahon, A Commentary on Proposals for Copyright Protection on the National Information Infrastructure, 22 *Rutgers Computer & Tech. L. J.* 233 (1996); Fred H. Cate, Law in Cyberspace, 39 *How. L. J.* 565 (1996).

4.2 Fixation in Tangible Expression

The definition of "copy" is directly related to the notion of "fixation" of the information in a tangible medium of expression.[45] As the information on the Internet lacks the physical form it is embodied upon, the copyright model that is based on physically defining the expression of ideas does not ensure reliable payment for mental works. Attempting to fix this problem, broader and more vigorous enforcement has been suggested. However, broader enforcement would result in prohibiting or stifling virtually all the activities of receiving and transmitting information on the Internet.

Another related doctrine is the provision that it is not "ideas" but a particular "expression" of ideas that is protected by copyright law.[46] The property rights of copyright have been focused not on their ideas but on the expression of those ideas. [47] The rationale behind this is the assumption that ideas and facts are collective property of mankind. But the difficulty in separating expression from idea has puzzled many courts and commentators. The difficulty is compounded in the case of digitized information which may not necessarily be fixed in any tangible medium, and in the Internet environment which makes modifying the information extremely easy. Works in digital form move away from printing presses, and the traditional copyright paradigm will be less useful in this area. In order to suggest an appropriate framework, it seems to be critical to ask where the value of information lies, especially when it's free of physical form and packaging.

4.3 Authorship

The Copyright Act protects the "original works of authorship".[48] As suggested in the discussion of the ease of modification, works in digital form on the Internet generate complicating problems of defining what is original and who is the author. An example can be drawn from the hypertext function. When an encyclopedia is converted from a text version to a hypertext version, the problem arises as to which aspects of the extra intellectual labor that has gone into the creation of the hypertext version of this encyclopedia are instances of "authorship".[49]

Another example is hypertext fiction, which can be more readily available in the context of the Internet. A scenario of hypertext fiction is as follows. An author writes the beginning of a short story, and posts it on the electronic bulletin board, inviting any and all partici-

---

[45] 17 U.S.C. section 102 (a) provides that copyright protection "subsists in original works of authorship fixed in any tangible medium of expression, ..."

[46] 17 U.S.C. section 102 (b) states that "In no case may copyright protection for an original work of authorship extend to any idea, procedure, process, ..."

[47] The Supreme Court in Mazer v. Stein explicitly reiterated that "copyright may not protect 'ideas' but protects the particular 'expression' of the ideas." 347 U.S. 210 (1954). Justice Sandra Day O'Connor commented that "The primary objective of copyright is not to reward the labor of authors, but '[t]o promote the Progress of Science and useful Arts.' To this end, copyright assures authors the right to their original expression, but encourages others to build freely upon the ideas and information conveyed by a work. This result is neither unfair nor unfortunate. It is the means by which copyright advances the progress of science and art." Feist Publications, Inc. v. Rural Telephone Services Co., 499 U.S. 340, 349 (1991).

[48] 17 U.S.C. section 101.

[49] Samuelson, *supra* note 25.

pants to compose endings for the story.[50] Hundreds of writers respond. When a hypertext system program is intended to permit authors of fiction to create hypertext stories, the hypertext stories can change with every reader and every reading, because they need not be read in the same linear fashion that has characterized printed text.[51] In this case, not only is the person who writes a series of episodes and decision points for linking episodes together involved with some kind of authorship, but the reader is also making choices regarding the order and range of components, in a way that will significantly change the story. This might be a case in which new kinds of authorship can be made possible by developments in computer technologies. Although copyright law has not acknowledged the "creative use" of readers as constituting "originality," the hypertext function provided in the Internet may extend the concept of authorship.

The characteristics of digital technology and communications networks are likely to make creators of information reluctant to put their information products on the Internet. But if the Internet threatens authors' ability to control the use of their works, it also offers them new opportunities for creation. By facilitating communication among creators and enhancing their ability to disseminate the fruits of their labors, the Internet may promote new modes of authorship, particularly a collective kind. In the example of hypertext fiction, if an initiating author tries to publish her beginning, as joined to her favorite fifty conclusions, which she has assembled and edited, who owns what rights under copyright in the story and its components?

This question is related to the issue of joint authorship. The Copyright Act defines a "joint work" as "a work prepared by two or more authors with the intention that their contributions be merged into inseparable or interdependent parts of a unitary whole."[52] The legislative history of the Act in the House and Senate Reports envisions two situations giving rise to a joint work: (1) "the authors collaborated with each other," or (2) "each of the authors prepared his contribution with the knowledge and inten[t]ion that it would be merged with the contributions of other authors as 'inseparable or interdependent parts of a unitary whole.'"[53] The legislative history suggests that the co-authors must not only intend, but also be aware of each other's contributions. The initiating author and the other authors did collaborate with each other and the contributions were merged to make a unitary whole. Thus all the authors may become joint authors of the story. However, although a co-owner of a "joint" work is entitled to exploit the work on a nonexclusive basis without obtaining her co-author's permissions, she cannot grant third parties exclusive rights without all the co-authors' permission.[54]

In the Internet, it is easy to make various degrees of original contribution to the existing works of authorship. A reader can easily edit and modify. When she carefully chooses different information, criticizes, and adds her own thoughts, would the resulting work product

---

[50] Jane C. Ginsburg, Putting Cars on the "Information Superhighway": Authors, Exploiters, and Copyright in Cyberspace, 95 *Colum. L. Rev.* 1466 (1995). She notes that this hypothetical scenario is inspired by Mark Twain's proposal. *Id.*, note 11.

[51] Samuelson shows an example of some English professors who wrote this program, described in Jay David Bolter & Michael Joyce, Hypertext and Creative Writing, *Hypertext '87 Papers* 41 (1987). *Supra* note 25.

[52] 17 U.S.C. section 101.

[53] S. Rep. No. 473, 94th Cong., 1st Sess. 103-04 (1975); H.R. Rep. No. 1476, 94th Cong., 2d Sess. 120 (1976).

[54] See, e.g., Werbungs und Commerz Union Austalt v. LeShufy, 6 U.S.P.Q.2d (BNA) 1153, 1154 (S.D.N.Y. 1987) (citing M. Nimmer, Copyright section 6.10 (1987)).

be another work of authorship? According to copyright law, it depends on the degree of creativity exercised in making the new work. If she has engaged in a minimally original selection and arrangement of the existing materials, the resulting collection enjoys its own copyright as "compilation".[55] The owner of the copyright in a compilation holds exclusive rights in the reproduction and public display (among other rights) of the work, as edited by its compiler.[56] But the compilation copyright does not extend to the compiled elements, but only to their collective presentation. This means that the new author may be a copyright owner, but her legal ownership of certain exclusive rights in the compilation is not so meaningful in a networked environment.[57]

When an author has put more creative effort in making the new work product than compiling, how much creativity is sufficient to constitute "original" contribution? This is not a new question in copyright. But the significance of this question in the context of the Internet lies in the fact that creative efforts become more central and routine activities involved in reading and using the information. This phenomenon not only demonstrates the importance of discerning the level of "creativity" required by copyright law, but also shows the difficulty of clarifying the distinction between creators and users that becomes blurred in the Internet environment.

4.4 Fair Use

The fair use defense is considered to reflect the basic goal of copyright law as it balances the individual's claim to her work against the public's right to make the most beneficial use of the work. The concept of fair use is given statutory recognition in section 107 of the 1976 Copyright Act which states that in determining whether the use made of a work in any particular case is a fair use, the following four factors shall be considered:

1. the purpose and character of the use, including whether such use is of a commercial nature or is for nonprofit educational purposes;
2. the nature of the copyrighted work;
3. the amount and substantiality of the portion used in relation to the copyrighted work as a whole; and
4. the effect of the use upon the potential market for or value of the copyrighted work.[58]

A recent case that dealt with the issue of fair use in the context of the Internet is *Religious Technology v. Netcom*,[59] in which Religious Technology, the copyright holder of the works of the late founder of the Church of Scientology, claimed that Netcom, a provider of Internet service whose client includes BBS that operates Usenet service, was a contributory infringer of the copies of their works that are made and stored on its computer while servicing BBS. Through the analysis of the four statutory factors, the court held that Netcom's

[55] 17 U.S.C. sections 101, 103; Feist Publications v. Rural Tel. Serv., 499 U.S. 340 (1991).
[56] 17 U.S.C. sections 103, 106.
[57] Ginsburg, *supra* note 11.
[58] 17 U.S.C. section 107.
[59] 907 F. Supp. 1361 (1995).

posting of the plaintiff's works may not be found a fair use, in denying Netcom's motion for summary judgment.[60]

The court first decided that Netcom's use of copyrighted materials served a completely different function than that of the plaintiffs. Thus, the first factor weighed in favor of Netcom. The court also argued that although some of the works transmitted by Netcom were unpublished and could be considered original and creative, which usually makes the work qualified for stronger protection, the purpose of the use is more important in this case. For the amount and substantiality of the portion used, the court held that even though Netcom made available to the Usenet exactly what was posted by BBS, Netcom still copied no more of the plaintiff's works than necessary to function as a Usenet server. Only the fourth factor of the effect of the use upon the potential market for the plaintiff's works weighed against Netcom, because the court acknowledged the possibility that the postings of the works, made available over the Internet by Netcom, could hurt the market for the plaintiff's works.

It is interesting to note that although the court stated that the analysis of the other three factors weighed in favor of Netcom, it still held that Netcom's posting of the Religious Technology's works may not be considered a fair use, solely based on the possible effect on the market. Giving more weight to one factor than others, particularly the fourth factor, is not unique in this case. However, the court's reasoning in the analysis of the fourth factor raises a concern regarding the applicability of the fair use defense on the Internet. A critical question in assessing the possible effect on the market is market substitution, whether consumers would not purchase the plaintiff's product because of the availability of the defendant's product. Netcom argued that the relevant question is whether the postings fulfill the demand of an individual who seeks to follow the religion's teachings, and not whether they suppress the desire of an individual who is affected by the criticism posted on the Usenet. But the court accepted Religious Technology's argument that the Internet's extremely widespread distribution multiplies the effects of market substitution, and held that this factor weighs in favor of Religious Technology.

In this case, therefore, the potential effect on the market becomes the most important factor in determining whether posting a copyrighted work in the Internet is a violation of copyright, and the copyright holder could use the Internet's widespread distribution system to argue for the case of market substitution. Consequently the very characteristics of the Internet, with its efficient and widely available electronic delivery system, will render any transmission of information on the Internet devoid of privilege as a fair use. James V. Mahon raises a similar concern in his analysis of the White Paper, arguing that the White Paper addresses the needs of commercial providers of copyrighted information while providing little support for the preservation of fair use and other public benefits.[61]

While the fair use defense is designed to moderate the possibility of overprotection, which retards the flow of ideas through society, the defense may not turn out to be useful in

[60] A summary judgment is granted when the court holds that there is no genuine issue of fact that requires a trial. When there is no genuine issue of fact remaining in a case, the court decides the case as a matter of law. In Religious Technology v. Netcom, the judgment was denied because the court held that a genuine issue of fact was raised regarding whether Netcom substantially participated in the infringement, and whether Netcom has a valid fair use defense. Accordingly, the case could not be decided as a matter of law. Being denied a summary judgment is different, however, from a judgment finding that there was a valid defense of fair use. If the parties choose to bring the case to a trial, the court may or may not find a valid fair use defense at the trial.

[61] Mahon, *supra* note 44.

the case of the Internet. An approach that seems to be useful in the Internet can be drawn from a case that did not involve issues related to computer technology. In *Campbell v. Acuff-Rose*,[62] the Supreme Court applied the fair use doctrine to works that would be otherwise infringing derivative works. *Campbell* concerned a parody of the song "Oh Pretty Woman." Justice Souter focused not on the market effects or on the commercial purpose but on the creative relationship between the infringed work and the infringing work. He stated that "The central purpose of this investigation is ...; it asks, in other words, whether and to what extent the new work is 'transformative'."[63] If this rule of "transformativeness" is applied to works of authorship on the Internet, it can provide a helpful guideline about what kind of use of the existing copyright materials may constitute creative enough efforts to make new works of authorship, that is, "transformative use".

4.5 Access

The issue of access is often discussed in the context of National Information Infrastructure. Niva Elkin-Koren contends that the overall effect of enforcing copyright law in cyberspace is to restrict access to information in an unprecedented way, and prohibiting users from using and acting upon copyrighted works is dangerous from the perspective of social dialogue and political participation.[64] Fred H. Cate also argues that the broad application of the exclusive rights to reproduce and to display and perform publicly prohibits public access to expression in the digital context of the Internet.[65]

Margaret Chon has conducted an extensive analysis of the term "Progress of Science and useful Arts."[66] She argues that under the "incentive v. monopoly" test, the public's interest in the "free flow of ideas, information, and commerce"[67] is considered to be competing with the incentive to authors and inventors, rather than to be more fundamental. In the term "Progress" she tries to find an "Enlightenment faith in knowledge" that is inferred - that the encouragement of knowledge production and knowledge deployment via the copyright power is in the nature of a trust, which all persons (not only authors and inventors) have a stake in, and a fundamental right of access to.[68] This right of access is considered more basic as a constitutional matter than the provision of incentives to authors and inventors.[69] According to Chon, increasing knowledge does not necessarily correspond to increasing control of our natural and social world through the acquisition and manipulation of knowledge,[70] but could be merely the reflexivity of knowledge.[71] Therefore, the knowledge could

---

62 510 U.S. 569, 114 S. Ct. 1164, 127 L.Ed.2d 500 (1994).
63 114 S.Ct. at 1164.
64 Niva Elkin-Koren, Cyberlaw and Social Change: A Democratic Approach to Copyright Law in Cyberspace, 14 *Cardozo Art & Ent. L.* J. 215 (1996).
65 Cate, *supra* note 44.
66 Margaret Chon, Postmodern "Progress": Reconsidering the Copyright and Patent Power, 43 *DePaul L. Rev.* 97 (1993).
67 Harper & Row v. National Enterprises, 471 U.S. 539, 546 (1985) (quoting Sony Corp. of America v. University City Studios, 464 U.S. 417, 429 (1984)).
68 Chon, supra note 66, at 99-101, 103.
69 *Id.* at 103.
70 *Id.* at 103.
71 *Id.* at 98. The reflexivity of knowledge is defined as its capacity to influence human social practices, which in turn influence knowledge production. Also see Giddens, The Consequences of Modernity 39, 43-44 (1990).

not only improve socioeconomic conditions but also could provide an opportunity for political participation. Here the accumulation of knowledge achieved by providing incentives to authors and inventors is distinguished from the increased access to knowledge by all, because "the value of knowledge lies not in its mere possession, but in the range of possible uses and users for it."[72]

5. Summary and Conclusions

Applying copyright law in the Internet creates substantial and conceptual problems. Technological developments and subsequent changes influence changes in the value of information products and services. But assessing creative value in the copyright regime is a complicated matter, especially in the changing environment of the Internet. It has been suggested that the value of information in the Internet's environment lies in personal effort, process and services, rather than in ownership of assets. Whether the personal effort in this case means labor or creativity is not clear. It seems that the personal effort and services often require being creative, but at the same time, in the absence of authorship, interactivity may generate important value. Further, the Internet expands the possibility that receiving and using information can be as creative as generating it. Therefore, the changing environment in the Internet seems to suggest that there would be many different ways to generate value, both by being creative in the "production" of work in the traditional sense and by being creative in other areas such as servicing, distributing, performing, receiving, and communicating.

The new possibilities of value are not well encompassed by current copyright doctrines. Rather, the acknowledgment of the newly found importance of relationship and interactivity among the creators, sellers, and receivers of information products and services seems to be reflected in the debate about the notion of access. A few scholars focus on the term "Progress of Science and useful Arts" more closely in their attempt to discern what is ultimately designed to be promoted by copyright law, which may or may not coincide with the current scope and degree of rewards to the authors. The essential problem regarding access is not so much about having access to a particular product as it is about creating the general mechanism of ensuring access to the existing products and services. The scholars who focus on the concept of access seem to ask, what good does innovation do if the public does not have access to it?

Therefore, we need to ask what it is that copyright law is designed to promote, and what it is that we really want to achieve from enforcing copyright law. As shown in this paper, some argue that democracy is the ultimate goal, while others perceive that it is irrelevant to copyright law. These are some of the concerns we should first resolve before we can have meaningful dialogue about how the copyright law should be applied to contents on the Internet, and what kind of role copyright law could play in maximizing the benefits of the Internet.

Legal frameworks shape the ways in which technology is developed and its applications are adopted in the society. At the same time, technological developments and subsequent changes in the economic and cultural environment may also have important implications for

---

[72] *Id.* at 145. Quoting Steve Fuller, Studying the Proprietary Grounds of Knowledge, in To Have Processions: A Handbook on Ownership and Property 105, 109 (Floyd W. Rudmin ed., 1991).

the law, by challenging our assumptions that are relevant to the legal framework. The advent of computer technology, the merging of computer and communications technologies, and the network system of the Internet have presented a considerable difficulty in applying the existing framework of copyright law, which leads us to ask fundamental questions that have been missing in the debates about intellectual property, such as: Why do we need innovation? Is it always desirable? What does it mean to innovate? More production of information goods? Further production of goods of better value? Where does value come from? Creativity? Effort? Communication? Relationships?

Applying the conceptual framework of copyright law to the Internet requires translation of copyright principles, rather than copyright doctrines, to the digital technology and changes generated by the technology. The conceptual problems reflect the fact that copyright law adapted itself to address the special needs of print technology.[73] Legal rules that are based on one kind of social reality may no longer be a coherent standard in the face of changes in the social reality. The ways of thinking about laws governing innovation may be guided by the changes in the culture and economics of works of authorship. The challenge of the Internet to copyright law demonstrates that innovation does not have to be in terms of more production of information. When more value is assigned to the communication of information than the ownership of information, encouraging more communication would be an innovative practice.

[73] For the view of copyright law as a product of print technology, see Elizabeth Eisenstein, The Printing Press as an Agent of Change (1976); Ethiel de Sola Pool, Technologies of Freedom, On Free Speech in an Electronic Age (1983); Ethan Katch, The Electronic Media and the Transformation of Law (1989).

Telecommunications Transformation: Technology, Strategy and Policy
E. Bohlin and S.L. Levin (Eds.)
IOS Press, 1998

CHAPTER 18

The Internet and the Future of Security: The Globalization of Space, Time and Image

Knut Erik Solem

*Department of Political Science and Sociology,
Norwegian University of Science and Technology, N-7055 Dragvoll (Trondheim), Norway*

Abstract. The Internet, and its increased usage, have been the subject of much debate. It has had a profound impact on many sectors of society, on ways in which we work and interact, and on our understanding of social forces around us, the evolution of society and of ourselves. Like numerous technologies, it is a «two-edged sword», which on balance – when understood and used correctly – contains great promise for the furtherance of human development in the areas of truth maintenance and freedom of expression, hence for human long-term security and well-being.

1. Introduction

The Internet, which has been described – and quite correctly it seems – as an anarchic but liberating system, has grown at an explosive speed. By the beginning of 1996 some 9.5 million computers were directly connected to this global network, which is nearly a 200% increase from last year alone. However, the number of users is, of course, even much larger. In addition to these 9.5 million computers, there are also machines which are jointly connected to networks with common access to the Internet. Since the majority of these machines are operated by several individuals, only a very rough estimate of actual users can be made. However, with – say – somewhere between 3 and 5 users per direct Internet access, this would give us an estimated figure of from nearly 29 million to 48 million users globally. IBM believes that by the year 2000 some 400 – 800 million users will be connected to the Internet.

The implications of this are staggering.

For International Relations (I.R.), or International Organisation (I.O.) for example, some of the key questions are: What does this mean for such central concepts as security and foreign policy-making? Will there be new forms of warfare and conflict resolution and, if so, which shape will these take? If previous notions of censorship and control are on the way out, as it seems they may be, what will – or should – take their place? Or is the question in fact redundant? This chapter provides a conceptual framework and is a first attempt to explore the implications of these concerns.

2. Growth and Development of the Internet

The growth and development of the Internet have been both quantitative ***and*** qualitative. According to Vint Cerf, it has gone from "near-invisibility to near ubiquity" in a little more than a year [1]. And if we look back somewhat further in time, the near-explosive development of the Internet is evident. The following charts and figures may illustrate the system's quantitative success [2]:

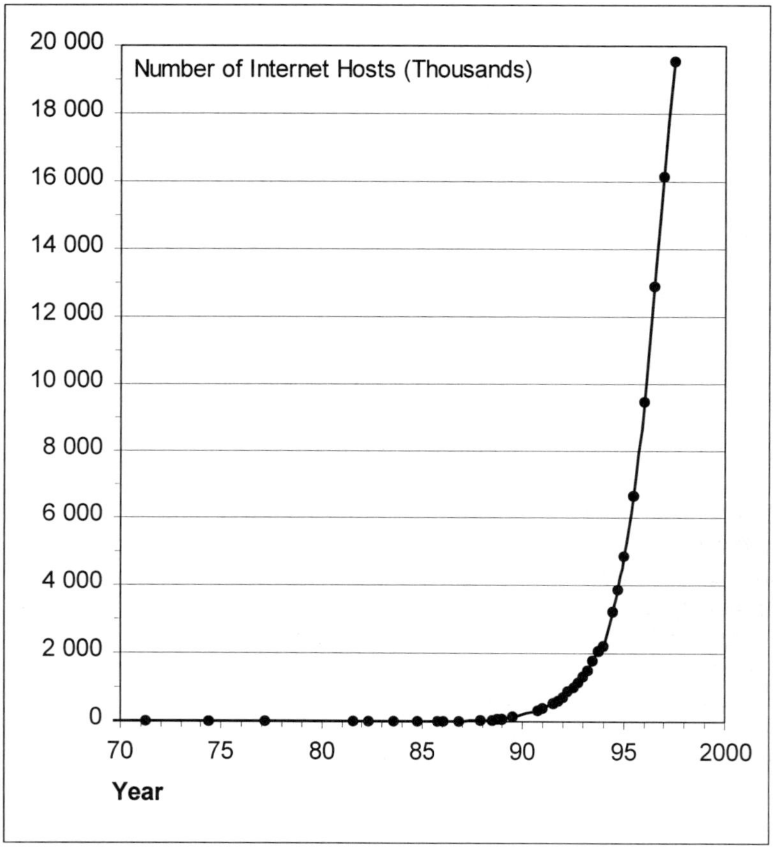

Figure 1. Growth of Internet

As of January 1996 the «Net» domain was more than twice as large as it had been some six months earlier. Some 76,000 systems are now named www, up from only 17,000 six months ago.

Growth of Internet Hosts and Networks is indicated by the fact that from July 1996 until July 1997 Hosts increased from 12,881,000 to 19,540,000, whereas Networks within the period from July 1995 until July 1996 went up from 61,538 to 134,365 [3].

These are staggering statistics.

Not only in terms of sheer numbers and their implications, but also with respect to quantitative changes taking place *within* the user constellation of the Internet, some specific recent developments are well worth taking into account. First, the *commercial* part of the Internet has now passed that of the *university* and *public* sector usage of the system. According to reports from Network Wizard, by the end of 1995 there were 2.4 million users directly connected to the Internet with the registration *com*, which in the U.S. signifies commercial services. This number represents an increase of some 45% compared to 1994. In comparison, the number of direct connections within the U.S. educational sector within the same period of time amounted to some 1.8 million, representing an increase of only 8%. Since U.S. trends in this domain, as well as in others, frequently signify larger, often global, trends, this observation is striking.

A major implication of this trend is that we may now well be entering a *new period* in the Internet saga. If this is the case, no longer will «the Net» remain the domain of principally the computer-literate and "net-initiated" *aficionados*. From now on it could in fact turn into the full-fledged open marketplace for trade and commerce, including the trade and exchange of ideas, as contained in its initial promise. And as I.R. or I.O. theorists, practitioners, or observers, this development should – and hopefully it does – concern us. Conflicting claims about the Internet, such as that it is the "greatest free marketplace of ideas that has ever existed" and/or that the "most remarkable thing about it is the banality of the material that can be found on it" [4], may now be observed more closely and tested. Although Keith Webb is no doubt right is stating that the truth, in all likelihood, lies somewhere in between these two opposing statements, we will no doubt from now on be treated to a different flavour with regard to the Internet, its usage and the – perceived and de facto – roles and impact it will have regarding society in the near future.

Because of the quantitative and qualitative changes in the Internet's user constellation, because of the massive increase in usage, and because of the need for new services to stem the information overload, new services will probably materialize. Tailor-made information packages and subscription services are already seeing the light of day, or are on the horizon. This process will doubtless increase substantially and – only after a while – stabilise somewhat. Consequently, the list of potential services on the Internet may prove to be almost limitless. Already, the entertainment industry, for example, is experimenting with ways of pumping movies and music through the Web directly into consumers' homes, and – in the process – bypassing a series of middlemen, such as retail stores or video-rental stores. It could, and most likely also will, bypass a variety of regulatory and/or controlling mechanisms. Some of these issues will be discussed later on.

The counterargument to the above "rapid-growth/development" scenario is the opinion held by some observers that the Internet is merely a passing fad, which will fail because 1) it is incapable of keeping up with the growing traffic and 2) businesses cannot find a reliable method of selling their products over the network. It should perhaps be noted here that similar scepticism did greet the introduction of a whole host of other technologies, which are now taken for granted by almost everyone. And – everything else being equal – the useful parts of new technology have often tended to become imbedded in our daily life.

Human history is, in fact, full of miscomprehension and miscalculations with respect to the introduction of new technology. The telephone and the car, for example, were at the time of their introduction both denounced by critics as nothing more than "toys for the rich". The telephone and the radio were seen by some as the "destroyers of family life", and television was condemned outright by some observers. No less a "reliable source" than *The New York Times* reported after a prototype viewing of TV in 1939 that «the problem...is that people must sit and keep their eyes glued on a screen; the average American family hasn't time for it».

The problem we encounter here is an age-old one. It is that people have always had, and will surely continue to have, a tendency to view new technology in terms of what already exists. To state about the Internet that it is "like TV" or "like a newspaper but only better" is to totally miss the point. To try and *define* it in a precise manner may even be an exercise in futility, as the nature of the Web is both *dynamic* and *unstable*. We do not, as yet, know what exactly the Internet will evolve into. The only thing we *do* know, watching both quantitative and qualitative trends and changes, is the following: the present era, in which very large quantities of data of every sort can be replicated and transmitted at will, is going to create genuine upheavals in the domains where we work, learn, do business and interact with each other.

In the meantime the scramble is now on to try to meet a whole series of new services, which the industry itself anticipates will be wanted. Already by 1997, Compaq, IBM and other desk-top manufacturers plan on introducing low-cost Internet "appliances", i.e. stripped-down computers which can connect to a TV set and cost in the vicinity of $500 or less. Other, more advanced systems are on the horizon, and they will meet (and change) many of the consumers' preferences and choices.

3. Security and Foreign Policy-making

Whereas genuine security, in the final analysis, must be seen as more important than foreign policy-making per se, the latter activity has more often than not tended to take precedence in the ordering of the affairs of men. This ordering of priorities may invariably prove fallacious; however, it has not deterred foreign policy-making establishments from making and acting on such assumptions, for reasons internal to themselves. As we are interested in security and foreign policy-making in the future, in the "age of the Internet" so to speak, two fairly brief illustrations illustrate trends here.

Thomas Friedman of *The New York Times* reports an incident that caught his eye and attention. A front-page picture of the *Financial Times* shows Bill Gates, Chairman of Microsoft, holding talks with Jiang Zemin, president of China, with a caption stating that the two men have held "very cordial" talks, in contrast to their "frosty summit" of 18 months ago. Friedman then observes that, this being the case, Mr. Gates will have met the Chinese president twice the number of times as has Bill Clinton, the U.S. president within the same 18-month period. The journalist concludes – and he may very well be correct in this assumption – that the Chinese believe they need Bill G. more than they need Bill C. Friedman therefore poses the interesting, and for us quite relevant, question: does Microsoft have a foreign policy? [5]

The direct answer to this question is that Microsoft does not have a foreign policy per se. But what the company has is a global business agenda. And since Microsoft today possesses twice the market value of General Motors it may well be, argues Friedman, that it would seem to be bound to influence U.S. foreign policy. But does it? And does it have to?

Microsoft is a $50 billion company, yet it has only fairly recently opened a Washington office. With an army of PR experts and lawyers to defend it against antitrust suits and copyright violations, it is unlikely that the company would need to look to Washington for any specific help. So, Friedman is right, Microsoft does not *need* Washington to open doors for it, since – it seems – foreign governments are more or less begging the company to enter their domains. And the general importance of China is not only due to the fact that the country has 1.2 billion inhabitants. It is also, as far as Microsoft is concerned, due to the equally important fact of China's restrictive birth-control policies. With every family being restricted to one child, this means that there are often two sets of grandparents and two parents – a total of six adults – with the propensity of saving to buy a computer and software for each child.

Other countries which will be "up and running" in this process, as far as Microsoft is concerned, are Japan and Israel. The company's hottest market in the Middle East is Saudi Arabia. Does Microsoft, therefore, "influence" U.S. foreign policy? Is it, in fact, an "American" company? The answer to this last question, coming from Microsoft itself, is that it is a company based in the U.S., but that it is a global company. Hence there is Microsoft Japan, Microsoft Italy...and so on [6]. This does not mean, of course, that it is indifferent to the U.S., for the company needs a cutting-edge U.S. technology market to design, test and perfect its products prior to global market penetration. With a virtual worldwide monopoly – Microsoft's operating system runs 85% of the world's computers, and has access to most, if not all, of the modern lanes of communications – the company resembles, in Friedman's words, one of the great sea powers of old.

Is what is good for Microsoft good for America? In one sense the answer is yes, and the explanation is as follows. In order to take advantage of the technology of which we speak, and which is desired by a whole host of recipient countries, societies have to be more open, deregulated and interactive. This is, by and large, good for democracy.

Still, software technologies which enable individuals to communicate horizontally through the Internet, across national boundaries, hence which allow them to set up groups and information pools outside the controlling power of governments, *must* surely upset or frighten someone, one would think, and that someone is likely to be an individual or group of individuals "in authority". Apart from the well-known general type of «power-mongering» found here and there, and the somewhat odious but persistent desire in some individuals for control over other people, there are in fact a few legitimate concerns. However, these tend, in general, to be related to crime and address themselves to quite specific issues, such as child pornography, which – at least until now – have been seen as largely outside the confines of foreign policy-making.

In the final analysis, it is difficult to predict what the full impact of the Internet will be for the full and proper process of foreign policy-making. Chances are that, if understood and used properly, it could provide great opportunities as an indispensable aid to decision-making.

As far as security is concerned, it is perhaps not all that easy to see that there should be a direct link between the operation of what – until now – has primarily been considered an information technology tool (some sort of super-library-cum-TV with extras) on the one hand, and important aspects of national or military security on the other hand. However, considering a recent article on Cyber Wars from the Science and Technology section of *The Economist* may make us change our minds. «Logic bombs may soon replace more conventional munitions» [7]. This article, based on a scenario development from the U.S. Pentagon military planners, may well be of particular interest for future International Relations theory and practice. In this scenario (future history) several local computer users, when trying to log on to the global communications network, find it impossible to connect. In horror they watch their computer hard-drives becoming clogged with streams of incomprehensive e-mail messages containing tens of thousands of lines at a time. Concurrently, managers of large w.w.w. (World Wide Web) information sites watch helplessly while their large server computers hopelessly grind to a halt after being overwhelmed by tens of thousands of simultaneous requests for data. It is several hours later that engineers at telephone companies, defence labs and universities start to realise that the streams of data traffic swamping the Internet connections are in fact in no way accidentally caused. Rather, according to this scenario, they originate from outside U.S. territorial grounds. One clue is the way that the packets of data flooding the network keep changing their addresses of origin; they seem – in this story – to originate somewhere in Eastern Europe.

A few days later, the lights have gone out, telephone lines are jammed solid, trading on the New York Stock Exchange has stopped, automated teller machines have started indiscriminatingly crediting and debiting customers' accounts, airlines have lost their air-traffic controls, and the ghastly realisation dawns: America is under attack, the victim of a cyber war [8].

This is of course fiction, but such scenarios are now being developed in the U.S. and in defence departments elsewhere, as well as within security planning organisations. According to at least one Silicon Valley technology guru, Geoffrey Baehr, referred to in the above article, it is quite clear that, by now, a network war could halt a country's economy as effectively as would an electromagnetic pulse following a nuclear detonation. Several similar scenarios have been developed by, for example, researchers at the Rand Corporation in California. Some of these – assuming they become unclassified – may be available to I. R. experts and other researchers in the university world in the near future.

Scenarios and strategic games are now being drawn up, and it seems certain that the question of «information warfare» is taken seriously by top military authorities in many countries. Roger Molander of the Rand Corporation's Washington office, for example, referred to in the same article, has developed several such security-related exercises. They have tended to raise as many questions as answers, but at the very least these "serious games" have alerted military and strategic planners alike to some of the information-based security threats in the future.

By *security* we may now have to understand a much broader concept than that traditionally applied by national defence planners. Security is closely related to strategy, previously a somewhat narrowly defined concept. Until fairly recently in many circles, the latter concept has tended to be based primarily upon military power per se. We will therefore propose an alternative working definition for present purposes. It seems to us that for very good reasons a (strictly speaking) "military strategy" for a nation is no longer sufficient, if indeed possible. In many instances, for example, the line of demarcation between military, economic and political matters is no longer clear-cut. Hence, the development of a long-term strategy for national survival, even with its military meaning intact, will of necessity have to incorporate political, economic and even social factors. Conversely, political strategy is by now increasingly seen as firmly based upon military and economic power realities. Does foreign policy-making, then, base itself upon or does it determine security? To arrive at this point of conceptual agreement, with subsequent prescriptions for action, a fair bit of work has to be done by future defence and foreign policy planners alike.

As we have seen, the state is no longer the only actor in the field of International Relations and the arena of foreign policy-making. But to the degree that the state remains the primary actor and guarantor of national security, the following major assumptions should be kept firmly in mind. For *security* to be achieved, strategy – as applied here – must be seen in its broadest of terms. As such it amounts to the art and science of employing all elements of power of a nation to accomplish the objectives needed in peace or war, and – increasingly – within so-called "peace crises". Strategy in this sense, therefore, involves the use and close integration of the economic, political, cultural, social, psychological, moral, and even spiritual power available. According to this line of thought, strategy can be formulated only after the objectives to be accomplished have been determined. Hence national objectives and national power are irreducible elements of national strategy. When national objectives have been determined, all aspects of the problems confronting the nation should ideally be thoroughly examined, and accurate evaluations made of the character, size and capabilities of the various elements of national power available for the construction of an optimum strategy. This type of strategy, then, should be flexible enough to counter unexpected moves by any adversary. Hence strategic options – or contingency plans – are necessary ingredients in the overall scheme [9].

If this sounds like a good counter-conventional scheme for meeting some of the challenges of the future, such as they present themselves for purposes of proper foreign policy-making or for the maintenance of national security, it probably is. And will new forms of technological, informational and analytical tools based on the bridging of gaps and on technology break-throughs – such as the Internet – be of assistance for us in the construction of such a scheme? The answer seems to be an unmitigated yes.

4. Censorship and Control

At the outset it was stated that perhaps some of the previous notions of "censorship" and "control" were outdated, and that, consequently, a remaining question might be: what would – or should – take their place? As a corollary, is that question itself redundant?

According to the past Norwegian Prime Minister, Mrs. Gro Harlem Brundtland, new technology will demand new forms of co-operation among countries. Censorship and control – in the context of the *information society*, which is the direction in which we are now collectively steering – become impossible for that reason alone. The informational and biotechnological revolutions, within which it seems we now find ourselves, possess such characteristics that, according to Mrs. Brundtland, the "boundaries of censorship are being exploded" [10]. In the light of recent events, this claim seems to be a visionary, perhaps even revolutionary, but essentially correct way of viewing these developments. Attempts by individual countries to use national legislation in order to control or suppress the content of the Internet are bound to fail, according to the past Prime Minister of Norway. Instead of applying such traditional methods as censorship, control and regulation, according to her, work should be undertaken to try to influence attitudes. International co-operation, therefore, becomes important and hence takes on renewed relevance. Furthermore, since the private sector seems to be ahead in the game, the public sector ought to follow suit, the argument goes.

However, this is not always and everywhere the case. The world's most populous country, China, instead recently embarked on the process of censoring the Internet. It has been blocking some 100 Web sites, mostly but not exclusively of U.S. news organisations, the Taiwan government, dissidents and other groups deemed to be politically sensitive. All information is tightly regulated in China, and the new restrictions are therefore the latest way to attempt to control information entering the country, and to keep out criticism of the government. The link between information censorship and general security is clear. The attempt to stem the onrush of information in order to protect, in the short term, political correctness, and in the long run the survivability of the (particular) regime cuts both ways. Perhaps a more fine-tuned handling of the issue of Internet information vs. perceived political fear(s) expressed by many of the authorities could – in the longer run – provide some sort of improved security, albeit probably only in the short to medium run, for the country as a whole.

Elsewhere in the world some of the content on the Internet has caused problems for the authorities. Germany, for example, imposed strict censorship on the Internet, forcing the on-line service CompuServe Inc. to ban worldwide access to about 200 bulletin boards that contained material with sexual content. According to the company, which has some 4 million members around the world, Germany had identified 200 Internet newsgroups with content that was illegal under German criminal law. Since it was unable to block out access only in Germany, the company «temporarily» locked out all its members. According to CompuServe, the German Government was investigating newsgroups and other Internet content that might contain child pornography, pornographic material that is illegal for adults to see, and pornography that is legal for adults, but too explicit for children [11].

Censorship and security are interlinked in other ways as well. The Internet has, so far, often also played the role of the defender of the politically persecuted, the underdogs of world politics. Human rights is one such example. Aung Suu Kyi, the symbol of resistance and struggle for freedom in Burma, benefits from the use of the Internet by groups who defend her, and that country's, civil liberties and human rights. «Free Burma» is a comprehensive Internet site with full coverage of history, pictures and other illustrations of the suppression of civil liberties and freedom of expression in Burma. Amnesty International, Human Rights Watch and the Centre for Human Rights are examples of organisations which make use of Internet as a tool. Not only Asia but also Latin America are examples where the Internet has served as a useful tool for furthering human rights, freedom of expression, hence long-term global security. Concrete cases exist, as for example in Honduras and Guatemala, where victims have been rounded up and tortured and sometimes disappeared due to semi-military groups. These activities have been observed, noted and made known worldwide by some Internet users to others, with resultant pressures being brought to bear on respective authorities in the countries concerned. It seems that long-range security is best served by an open system with free information.

5. The Remaining Problem of Inequalities

However, the picture is not all rosy. Information technology (I.T.), in which the Internet becomes the central component, will doubtless create new lines of demarcation and perhaps even new classes of people, and also here the impact upon security seems important enough. The old, unattractive-sounding concept of "gods vs. clods", which was used in the literature a decade or so ago, could in fact take on a new relevance. One may hope that this will not be the case. The scenario of a world consisting of those who control and can operate the technology in question vs. those without knowledge and access is far from pleasant. It seems in some way offensive to the egalitarian element of the human spirit. Should, therefore, the notion of full and equal access to and development of I.T. be made the principal goal of action in this arena? The issue is somewhat broader than the Internet and global security, and the jury is still out on this one. But the link between inequality and security is patently both clear and pressing.

It is perhaps useful to consider some related important and possibly intervening trends. By the year 2000 one half of the poor population of the world's less developed countries (LDC) will live in cities. Increasingly, rural populations shrink – due to lack of employment opportunities – whereas the urban populations grow. By the early part of the next century the number of mega-cities (more than ten million inhabitants) will rise sharply. Mexico City – by now the world's largest city – is soon approaching 25 million inhabitants, followed by Sao Paulo, Brazil, with 22 million. Calcutta, India has nearly 16 million while Shanghai, China and Bombay, India have populations of 15 million. According to recent United Nations statistics, whereas there were 1.4 billion city dwellers in 1970, that figure had increased to 2.4 billion in 1990 and by the year 2000 will have reached 3.2 billion. Further out, according to calculations made by the United Nations (UN) and the United Nations Development Programme (UNDP), the urban populations could reach 5.5 billion by the year 2025.

Add to this the rather deplorable conditions under which these individuals will try to eke out their existence. Some 300 million, which is nearly a quarter of all the Less Developed Countries' (LDC) city dwellers, live under what the UN describes as life-threatening conditions and extreme poverty. The lack of water, absence of proper renovation, massive pollution, very high unemployment and an increasing level of homelessness will contribute to make this situation more dangerous, perhaps even explosive.

Information, education and the development of democracy may not be the sole panacea for salvation. But anything that tries to stop or in some way stultify mankind's knowledge of what is actually going on seems to us to be a step in the wrong direction. For that reason alone, and there are others as well, the attempt to try to limit or restrict the content and/or use of the net would probably not succeed, nor should it. This of course does not mean an absence of some sort of user's code of ethics, which is an entirely different thing. We know of misuses and misapplications of the Internet, such as harassment cases, or stalking, but these appear in other avenues of life also. And, needless to say, the important legal task ahead with the Internet seems to be less the writing of new laws than the working out of proper parallels and analogies between Internet acts and real ones. Censorship and security, freedom of information, and subsequent freedom of action vs. power and control are in the final analysis not contradictions. In reality they are perhaps only different sides of the same coin. And so it will be seen.

6. Conclusion

It goes without saying, or so it should do, that these issues are in fact not contradictory. Rather, there are important elements of complementarity involved. Real security lies in the protection of human rights and in our ability to develop ourselves freely without undue constraint or real danger to others. This, of course, brings us in a sense full circle back to some very important and basic normative questions, such as a proper and expanded notion of security itself, such as the one developed through this chapter.

Truth is one very critical component, without which nothing is in fact possible. To quote Vinton Cerf: "Truth is a powerful solvent" [12]. The Internet is perhaps one of the most powerful agents of freedom of information that there is. Its informational ability to expose the truth to those who want it is nearly unmatched. And as long as we *know* that this tool, like many others, is a two-edged sword, we are still on a safe ground. There are, obviously, many contradictions still to be properly resolved, given the will and the means. The lack of censorship may in fact have an impact on «Pax Americana» as American values and language are spreading around the world (cf. engaging China instead of containing it). On the other hand, some large drawbacks, often caused by the process of «forced modernisation» which itself at times has been identified, and even directly (hence unjustly) related to certain ills of American society, must be observed and analysed closely so as to be mitigated in time.

As we have already seen, the increasing problem of the (informationally) rich vs. poor can also have security impacts, which may take new, unfamiliar and unexpected forms. For these types of future threats we shall need new, counter-conventional strategies.

Finally, what we must do – in this field of human activity as in others – is not to attempt blindly to stop technology as some seem to be urging us to do, but certainly to stop blind technology. The Internet, when understood and used properly, could help us understand this difference and make the right choice.

References

[1] V. Cerf, Computer Networking: Global Infrastructure for the 21 Century. Data Services Division, MCI Telecommunications Corporation, 1996.
[2] R. H. Zakon, Hobbes' Internet Timeline v3.1, http://www.isoc.org/zakon/Internet/History/HIT.html.
[3] Op.cit.
[4] K. Webb, The Internet as an Object of International Relations Interest, paper presented to the ECPR Planning Session on The International Relations of the Internet. Joint Sessions, Oslo 1996.
[5] T. L. Friedman, The Microsoft Foreign Policy Has to Do With Open Windows, *The New York Times*, February (1996).
[6] Friedman, op.cit.
[7] Science and Technology, Cyber Wars, Logic Bombs May Soon Replace More Conventional Weapons, *The Economist*, London January 13 (1996).
[8] *The Economist*, op.cit.
[9] K.E. Solem, Should Canada Have a Long Term Strategy for the Arctic? In: L'Arctique: Espace Stratégique Vital Pour Les Grandes Puissances. Collection Etudes Strategiques et Militaires. Le Centre Québécois de Relations Internationales (Québec), l'Université Laval, Québec 1986.
[10] G. Magnus, Sensur og kontroll blir umulig (Censorship and Control Becomes Impossible), *Aftenposten*, Oslo, February (1996).
[11] M. Martin, Germany Forces CompuServe to Censor Sex on the Internet, *International Herald Tribune*, Hague, December 29 (1995).
[12] V. Cerf, op.cit.

Telecommunications Transformation: Technology, Strategy and Policy
E. Bohlin and S.L. Levin (Eds.)
IOS Press, 1998

CHAPTER 19

The State Borderline In the Cyberspace

Zixiang (Alex) Tan
Center for Science and Technology, Syracuse University, Syracuse, NY 13244, USA

Ding Lu
Department of Economics and Statistics, National University of Singapore
10 Kent Ridge Crescent, Singapore 0511, Singapore

Abstract. Contrary to the claim that the cyberspace is "borderless" with regard to cross-country information flows, governments and regulators around the world have tried a variety of ways to impose national boundaries in the cyberspace. This paper categorizes these efforts into four models of Internet content regulation. Based on these models, we identify three main approaches Asian countries have so far taken to control information flows in the Internet. Two country cases, Singapore and China, are used to examine the effectiveness of government policies aiming to control access to information in the Internet.

1. Introduction

The success of the Internet as a communication medium is largely attributed to its open, decentralized architecture and to the absence of any centralized rule-making authority. From July 1995 to July 1996, the number of total host computers on the Internet grew from 7.2 million to 12.9 million, a 79% increase. During the past five years, the world's population of "Netizens" has increased at an explosive rate of about 90% to 100% annually.

Such phenomenal growth has occurred globally. The global network consists of numerous independently administrated networks, which are interconnected according to TCP/IP network protocols. It is these protocols that allow any network in the world to connect to the Internet. The unprecedented open and decentralized nature of the TCP/IP protocols and the Internet itself has enabled the Internet to expand into a global network that encompasses all national or regional networks. It has therefore functioned as a remarkable bridge across cultural and geographic boundaries and emerged as a prominent medium that facilitates communications amongst societies of different socioeconomic and political backgrounds.

This same open and decentralized nature of the global Internet has also brought in unprecedented challenges to the existing orders of the human society. The Internet is "creating a new realm of human activity and undermining the feasibility and the legitimacy of applying laws based on geographic boundaries" [1]. In a world of independent nations, different legal frameworks are instrumental to the maintenance of national sovereignty. The new realm of human activity in the cyberspace poses serious questions to the nation-based legal framework. What kind of laws or ideologies should be applied to this "borderless" global cyberspace created by the Internet? What kind of policies should national governments adhere to when they cope with issues such as the exposure of politically and culturally sensitive information or the access to objectionable materials? What kind of legal standards should be established to protect individual rights to privacy and intellectual property rights such as trademarks and copyrights?

Before a global consensus can be reached on these controversial issues, some sovereign countries have already pioneered in developing laws to regulate and control the information flows in the cyberspace. They have also taken various administrative measures to block and filter content accessible in the Internet. On this front, the practice in some East Asian countries such as Singapore and China is particularly remarkable. The purpose of this paper is to review the policy alternatives regarding Internet content access. By looking into two country cases, China and Singapore, the paper examines the effectiveness of the policies to control Internet content. In the following sections, we first introduce four models of state policies toward Internet content regulation. We then take an overview of the three main approaches Asian countries have taken so far in regard to the regulation of Internet content. Two country cases, Singapore and China, are presented in Section 4 and 5. The last section summarizes the findings.

2. Governing Internet Content – Four Models

Ever since the Internet first emerged as part of its trial operation in 1969, passwords and other security methods have been employed to prevent access to private information. There has been, however, little restriction on access to public content on the Internet. This was because for many years the Internet was typically a means for communication well contained within the research and education communities. This nature of non-control and non-regulation in regard to content access did not become a serious issue until about four or five years ago when it first attracted public interest.

In the past few years the Internet has rapidly expanded out of the research and education community's boundary. The exponential growth of the Internet has shaped it into a crucial part of the global communication system for various communities. The Internet has evolved into the most powerful medium the world has ever known. It performs a more comprehensive and sophisticated function than any other communication medium in its capable provision of content publishing, distributing, and broadcasting, while not contained effectively by any geographic boundary. The information flow in the Internet relates to all aspects of human beings' social, political, and economic life. Given the fact that all other forms of communication media are not spared from some kind of regulation anywhere, it would be naive to believe that sovereign states would not try to exercise their governance on the content in the Internet.

In fact, various content regulations, taking different formats, are presently enforced to some degree in many nations. Bill Gates of Microsoft noted at the 1996 World Economic Forum that, "All countries, from China and Singapore to Europe and the United States, [want] to control what they perceived to be undesirable materials" [2]. Nevertheless, while such content regulations certainly challenge the Internet's global non-control nature, governments have to be cautious that their restrictions do not take away the benefits brought by the Internet to their citizens and to their nations as a whole.

From a global perspective, four models might best categorize current government policies toward content access in the cyberspace.

1) The model of non-regulation policy, which is based on the principle of "Free Speech" and supplemented by self-blocking and self-filtering principles individually chosen by Internet users.
2) The model of territorial regulations, which establishes state regulations and enforces regulatory jurisdiction within the territories of sovereign states.
3) The model of extended territorial regulations, which extends state regulatory jurisdiction beyond the physical borders of sovereign states either passively or actively.
4) The model of "Global Policing", which creates international authorities to set up basic rules for the global Internet regulation.

The first model is a logical extension of the Internet's current self-governing principle. This model calls for a laissez–faire attitude of government regulations and a reservation of free speech rights for Internet users to publish and to access to any content. It decentralizes content control to users themselves who determine what is the appropriate content to publish and to access.

However, this widely open model is certainly not capable of satisfying those who desire not to be exposed to offensive content. Nor is the model capable of protecting minors from "inappropriate" materials globally. Some technical solutions have been proposed to give users, especially concerned parents, a certain level of control in the selection of content. One technique is to install software that can analyze content and identify user-designated words. Access to any material containing those words can then be filtered and/or blocked.

Another technique is to rate on-line contents in order to better guide users to desired materials. One popular rating technique is PICS (Platform for Internet Content Selection). PICS was developed by MIT's WWW Consortium in the United States and works like a movie rating system [3]. Its labels have several numbers which indicate levels of violence, nudity, sex, offensive language, or other concerns. These labels are then attached to materials on the Internet. Users are thus able to determine the filtering or blocking of materials by designating access only to those sites above a given PICS rating.

While these techniques may achieve their desired goals, there are many factors to consider before implementing them. For example, the technique of sensing words has a great chance of filtering many innocent materials in which "undesired" words might appear. Also, by following a PICS-type rating system, users have to rely on a rating party or the software's authors to determine what is offensive. Therefore, a significant question of whose criteria should be applied to rate these materials also becomes an issue. Nevertheless, these techniques could, to some extent, guide our web surfing and protect our children from being exposed to undesired materials.

This policy may thus not be feasible when considered for adoption by the more interventionist governments who believe that the governments should define what materials are desirable and undesirable for their citizens. It may be feasible, however, for some democratic societies where free speech is a constitutional right.

Many governments therefore tend to implement the second model of imposing content regulations on the Internet within their own territories, though the degree of regulation varies. These governments have passed laws to specify the range of materials which may not be distributed. For instance, the Singapore government licenses its Internet Service Providers as broadcasters who are subject to strict content regulation. Some of these content regulations are even stricter than are restrictions placed upon traditional broadcasters.

In the United States, the second model seemed to be working in 1996 when President Clinton signed into law the Communications Decency Act (CDA) passed earlier by the Congress. The law prohibits the distribution of certain pornography or hate materials over the Internet. The trend has been set back since June 1997 when the US Supreme Court shot down the anti-smut law by declaring it unconstitutional in violating citizens' free speech rights [4].

For nations that prefer some kind of centralized regulatory instruments, the debate centers around what kind of regulations should be applied. Since the Internet has the triple function of being a content publisher and a distributor as well as a broadcaster, it often turns out that none of the existing laws for regulating classical media can cope with the Internet. Such complexity often initiates the process of revamping the regulatory framework as a whole. This is because too little regulation will not have sufficient impact on users and too much regulation may violate the users' basic constitutional rights. The CDA case in the US vividly suggests this dilemma [5]. In addition to legal complexity, the potential economic impact of those content regulations is still largely unknown.

The third model proposes that existing states extend their "jurisdictions" beyond their physical boundaries since the regulated contents are often stored outside their territories. Proponents believe that in this way, states might truly enforce their domestic regulations. However, this model may be difficult to achieve because of the fact that the content access has long been subject to totally different regulations in different state territories. It is very likely that what is unlawful in one nation might be perfectly legal in another.

For example, in 1996, the French government banned a book on the health history of François Mitterand, the former French President. The book was, however, soon published on a web site in Canada. In this same way, much of what is considered politically and culturally sensitive information by the Singapore and Chinese governments often resides on Internet servers outside their territories. Therefore, if they are serious about enforcing their regulations, these governments must "extend their jurisdictions". To best accomplish this, these states would have to actively extend their authorities to others' territories. According to this model, to ensure that its citizens are not capable of retrieving the book on Mitterand from the Canadian site, the French government must attempt to convince the Canadian government to enforce French laws. This may well be considered by many states as impractical. Along this line of thought, Johnson and Post argue that "the US will surely resist assertions of jurisdiction by other countries over servers within its own territory. And other countries would just as surely reject any effort to impose US laws on the rest of the world" [6].

A more passive approach of instituting this model is for the states to create a national firewall capable of filtering and blocking any undesired material from entering their territories. In this way, states create what may be termed "virtual boundaries" and enforce their own laws and regulations within their "virtual territories". This approach has been embraced by many governments including Singapore and China. The feasibility of enforcing such country-based regulations depends on the policy's technological effectiveness, which has been challenged in practice. On top of that, the resulting political, economic, and social costs need to be carefully examined.

The fourth model is to sign international treaties and to form international organizations to work as "global policemen". Currently, there are some organizations, such as the World Intellectual Property Organization (WIPO), which have fulfilled this kind of global coordination on certain issues. Therefore, perhaps a global organization should be established to regulate and enforce basic rules governing content access and other issues related to the Internet.

So far, however, no existing international organization has volunteered to take such a responsibility. Some associations, such as the Internet Freedom Network, have been organized to discuss and promote international regulatory criteria on governing the Internet. In many cases, however, they have been doing nothing more than preserving the Internet's founding nature – self-governance. Given the vast variance of cultural values and ideologies around the world, it is very challenging and difficult, if not impossible, to find a global common ground for building basic rules of content regulations.

3. Internet to Asian Governments: Economic Benefits vs. Political and Social Costs

The importance of the Internet in a modern global economy has been recognized by most Asian governments. Information technology has been noted as a crucial factor for sustaining economic growth, promoting international business, and enhancing education and R&D capacities. Many government-initiated information technology projects have been launched to nurture these potential benefits throughout Asia.

For example, the Singapore government initiated its IT2000 National Information Infrastructure (NII) program in 1992, almost two years ahead of a similar initiative in the US. The goal for this program is to build Singapore into an "Intelligent Island" by the year 2000 through broadband networking on fiber optical technologies [7]. In addition to infrastructure hardware, the city-state launched the Electronic Commerce Hotbed program in August 1996 to jump-start electronic commerce in Singapore by identifying new applications with economic potential and in demonstrating the technical feasibility of using the NII for commercial purposes. The rationale is to capture the opportunity emerging in the Information Era for Singapore to sustain its traditional role of an "entrepot" for East Asia, albeit in an electronic trading sense [8].

Table 1. The Number of Internet Users in Some Asia-Pacific Nations

Country	1996	1997 (estimated)	2001 (estimated)
Australia	1,000,000	1,600,000	5,300,000
Taiwan	590,000	870,000	3,350,000
Korea	155,000	325,000	2,900,000
China	70,000	250,000	2,700,000
Malaysia	90,000	210,000	1,750,000
India	40,000	200,000	1,800,000
Singapore	150,000	255,000	900,000
New Zealand	210,000	310,000	740,000
Thailand	80,000	150,000	630,000
Hong Kong	200,000	275,000	700,000
Indonesia	60,000	115,000	550,000
Philippines	40,000	70,000	420,000
Total	**2,685,000**	**4,630,000**	**21,740,000**

Source: The New Century Group.

The Chinese government has launched a series of "Golden Projects" to wire the nation and to interconnect government agencies, firms, and consumers through electronic networks. These projects aim to transform the Chinese economy into an integral part of the "information age" [9].

In the Asia-Pacific region, governmental promotion of information technology has led to a strong expansion of Internet users. According to a market researcher, the New Century Group, the total users increased 86% from 1.44 million in 1995 to 2.69 million in 1996 in the major Internet markets in the Asia-Pacific region except Japan [10]. It is predicted that the user population will keep growing at a 52% annual rate until it reaches 21 million users in 2001 (see Table 1 for details).

On the other hand, the expansion of information technology, and the Internet popularity in particular, raise many concerns to Asian governments. Asian countries often view their cultural values as quite different from those of Western nations. Compared to Western countries, most Asian societies place more emphasis on social order, respect for authorities, and harmony between individuals and the community. Traditionally these cultures also have more restrictive attitudes toward obscene materials both for adults and for minors. In some multi-racial, multi-cultural countries and regions, sensitive issues related to religions and races are subject to censorship because they could stir up interracial or interreligious hates and riots. The free-expression nature of the Internet is believed to threaten these societies' political and social stability by undermining these values and orders. Free expression could also challenge the authority of various political regimes and the cultural identification of those nations. These concerns have caused most Asian countries to act more resolutely and actively in Internet content control and regulations than the nations in North America and Europe.

Generally speaking, Asian countries are adopting three approaches to deal with the content control. An often seen, but rather passive and inefficient, approach by these countries is to delay the expansion of their IT infrastructures in a bid to slow down the diffusion of the technology [11]. Governments taking this approach tend to limit its user base mainly to selected academic and business communities. Some even restrict the Internet services to limited basic services such as e-mail or very limited WWW browsing functions. Their belief is that information benefits to economic growth, education development and R&D activities can be derived from provisions to these users alone. They perceive that the political and social risks of such usage could be minimized by the restriction on the range of users. Countries pursuing this approach include Iran, Iraq, and several others.

When governments start to fully exploit the economic benefits of the Internet by allowing its provision to a wider range of users, they often seek alternative means of control. Laws are passed to prohibit the distribution of certain materials and the conduct of anti-government communications. Meanwhile a national firewall is often instituted to filter and block undesired information from the outside world.

Under this second approach, Asian countries choose to institute tight internal laws and build a national firewall. Currently, the most active practitioners of this approach are Singapore and China. Several other Asian countries are also expected to adopt similar policies in the near future. In September 1996, the broadcasters and officials of the Association of South-East Asian Nations collectively announced that they would police the Internet and block access to sites that offend Asian tastes and values [12].

In the process of following the approach, Asian countries usually extend regulations currently in place for other forms of media such as publications, movies, and broadcasting to the emerging Internet. The concept of a national firewall, although involving a new technology, is actually a rather old idea, similar to the existing customs systems which guarantee the implementation of their content regulations within national territories. The effectiveness of this technology has been widely questioned and nations pursuing it have encountered international criticisms. It remains to be seen whether such an approach has negative side-effects on international transactions and global information exchanges through the Internet.

Several countries and regions in Asia have adopted a third approach, i.e., the self-governance approach. For example, governments in Hong Kong, Taiwan and Japan have done little to block content access domestically or internationally. In Hong Kong, as said by the president of the second-largest ISP in Hong Kong – Hong Kong Star Internet – "the problem lies more in self-censorship, because people don't want to do anything wrong" [13]. The Hong Kong Internet Service Providers Association (HKISPA) has published a Code of Practice to guide its members on content regulation and other issues, while the Hong Kong government has not formulated any content regulation yet [14]. The Code states that "Members will not knowingly host any illegal content and facilitate any illegal activities."

The self-governance seems to work well in those nations so far. Typically, this is because those countries are already a developed social environment of free expression, or else because political and cultural stakes are considered to be of less importance than economic concerns. It is generally believed that few other Asian nations will pursue such an approach in the near future because of the uncertainties in its long-term outcomes.

4. Internet Content Control in Singapore

There are mainly three instruments used by the Singapore government to control Internet content. The first is the regulation of domestic content provision. The second is the imposition of proxy servers to filter undesirable information from abroad. The third is the application of existing laws regarding information distribution to the net users.

The government regulates the provision of Internet content through the Singapore Broadcasting Authority (SBA). On 11 July 1996, the SBA announced the Class License Scheme, which classifies the regulated entities into two groups, the Internet service providers and the Internet content providers. The former group includes the three primary access providers in Singapore, namely the CyberWay, Pacific Internet, and SingNet. It also includes those which give secondary access such as cybercafés and schools. All Internet service providers must register and pay license fees to SBA. The second group includes all institutions or individuals who provide content to the Internet, such as information on a web page. Of those in the second group, only the ones which discuss mainly politics or religions in Singapore, and on-line newspapers which target local subscribers, will have to register when notified by the SBA. Institutions or individuals who merely present factual information about a particular religion or whose sites contain occasional political material need not register unless told to do so by the SBA. No licensing fee is payable for anyone belonging to the second group. Table 2 presents a summary of the license scheme.

Table 3 lists the legal responsibilities of the licensed Internet firms. All licensees must ensure that they do not allow their services to conduct gambling games and lotteries, or to broadcast horse-racing related information. Neither should they advertise or promote any types of fortune-telling such as astrology, geomancy, or palmistry. They are also not allowed to use their service to solicit prostitution or any other immoral activity. Only qualified persons recognized by relevant professional bodies in Singapore can offer professional advice or specialist consultant service through the Internet. Following these responsibility guidelines, all of the three Internet Access Service Providers obtain written guarantees from their subscribers not to abuse the facilities or to violate a code of laws [15].

Table 2. Singapore's Class License Scheme

Categories	Examples	Registration	License Fee (per annum)
Service Providers			
Internet Access Service Providers	CyberWay, Pacific Internet, SingNet	Yes	S$ 1,000
Internet Service Resellers (non-localized)	Singapore Network Service, National Computer Systems	Yes	S$ 1,000 (> 500) or S$ 100 otherwise
Internet Service Resellers (localized)	Cybercafés, libraries, schools	Yes	$100 per location
Content Providers			
Political parties' web pages	National Solidarity Party	Yes	
Groups' web pages that mainly discuss Singapore politics	Socratic Circle, Sintercom	Yes	
Individuals' web pages which discuss Singapore politics	A parliament member's homepage	Yes	
Groups' web pages about a particular religious group	Bible Society of Singapore, KampungNet	No	
Groups' web pages discussing or promoting religion in Singapore	Catholic Archdiocese of Singapore, NUS Buddhist Society	Yes	
Individuals' web pages which discuss religions in Singapore	Roland Yap's Homepage	when notified by SBA	
On–line newspapers targeting local subscribers	*Straits Times, Lianhe Zaobao*	when notified by SBA	
Other group web pages	A professional society's web page	No	
Individuals providing web pages for business		No	
Other broadcasting services, broadcast data and value-added networks, computer on-line services	Audiotext, Videotext, Teletext	No	

Source: Singapore Broadcasting Authority.

In Table 3 we can observe the second instrument to control Internet content, that is, the imposition of proxy servers to filter undesirable information from abroad. The three Internet Access Service Providers must use proxy servers to deny access to sites blacklisted by the SBA as objectionable. All the resellers, including cybercafés, libraries, and schools, must connect to the proxy servers. The blacklisting is based on the principle to bar the following types of information:

Table 3. Responsibilities of Licensed Internet Firms

Requirements	Internet Access Service Providers	Localized Resellers	Non–localized Resellers	Web Publishers and Server Administrators
For the World Wide Web: Deny access to sites blacklisted by SBA	Yes			
Connect to provider, which has blocked access to blacklisted sites		Yes	Yes	
Remove web sites as directed by SBA	Yes	Yes	Yes	Yes
For newsgroups Subscribe according to SBA guidelines	Yes			
Deny access to newsgroups except those brought in by access providers		Yes	Yes	
Unsubscribe from newsgroups and remove articles as directed by SBA	Yes	Yes	Yes	
General Ensure proper use of Net by account holders	Yes	Yes	Yes	
Keep audit trails, help SBA to investigate breach of - license, and sanction errant users	Yes	Yes	Yes	

Source: Singapore Broadcasting Authority.

- Information that affects public security and national defense. This includes content that undermines confidence in the justice system and incites hatred or disaffection against the government.
- Information that affects racial or religious harmony, including information that denigrates or satirizes a racial or religious group, inflames racial or religious hatred, or promotes religious deviations or occult practices such as Satanism.

- Information that affects public morals. This category includes pornography, content that encourages promiscuity or depicts gross violence, nudity, sex or horror, and items about homosexuality, lesbianism and paedophilia.

The date 15 September 1996 was the deadline for all Internet resellers to link to proxy servers. From that date on, only filtered access to Internet has been available in Singapore. Whenever a user tries to access to a web site, the proxy server will check a user's request against the SBA blacklist before granting the access [16].

The third instrument used to control flow of information through Internet is the application of existing laws regarding information distribution to the net users. The SBA's Class License Scheme stipulates that sound recordings broadcast through the net service should have been cleared by the censorship section of the Ministry of Information and the Arts. Films or video broadcasts should only include those approved by the Board of Film Censors unless the material is one to which the Films Act does not apply or which is exempted from it [17]. Laws on advertising also apply to advertisements in the net. For instance, in April 1997, the Ministry of Health fined two companies for placing advertisements on the Internet which flout Singapore's medical regulations [18].

The most well-known case so far has been the prosecution of a 41-year-old net subscriber who had downloaded 76 obscene files from the net and stored them on his personal computer's hard disk. He was convicted in court and was totally fined S$ 61,500 (US$ 1.00 is approximately equivalent to S$ 1.42) in September 1996. It was reported that, before the man's arrest, the police had secretly watched the net user's surfing conduct for three weeks [19]. Such police investigation is possible since the law requires that Internet firms "keep audit trails and help the SBA to investigate breach of license and sanction errant users" (see Table 3).

5. Internet Content Control in China

China has attempted to develop and adopt cutting-edge information technology for years. Recently, it has launched its own "National Information Infrastructure" projects, entitled the "Golden Projects". These projects are designed to transform China's current economy into an integral part of the "information age". Further, perhaps at the forefront of China's Golden Projects is its strong desire to build up its own ubiquitous Internet.

China's Internet efforts first began in its research and education networks. The CAnet (China Academic Network), established in 1987, is generally cited as the first computer network in China [20]. A few other academic and research networks soon followed, including the China Research Network (CRnet), the Institute of High Energy Physics' (IHEP), and the National Computer Networking Facilities of China's (NCFC). However, these networks cover a very limited geographic area and interconnect only a handful of research institutions and universities. Additionally, they often perform with low transmission speeds and only function basic global Internet connections.

With the global Internet "fever" of the early 1990s, China's Internet has recently undergone an extensive national expansion and commercialization. The early, academic networks have been consolidated into the China Education and Research Network (CERNET) and the China Science and Technology Net (CSTNet). They have emerged as two national networks interconnecting all major universities and research institutions. The first commercial network, ChinaNET, has already extended its services to all the provinces. Additionally, the competing commercial network, China Golden Bridge Net (ChinaGBN), launched the provision of its services in 1996. Further, other commercial Internet Service Providers (ISPs) have since been launched independently on both local and regional bases throughout the nation. In early 1997, 603 networks registered their sub-domain names under China's Country Domain (CN). In 1994, a conservative estimate of total Internet users was approximated to be around 1,600. In 1995 this user base increased to a total of 6,400 users. Currently, official statistics for early 1997 estimated about 540 access providers and campus networks composing 90,000 to 100,000 Internet users nationally.

China's current embrace of the Internet is guided by several theories which center on the role of information technology and its applications to national economic development. Both China's top leaders and the academics have endorsed the so-called thesis of "Economic Informatization", which regards Information Technology a major driving force to sustain China's economic growth into the 21st century [21].

While the Chinese government has committed itself to the promotion of Internet expansion, it is also fully aware of the potential threats posed by the technology. With the increasingly pervasive availability of Internet services, there arise concerns that the central Chinese government may lose its control over the flow of information. Two facts justify the concerns. First, the Internet may enable organizations and individuals to collect, process, and/or distribute information that the government finds objectionable. Second, the Internet allows Chinese citizens to communicate easily and inexpensively with individuals who are not subject to China's regulations and laws. Fears are that the Internet may facilitate communications between domestic dissidents and overseas anti-government groups. The government is therefore worried that the Internet may turn into an effective and efficient means of conducting anti-government activities. It is also afraid that other unlawful activities such as distribution of objectionable materials may become rampant in the Internet.

The principle of China's regulation of Internet content contrasts sharply with the policies practiced in the Western countries. For example, in the United States, access to the content of adult-oriented sites is generally tolerable. In China, however, web users are prohibited from viewing as well as distributing such content. Additionally, many Western media's news reports about China are considered to be "distorted" and "biased" by the Chinese government and are therefore blocked.

Notwithstanding that, the Chinese government has decided to enforce strong and serious content regulations rather than suppress the Internet's expansion. The rationale is to realize gains economically while minimizing the potential political and social risks brought by the Internet. The government has promulgated and implemented a series of regulations to address the situation on three levels, namely the level of the end users, the level of ISPs and national interconnection networks, and the level of international access.

While the Singapore government has established laws and implementation measures, China is still in the stage of formulating its laws and their implementation. In 1994, China's State Council issued its Executive Order No. 147, "Security Regulations for Computer Information Systems in the P.R. China", which claimed that any computer information system should not put national security in danger [22]. The regulations require owners of any computer information system to implement access controls, administrative controls, and personnel controls.

Based on this order, the Ministry of Public Security (MPS) issued a decree in February 1996 to regulate the bottom level – end users. The decree requires that all Internet users register with the MPS bureau locally. Further, it also prohibits the searching for and distribution of anti-government information, pornographic materials, or any other information which might be deemed to violate national interests or other citizens' rights. The required registration also informs users of the possible monitoring of their Internet-related activities by the police. Finally, this vague and broadly defined content regulation allows the government an extensive degree of flexibility to implement further content control on the Internet users.

On the other hand, China's State Council finally published China's "Regulations on International Interconnection of Computer Information Networks in the PRC" in 1997, which is derived from its 1996 interim regulations [23]. At the ISP and national interconnection network level, the government has licensed only four national interconnection networks. Each of the four networks is under the administration of a government agency. ISPs are required to register with one of the four national interconnection networks and to allow themselves to be subject to the monitoring of the four government agencies. While the number of licenses for the ISPs is not regulated, failure to implement content regulations may lead to the suspension of an ISP's operation license.

At the international interconnection level, only the four licensed national interconnection networks are permitted to set up direct interconnections with the global Internet. In addition, all of the global interconnections are required to use the international gateway facility provided by the Ministry of Posts & Telecommunications. In this way, they can be more easily monitored and accounted for. With the limited gateways tightly controlled by the central government, a national firewall becomes a technically feasible option.

This administrative and networking hierarchy is very unique and bears some Chinese legacies. It is fully examined in the paper by Tan, Mueller and Foster [24]. Figure 1 has illustrated the hierarchy.

China's national firewall started to attract domestic and international attention in 1996 when blocking some web sites became a real policy practice. Blocking certain Western web sites appears to be instrumental in successful enforcement of domestic content regulations and minimizing access to undesired information. This policy has, however, not yet been officially and publicly announced. In September 1996, some non-Chinese media reported that about 100 Internet sites "suspected of carrying spiritual pollution information" were being blocked by ChinaNET [25]. These sites include major US news agencies, overseas dissident groups, and some adult sites.

Still, this seems more like a trial test of the technology and of its political and economic costs. This approach maintains China's goals for Internet expansion and also satisfies the government's desire to minimize the political and social risks brought by undesired Western information. While most Western pundits do not agree with this approach, many Chinese users and Internet Service Providers see it as a positive policy. In this way, the existence of China's Internet has been legalized by the government and thus the possibility of it being banned altogether is effectively ruled out. As one major Chinese ISP manager based in Hong Kong explained, "China is not closing its door to all information. It's just requiring that all information coming in has to follow Chinese laws" [26]. The Chinese laws are, of course, very different from the laws in other nations.

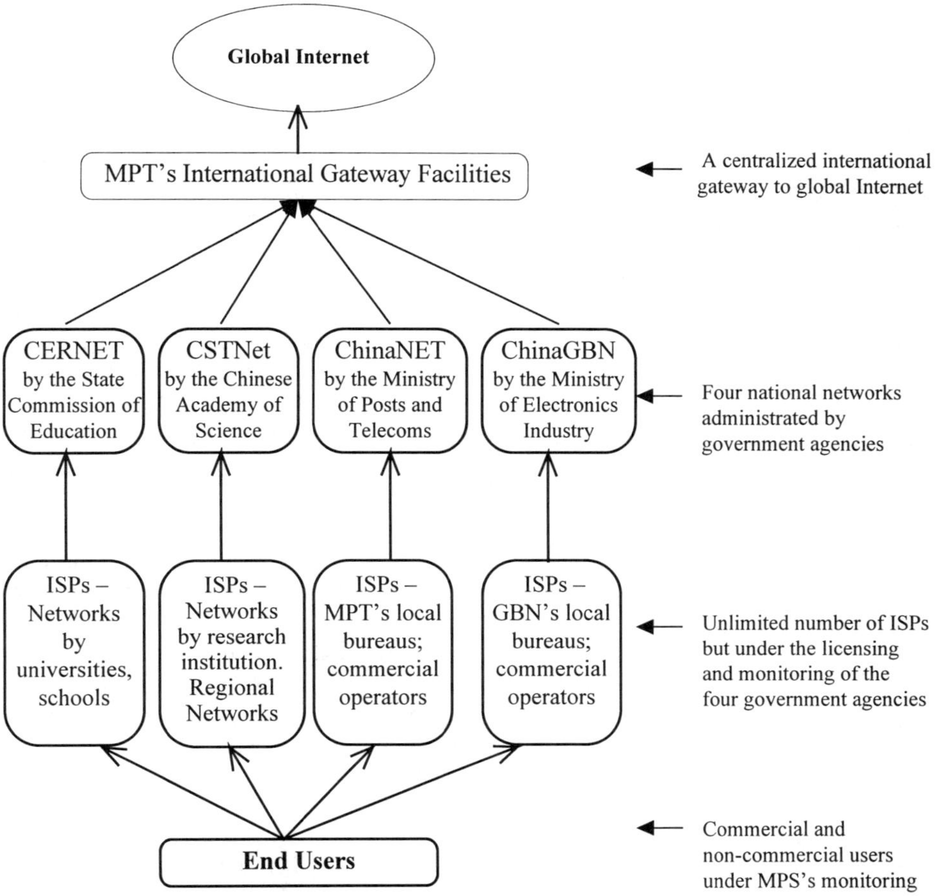

Source: Compiled and copyrighted by authors.
MPS: Ministry of Public Safety
GBN: Golden Bridge Network

Figure 1. The Administrative and Networking Hierarchy of China's Internet

Therefore, despite the fact that the actual political and economic implications are still unknown, such filtering and blocking regulations will certainly not be gone in the near future. As China becomes more familiar with advanced technologies, it may deploy more efficient means, such as sophisticated content-filtering and user-selected blocking, to allow some users full access but to deny such access to others [27]. In addition, the extent of China's blocking and censorship practice is likely to fluctuate as China's domestic political environment and general reform process undergo modifications. More stable and smooth economic and political transition will certainly make filtering and blocking less necessary. While recent news reports have recognized China's un-blocking of several news web sites including CNN, most politically labeled sites remain on the blocking list. Nevertheless, the

Internet's economic allurement, together with the available technology of filtering and blocking, is bound to keep China's Internet on line with the global Internet.

6. Concluding Remarks

Experience of Singapore and China shows that active state regulation of Internet content is feasible. Cyberspace is "borderless" only to a very limited extent.

A balance, however, must be kept between effective content control and economic benefits arising from easy and fast access to the Internet. In China the content control policy has so far safeguarded the legal existence of China's Internet. Technologically, however, the content control firewall may slow access to all the web sites. When Singapore started Internet filtering in September 1996, a major concern arising among users was the possibility of slower access due to proxy server connections. Some users claimed that they found access through proxy servers significantly slower than access through direct links. Even the manager of one of the primary access providers, CyberWay, agreed that access time may rise if the proxy server must check a long blacklist for every access. The SBA, which declines to disclose the content or the size of its blacklist, merely stated that "our objective is not to slow down or impede the performance of the proxy servers." It made the promise to hold regular discussions with the Internet service providers and to work with them to monitor the performance of the servers [28].

The economic costs of longer access time due to filtering technology remain unknown without further studies. One sign is the report that the three primary IASPs in Singapore announced an increase of their equipment investment after a main server of SingNet crashed in October 1996, shortly after the introduction of the proxy server system [29]. According to one of the providers' general manager, the IASP can accommodate all its users using half the number of proxy servers it has, but it must maintain back-up servers "in case there is a surge in the number accessing our service". From July 1994 to March 1996, SingNet had invested S$ 16 million while CyberWay had invested S$ 10 million. To speed up Internet access, the three IASPs introduced dial-up access for 56 kilobits-per-second modems in April 1997. To accommodate the anticipated load from the faster modem, CyberWay had to spoon out an additional S$ 750,000 investment [30]. Possibly on cost consideration, the Malaysian government has said that it will not impose content regulation on the proposed "Multimedia Supper Corridor" project, for which the government has been actively marketing [31]. Being afraid of a slow accessing speed to the global Internet is certainly a concern.

Notwithstanding the costs, the effectiveness of the state-of-art filtering technology is also questionable. A recent article in Singapore's newspaper pointed out that the existing proxy servers could only block access to a limited number of sites and were virtually ineffective in dealing with new sites emerging daily in hundreds or even thousands. The writer told of his personal experience of trying "easy access" in Singapore to many pornographic web pages [32]. Therefore, unless there is a major technological breakthrough, the firewall technology can best serve as a nominal warning to the web surfers.

A possible solution to this problem is the state surveillance to detect errant net users. As warned by Yeong [33], in Singapore, "you are never alone out there on the Net." According to him, the cache data kept by most browsers gives away the user's visiting records such as visited sites and visiting frequencies. Anyone surfing on the net will inevitably leave behind traceable digital footprints. With the cooperation of the service providers, monitor and investigation of users' conduct are feasible and effective, as shown in the legal case illustrated in Section 4. The prosecution of one errant user sends a strong signal to all net users.

Here again exists a policy dilemma between keeping such surveillance and protecting individuals' constitutional right of privacy. After publicizing the legal case of the errant user in October 1996, Singapore's SBA chief executive officer was quick to reassure all net users of their privacy in using the net. According to him, browsing objectionable content at home is not illegal. Users are free to visit whatever sites they like as often as they want to. The emphasis is on voluntary good conduct and self-imposed restraint. To prove its sincerity, the SBA has suggested to the three Internet Access Service Providers to insert a clause in their service contracts with the users, promising not to keep tabs on the users' activities. A few days later an editorial in the government-linked newspaper *Straits Times* also assured "no Big Brother in cyberspace" [34].

Explicitly in Singapore and perhaps implicitly in China, the authorities have ruled out an all-out censorship that could stifle the growth of the Internet. As stated by the SBA, "Our aim is to ensure some regulation so that users, especially kids, will be prevented from accessing sites too easily with a single click [35]." The Singapore government has said that, in consideration of costs, it would not impose censorship and filtering on information even in the local political news groups, such as the "soc.culture.singapore" site well-known for its dissident views.

Entering the new century, most Asian governments will continue to develop their cyberspace regulatory policies, which will remain remarkably different from counterparts in the Western democracies. While embracing the opportunities provided by the Internet and other information technologies, all countries have to fine-tune their policies toward information flows in the cyberspace with eyes on cost-effectiveness and social-norm compatibility. Some kind of borderline will certainly be drawn by many nations in the cyberspace although it might be quite different from the existing physical national boundaries.

References

[1] D. Johnson and D. Post, And How Shall the Net be Governed? A Meditation on the Relative Virtues of Decentralized, Emergent Law. Cyberspace Law Institute Working Paper, 1997, http://www.cli.org/emdraft.html.

[2] M.H. Chua, Internet Data Cannot be Filtered Totally – Bill Gates. *Straits Times* (Singapore), Feb. 7 (1996) 3.

[3] P. Resnick, Filtering Information on the Internet. *Scientific American*, March (1997) 62-64.

[4] R. Vesely, Cyberspace Braces for Bay of PICS, *Wired News*, June 27, 1997, http://www.wired.com/news/news/politics/story/4794.html.

[5] D. Brekke and R. Vesely, CDA Struck Down, *Wired News*, June 26, 1997, http://www.wired.com/news/news/politics/story/4732.html.

[6] D. Johnson, op. cit., reference 1.

[7] National Computer Board Singapore, A Vision of an Intelligent Island: The IT2000 Report. NCBS, Singapore, 1992.

[8] G.C. Hian and O.B. Chin, Singapore Setting Itself up as Electronic Entrepot for East Asia, *Straits Times* (Singapore), March 6 (1997) 36.

[9] M. Mueller and Z. Tan, China in the Information Age: Telecommunications and the Dilemmas of Reform. CSIS/Praeger, Washington D.C./Westport CT, 1997.

[10] J. Skillings, Asian Internet Use Grows Despite High Hurdles, *IDG News Service,* June 9, 1997, http://www.IDGchina.COM/NEWS/Jun97/asian.htm.
[11] B. Gabbard and G. Park, The Information Revolution in the Arab World: Commercial, Cultural and Political Dimensions – The Middle East Meets the Internet. RAND Corporation, CA, 1995.
[12] *Straits Times* (Singapore), October 10 (1996) 44.
[13] G. Chang and S. Said, For Hong Kong, A Censorship Threat from Within, *Wired News*, June 23, 1997, http://www.wired.com/news/politics/story/4602.html.
[14] HKISPA, Code of Practice. April 12, 1997, http://www.hkispa.org.hk/cop-970410.htm.
[15] *Straits Times* (Singapore), October 24 (1996) 40
[16] *Straits Times* (Singapore), August 14 (1996) 41.
[17] *Straits Times* (Singapore), July 12 (1996) 43.
[18] *Business Times* (Singapore), April 11 (1997) 1.
[19] *Yazhou Zhoukan* [The International Chinese Newsweekly] (Hong Kong), October 7-13 (1996) 70-71.
[20] W. Wu, Great Leap or Long March: Some Policy Issues of the Development of the Internet in China, *Telecommunications Policy* **20** No. 9, November (1996) 699-711.
[21] Q. Hu (Minister of the MEI), Greeting Speech on the Re-establishment of China's MEI. MEI's annual book, 1993, pp. I1-I8.
[22] PRC State Council, The Regulations of Safety Protection for Computer Information Systems in the PRC. Order No. 147, Feb. 18, 1994. http://www.cernet.edu.cn/LAW/qry_law2.html.
[23] PRC State Council, Interim Regulations on International Interconnection of Computer Information Networks in the PRC. Order No. 195, Feb. 1, 1996. http://www.cernet.edu.cn/LAW/qry_law3.html.
[24] Z. Tan, M. Mueller and W. Foster, China's New Internet Regulations: Two Steps Forward, One Step Back. *Communications of The ACM*, September 1997. Forthcoming.
[25] R. Aronson, Net Breaches Chinese Wall, *The Wall Street Journal*, 8/19/96 A10. Also see K. Chen, China Bars Access to as Many as 100 Internet Web Sites, *The Wall Street Journal*, Sept. (1996) B12.
[26] R. Schoof, Entrepreneur Wants all China in Her Net, *Los Angeles Times* July 7 (1996) D7.
[27] W. Yurcik and Z. Tan, The Great (Fire)Wall of China: Internet Security and Information Policy Issues in the People's Republic of China. Proceedings of 24th TPRC conference, Solomon, MD. October 1996.
[28] *Straits Times* (Singapore), August 14 (1996) 41.
[29] *Straits Times* (Singapore), November 11 (1996) 34.
[30] *Straits Times* (Singapore), April 14 (1997) 38.
[31] *Yazhou Zhoukan* [The International Chinese Newsweekly] (Hong Kong), January 27 - February 2 (1997) 50-51.
[32] J. Gao, Unpreventable Pornagraph on Web (in Chinese), *Lianhe Zaobao* [United Mornings] (Singapore), April 6 (1997) 2.
[33] Y. A. Seng, You are Never Alone Out There on the Net, *Straits Times* (Singapore), November 6 (1996).
[34] *Straits Times* (Singapore), October 10 (1996) 44.
[35] *Straits Times* (Singapore), July 12 (1996) 43.

Part VI

Networks, Competition, and United States Policy

Telecommunications Transformation: Technology, Strategy and Policy
E. Bohlin and S.L. Levin (Eds.)
IOS Press, 1998

CHAPTER 20

The Telecommunications Act of 1996: A Policy Analysis Test Bed

Edward E. Zajac
Department of Economics, University of Arizona
Tucson, Arizona, 85721, USA

Abstract. The Telecommunications Act of 1996 (TA96) presents policy analysts with a marvelous opportunity to test their theories. Rapid technological change has compressed time scales from years to months, and analysts should quickly be able to tell whether or not their analyses are on target. TA96 contains two conflicting agendas – a free market agenda that bodes for less telecom regulation and a social agenda that bodes for more. It is not what a benevolent despot economist (BDE) seeking economic efficiency would decree, but it is the result of deal making and vote trading, facilitated by political entrepreneurs and justified by appeals to "fairness" and "level playing fields". TA96 violates seven principles of good policy design that a BDE would follow. As a result, political entrepreneurs can eventually be expected to turn waste and economic distortions into fairness arguments for reform. These opportunities for future political entrepreneurs represent time bombs that TA96 has planted.

"Theories which explain everything and predict nothing are useless."
– Anonymous

1. Introduction[1]

Telecommunications technology is changing at an unprecedented rapid rate, and telecommunications global competition is intense. Moreover, in the U.S., interest groups wishing to open up access to the "information superhighway" or "infobahn" have lobbied Congress to do so. The Telecommunications Act of 1996 (TA96) is the U.S. Congress's attempt to react to these pressures.

The result is an Act that contains two major agendas – a free market agenda and a social agenda. The free market agenda attempts to promote the attainment of economic efficiency; the social agenda to facilitate infobahn access by mandating new government subsidies and to perpetuate current rural and other subsidies. The two agendas conflict; the free market agenda envisions an eventual lessening of government intervention into telecom markets, the social agenda envisions a permanent government telecom role.

Because of the noted pressures, we can expect the Act's consequences to unfold quickly. This puts a new wrinkle on policy analysis. Typically, passage of major legislation signals

[1] This paper is partially based on a report prepared for the Goldwater Institute [1].

the end of a long road of political activity. Analysis consists of parsing into component parts what happened along the road. Since further changes are likely to be a generation off, analysts pay little attention to predicting the evolution of what the legislation has wrought.

However, we can expect that, unlike the 62 years between the Telecommunications Act of 1934 and TA96, we may have a new telecom act in 62 months. This presents a challenge to telecom industry analysts to predict TA96's consequences.

This paper attempts such an analysis. It starts by sketching an embryonic theory of institutional change, based partially on the results of laboratory economic experiments. The theory suggests that institutions that are not firmly market-based will inevitably contain time bombs. These will be in the form of opportunities for political entrepreneurs to push for legislative or regulatory "reform", changes that will "level the playing field" or otherwise cure perceived "unfairnesses". The U.S. telecommunications industry, including its regulators, was just such an institution. It had evolved over many decades as a result of deal making and vote trading, justified by appeals to level playing fields and like fairness arguments. However, TA96 is itself the result of such fairness-based deal making and vote trading. It is a typical "Christmas tree" bill containing provisions to buy off disparate interest group ands get them to join the coalition that finally got the bill passed.

The conflicting free market and social agendas contained in TA96 and in the FCC's actions to implement it plant the seeds for the cycle to start again. For, from an economic efficiency, market-targeted viewpoint, TA96 and FCC actions violate at least seven principles of good policy design. The paper spells these violations out in some detail. Finally, it lists some of the main time bombs that TA96 and the FCC have planted. All of the listed time bombs are potentially significant. However, probably the most significant time bomb stems from the combination of TA96's "affordability" and "geographic leveling" clauses of Section 254(b). Taken literally, these clauses imply nationwide price controls on *all* telecommunications *and* information services, basic *and* advanced. Thus far, the FCC has not faced implementation of these clauses, instead seeming to defer them to no later than January 1, 2001.

2. Theory

TA96 sketches massive mandates for institutional change in the U.S. telecom industry, with details to be filled in by the FCC and State regulatory commissions. In addition, the courts will undoubtedly play a significant role in how TA96 is implemented and how the institutional changes play out. Predictions of this evolution and comparisons with actual events that occur amount to the testing of some theory of institutional change.

Unfortunately, a robust, universally accepted theory of institutional change doesn't exist. We have North's [2] assertion that "fundamental changes in relative prices are the most important source of that [institutional] change." But his assertion doesn't carry us very far. Even if we are able to identify changes in relative prices, we still have the formidable task of determining the mechanisms by which economic agents mobilize relative price changes to effect institutional changes. In my view, such attempts have been only partially successful.

For example, in 1971, George Stigler published his famous economic theory of regulation [3], the main work cited by the Nobel committee in awarding him the Nobel prize. This was followed in the 1970s by several theories of regulation: Peltzman's elaboration of Stigler [4], Goldberg's implicit contract model [5], Fiorina and Noll's facilitator model [6], and the strategic game model of Owen and Braeutigam [7]. By and large, these models focused on explaining why regulation came into being and generally predicted an ever increasing amount of it. None predicted the massive wave of *deregulation* in transportation, telecommunication, financial markets, and agriculture begun under President Carter and continued under President Reagan.

Thus, our first task should be to develop a theory of institutional change suitable for the U.S. telecom industry. This is a big job that's beyond the scope of this paper. However, we observe that legislation is typically the result of vote trading and deal making. As I have argued elsewhere [8], the public debate leading to the legislation usually revolves around fairness issues – fair prices, fair regulation, "level playing fields" and the like. Economic efficiency is not on center stage. If it comes up at all, it is in the guise of a fairness item. The outcome is rarely what a benevolent despot economist seeking economic efficiency would decree.

In fact, this outcome is what the experimental economics literature would predict. In laboratory experiments involving a large number of subjects, interacting in impersonal, non-face-to-face situations (e.g. subjects situated in carrels and in front of computer terminals), and in a market-like setting, fairness issues almost *never* arise. However, in experiments with a small number of subjects in personal, face-to-face interactions, and a non-market-like setting, fairness issues almost *always* arise (see Isaac et al. [9] for a further discussion of this result).

Thus, if we wish to escape political strategizing based on fairness arguments, the experimental literature suggests that we must have economic entities that are firmly market-based. Needless to say, legislators interacting on committees to frame legislation are not acting in market-based institutions. Instead, they are acting within institutions that contain exactly the kind of small number of interacters and face-to-face situations where, in laboratory experiments, fairness almost always arises.

This leads to an hypothesis: *if legislation does not sufficiently deregulate and marketize, it will plant fairness time bombs – future opportunities for political entrepreneurs to exploit the propensity of the legislative process to seriously consider cries of "unfair" and demands for reform and counter-legislation.*

The more the legislation deviates for the benevolent despot economist's ideal, the more likely it will lead to waste and economic distortions, and the greater will be the opportunities for political entrepreneurs to mount fairness arguments.

This suggests the beginnings of a theory, namely: legislation means using the police power of the state to effect ends that some group desires. Of course, someone has to bear the costs of catering to the group. The classic case is what James Q. Wilson calls "client politics" [10]; the legislation's benefits are concentrated on the group, but its costs are diffused throughout society (the sugar and peanut subsidies are examples).

Eventually, political entrepreneurs may bring to the public's attention that, although the legislation's per capita costs may be minuscule, added up over the entire population they represent huge sums of money that are simply being wasted, sums that could be put to much better alternative uses. This phenomenon is what Wilson calls "entrepreneurial politics" – the outcries in the media against the sugar and peanut subsidies are again cases in point. New legislation is proposed and the cycle starts again.

Of course, the world of legislation is much more complicated than this sketch, and, for that matter, much the same story can be told of the evolution of institutions through court and regulatory actions, as well as through legislation. For example, we also have what Wilson calls "interest group politics", where coalitions of more or less equal strength battle it out in the public arena. The outcome is uncertain, and can result in no legislation (the Clinton health-care initiative), or "Christmas tree" legislation containing give-aways to many disparate interests to buy their participation in the grand coalition that gets a bill passed. *TA96 is a classic case of Christmas tree legislation.*

A proper development of the theory suggested takes us into the depths of both public choice and political theory, as well as the history of U.S. telecommunications. In this paper, I don't try to develop the theory in more depth. Instead, I assume that it is true and use it to predict probable future developments.

More specifically, I state seven normative principles of economic efficiency[2] that the benevolent despot economist would follow in making policy. Violation of these principles implies that there will be waste and economic distortions. As a consequence, opportunities are seeded for political entrepreneurs to flourish at a later date by bringing the waste and distortions to the public's attention; the time bombs are planted.

The seven principles are:

1. Policy should have a solid factual basis, and should not be based on mere good intentions.
2. Policy should be incentive-compatible. It should motivate economic agents who are following self-interest to attain the policy's goals.
3. Subsidies should be granted only to cure market failures.
4. Subsidies should be paid directly to beneficiaries, not indirectly through intermediaries.
5. Subsidies should be paid for by taxes that are as broadly based as possible.
6. Policy should distort the workings of the market as little as possible. It should avoid managed competition.
7. New policy should contain mechanisms that make its implementers accountable.

Following a brief summary of TA96, I point out that it fails to follow these benevolent despot economist's principles. According to the above simple theory, we can expect that TA96 contains time bombs. And indeed it does, as I discuss in Section 4.

---

[2] The term "economic efficiency" is used in many senses. For example, Owen and Braeutigam [7] argue that people value stability in their lives and are willing to accept a considerable amount of regulation and government market intervention in order to avoid risk and uncertainty. This means that the standard notion of economic efficiency must be expanded to incorporate the risk-regulation tradeoff. And, of course, public choice theorists argue that gains through vote trading are a normal part of the legislative process and should be incorporated into any concept of economic efficiency. However, in this paper, I use the term "economic efficiency" in the narrow sense of exploiting all gains through trade but ignoring risk and uncertainty and political deal making.

3. Summary of TA96 and FCC Actions

The reader can get a fuller, but still brief summary of TA96 from either [1] or [11]. Moreover, on May 7, 1997, the FCC issued the second and third in its trilogy of major orders to implement TA96, and summaries of these orders are available [12, 13, 14]. The FCC has not completely met the incredibly short deadlines that TA96 mandates for it to complete its work. For example, the details of the funding for rural subsidies have been left to the future. At the same time, the FCC's trilogy of orders on interconnection, access pricing, and universal service clearly spell out its basic philosophy in implementing TA96. Here, let me sketch the spirit of the combined effects of TA96 and the FCC's trilogy.

TA96's free market agenda is to create competition in all aspects of telecommunications. TA96's most notable philosophical leap is that local exchange plant can be opened to meaningful competition by requiring the present owners of local exchange plant to lease their facilities to rivals. It pays considerable attention to the details of how the leasing is to be done. However, the complete spelling out of the details is the job of the FCC and State commissions.

TA96's social agenda tries to facilitate access to the "infobahn" in several ways. It perpetuates the long-standing rural subsidy that guarantees that telecommunications rates for rural subscribers be comparable to those for urban subscribers. It goes out of its way to require such comparable rates for rural health care providers. In addition, it mandates that K-12 schools and libraries be subsidized by being given a discount on telecommunications services. Finally, TA96 sets up a Federal-State Joint Board on Universal Service to recommend to the FCC how universal service shall be supported, including universal service for advanced telecommunications services. The Joint Board delivered its report on schedule at the beginning of November 1996.

A notable feature of TA96 is that it does almost nothing to curb the regulators' power to control prices. In several places, TA96 repeats the standard regulatory mantra that prices should be "just, reasonable, and nondiscriminatory". The only specific mandate that traditional rate of return regulation is to be abandoned occurs in Section 252(d)(1), which states that determinations of a State commission of just and reasonable rates for interconnection of network elements shall be "based on cost (*determined without reference to rate-of-return or other rate-based proceedings*)" [my emphasis].

Indeed, TA96 goes beyond the standard mantra to introduce a whole new dimension to the "fair" setting of prices. Section 254(b)(1) states that "Quality services should be available at just, reasonable, and *affordable* rates" [my emphasis]. In its March 8, 1996 Notice of Proposed Rule Making, "Federal-State Joint Board on Universal Service" the FCC remarks [15]:

> While the Commission has often determined "just and reasonable" rates, we have not generally grappled with the notion of "affordable" in the context of universal service.

"Affordable" potentially leads to a whole new set of efficiency issues discussed under Principle P4 in the next section.

Briefly put, an industry regulator with goals to meet ideally wants three control instruments: control of prices, control of quality of outputs, and control of entry into the industry. TA96 has given telecom regulators only one of these instruments: control of prices. In fact, TA96 explicitly instructs regulators to abandon control of entry as quickly as possible and is largely mute on the control of quality. This is akin to a 747 pilot's being given control of the 747's rudder but not of its tail elevators or wing flaps. Moreover, using his limited control means, the pilot is to gradually turn control over to automatic pilot (the market). Needless to say, the regulators have a tough job.

4. Economic Efficiency Evaluation of TA96

With these preliminaries out of the way, I evaluate TA96 in terms of the seven benevolent economist despot's principles listed above.

P1. Policy should be based on solid facts, not mere good intentions.

TA96 is in fact a far-reaching experiment that is not solidly grounded on previous experience or data. Thus, the FCC and State commissions that have to implement TA96 are partially flying blind.

The FCC's NEWSReport [12] that announced and summarized the FCC Local Competition Order states, "Today's decision reflects the experiences of those states that have already endeavored to promote local competition." True, according to Huber et al. [11, p. 6], at the time of TA96's passage in February 1996, 19 states had already introduced local competition. The furthest along were probably Ameritech in Illinois and Rochester Telephone in New York State, and the FCC obviously has relied heavily on their experience. However, in both cases, state introduction of local competition was recent and neither state had gained extensive experience in ironing out details. Other states were further behind.

TA96's social agenda is perhaps based on even less data. Indeed, a generation of experiments with computerized instruction in grades K-12 gives scant hope for big educational breakthroughs. For example, the late 1960s were a period of a great optimism that computers could work educational wonders. The Plato project at the University of Illinois using Control Data Corporation mainframes and touch screen terminals was hailed as a breakthrough in interactive computerized instruction. Further, the Plato terminals were (and still are) linked nationwide through satellite and other transmission media. In addition, a number of publishers formed joint ventures with computer companies to wreak the expected wonders. For example, General Electric joined forces with Time-Life Books and the Silver Burdett textbook publishing company to form the General Learning Corporation. The efforts by-and-large have had minimal success, at least as measured by the rhetoric of the time, rhetoric that is similar to that of today.

Oettinger's 1969 book [16] analyzed this era. In spite of massive technological improvements since 1969 his basic point is still on target and shows us the difficulty of the task ahead: if computers and telecommunications are to have a major impact, all the actors in the educational drama – teachers, teacher unions, parents, PTAs, professors of education, school boards, etc. – must radically change their ways in order to cost-effectively fit modern telecommunications into their daily lives. What will be needed is not an appliqué of computer and telecom technology on present teaching methods. Rather, the entire educational system will have to change.

However, any radical change will generate winners and losers, and we can expect the losers to try to exert veto power with vigor and ingenuity.

P2. Policy should be designed so that economic agents following self-interest will be motivated to achieve the policy's goals.

This is what economists call the principle of incentive compatibility. Ideally, any regulatory system should fully follow it. TA96 does not. Instead it raises a number of questions about the adequacy of the incentives it sets up, namely:

> *Do the major industry players have the proper incentives to promote TA96's free market agenda?*

At the time of this writing, the RBOCs have yet to try to expand into either in-region or out-of-region long-distance service and ATT, MCI, and Sprint have done little to get into the local exchange market. Duane Ackerman, president and CEO of BellSouth, argues that the major long-distance carriers have little incentive to enter local residential markets because if they do, TA96 calls for the RBOCs to be allowed into *their* markets. In a recent Wall Street Journal article, he writes [17]:

> If they [ATT, MCI, and Sprint] move quickly to offer local residential service, regulators would open their long-distance markets to BellSouth and its counterparts. It profits AT&T to stay out of residential service, to the tune of $1.6 billion a year in the nine states BellSouth serves. That's the share of the long-distance market that economic analyses suggest BellSouth would take if it were allowed to enter the long-distance market.

> *Do the FCC and State commissions have the proper incentives to promote TA96's free market agenda?*

After the passage of TA96, a coalition of the AARP, the CFA, and the Consumers Union promptly raised the specter of deregulation of local services leading to what the coalition considers to be exorbitant local residential rate increases [18]. Its strategy of complaining swiftly seems to have worked; for example, in its access charge order, the FCC states [19, Section 38]:

> For primary residential and single-line business lines, however, we decline to implement this goal [realigning rates more closely with costs] by increasing the SLC ceiling above its existing $3.50 level as urged by many companies, including price-cap LECs and IXCs. We do not wish to see increases in the price of basic dial tone charged by local exchange carriers to their end users for fear that such increases might cause some customers to discontinue service, a result that would be contrary to our mandate to ensure universal service.

The FCC's action gives consumer groups further incentives to exert political pressure at the state and federal levels to keep local residential rates low. Of course, mandated local rates that are below cost discourage new entry into the local market and defeat TA96's free market agenda.

But dealing with the AARP-CFA-CU coalition and consumer groups will be only one aspect of regulatory management of incentives. As already stressed, to get into the in-region, interLATA market, an RBOC must satisfy a checklist of actions that ensure local competition. This gives State commissions leverage to get the RBOCs to invest in infrastructure. For, if an RBOC underinvests, the State commission can drag its feet in certifying that the RBOC has completed the checklist. It remains to be seen whether or not this or other leverage will be sufficient, and whether or not the State commissions will have the incentives to effectively manage this leverage.

Finally, TA96's mandate for "affordable" services is a great incentive for consumer groups to gear up for battle, as the above discussion of the FCC's decision to keep the residential SLC charge at $3.50 indicates. For example, in its Universal Service Order [20, Section 22], the FCC defined services to be supported by the universal service support mechanisms as: voice grade access to the switched network and calls on it; touch tone dialing; single-party service; enhanced emergency service (911); access to operator services, interexchange services, and directory assistance; and toll-limitation services for qualifying low-income consumers. This is a plain vanilla list. As we enter the world of fancier and fancier telecommunications services, consumer groups will have greater and greater incentives to invoke the "affordability" provisions of TA96 to campaign to expand it.

Do local exchange incumbents have the proper incentives to continue to develop their in-region plant?

All sorts of hurdles and restrictions are placed on the RBOCs for in-region operations. Fewer are placed on out-of region operations. Thus, U.S. West may be given an incentive to pour investment into Atlanta where it will probably get minimum regulatory flak, while BellSouth may have a similar incentive to invest in Denver. For an RBOC to invest in its own region is comparatively more expensive because of the regulatory hassle the RBOC must overcome.

Furthermore, in a fast-growing area of the country, new infrastructure must be planned for many months in advance. Forecasting what services would be demanded and in what quantities was a tough enough problem when the LECs had a monopoly. Now, incumbent LECs must forecast not only demand but how much of it they will provision and how much rivals will provision.

Will TA96's incentive structure preserve the high-quality, reliable telecommunications service that we are all accustomed to and take for granted?

The telecommunications network is a complex, fragile organism. A few equipment breakdowns or sudden demand increases can lead to instabilities. This is what happened in New York City in the early seventies. Consumers suddenly couldn't get dial tone, which made them only uncradle their handsets more often to seek it. If they got dial tone, they would get a busy signal when they dialed a number. This made them redial and redial. New York City's phone system spiraled out of control. It took a massive infusion of capital and personnel and months to cure the problems. It could happen again.

P3. Subsidies should be granted only to cure genuine market failures.

We now have a large literature on network externalities. Simply put, it tells us that the initial stages of an emerging telecommunications service are like the initial stages of an infectious disease epidemic, but an epidemic that has a positive rather than a negative effect. Too high prices for the emerging service may act like a vaccine – they may stop the positive epidemic in its tracks. Subsidies may be in order to keep prices low to promote the epidemic's running its course [21].

However, the epidemic eventually does run its course; the emerging telecom services mature and stabilize. Each member of the network will have arranged, one way or another, to pay for value received. For these reasons, the network externalities literature also tells us that the subsidy should exist only to a point [22]. When the emerging network service is mature and all externalities have been "internalized", the subsidy should be discontinued.

The danger is that any government subsidy to combat a possible network externality market failure will hang on. People who receive the subsidy will take it for granted. They will see it as an entitlement and will fight its removal. Thus, a subsidy to develop rural telephone service originally had an economic efficiency basis. But nowadays, rural telephone service is fully mature and there is no economic efficiency argument for preserving the rural subsidy. However, TA96 does nothing to curtail it. If anything, it makes the subsidy more permanent. We have the following explicit language in Sec. 254(g):

> ...the Commission shall adopt rules to require that the rates charged by providers of interexchange telecommunications services to subscribers in rural and high cost areas shall be no higher than the rates charged by each such provider to its subscribers in urban areas.

Likewise, the FCC seems to have bought the network externality argument regarding the Internet. It has not been receptive to telcos' demands that Internet providers be charged access fees and otherwise bear costs that they are causing. Further, recently, 160,000 Internet users bombarded the FCC with messages to protest phone rate increases that telcos have proposed [23]. Thus, a constituency of Internet users seems already to feel that cheap, flat rate Internet pricing is an entitlement.

P4. Subsidies should flow directly to beneficiaries, not indirectly through intermediaries that can siphon them off or distort the subsidies' intent.

Subsidies should not be indirectly conferred by being given to someone who will transfer them to someone who will transfer them to someone who will give them to the beneficiary. This allows the intermediate someones to bleed off a percentage of the subsidy.

Likewise, subsidies should not be indirectly conferred by distortion of prices. For example, in many countries price controls have subsidized the prices of "necessities"; the price of bread is controlled so that the "poor can afford it". But, because bread is artificially very cheap, farmers end up feeding bread rather than chicken feed to their chickens. Bread then becomes in short supply and more of it is produced than necessary.

Congress seems to be aware of the danger of misdirected subsidies. As noted in the FCC Universal Service Report and Order, the Congressional Conference Committee report makes clear that Congress intended that subsidies under TA96 "should be explicit" [20, Section 9], and the FCC's trilogy of orders has taken this intent seriously. However, making subsidies explicit is not the same as directing them to final beneficiaries, and TA96 leaves much room for misdirected subsidies.

For example, because of long local loops and old switching equipment, rural exchanges may be costlier than those in urban areas. Exchanges whose cost is greater than 115% of the national average are presently deemed to be "high-cost" and receive an explicit subsidy. The subsidy is funded by a tax levied on telecommunications providers.

Thus, subsidies are not directly sent to rural subscribers based on the subscribers' financial need. Instead, subsidies are sent to rural phone companies. These subsidies allow the rural companies to charge uniform rates, the same rates to poor rural subscribers as to millionaire gentlemen farmers and ranchers.

From the standpoint of economic efficiency, similar dangers exist in TA96's mandate that rates be "affordable". Over the last roughly 100 years, public utility regulatory and court decisions have given "just and reasonable" the connotation that rates should be high enough to properly reward a utility's stockholders for the risk they have assumed, but not so high as to award them exorbitant profits.

"Affordable" is something else, and raises a number of questions. Although terms like "affordable" and "necessities" were commonly used by economists a century ago, they no longer fit into the economist's lexicon. You will search in vain for "affordable" in a basic economics text or in *The MIT Dictionary of Modern Economics* [24]. The point, of course, is that nowadays economists assume that a person will maximize utility, with utility being limited by the amount of money the person has to spend. Except when the price of something exceeds your wealth or income, you can buy it by giving up other things.

"Affordable" implies that something is priced at or below some acceptable level. The acceptable level must be below the market level, otherwise there would be no need to worry about "affordability." Then, to whom should "affordable" apply? To classes of customers (e.g. libraries, K-12, health care providers), worthy persons, or to everyone? How should prices be held to "affordable" levels? By direct subsidies to worthy classes or persons or by price controls on services?

If one takes a broad view of "affordable," that it should apply to everyone and not through means-test-based subsidies, it means price controls. At the time of this writing, the Universal Joint Board and the FCC have refused to go this far. After considering a plethora of candidate definitions of "affordable" submitted by numerous commentators, the Joint Board recommended that "the states exercise primary responsibility, consistent with the standard enumerated above, for determining the affordability of rates" [25]. The FCC has almost entirely accepted the Joint Board's recommendations. Further, in its Universal Service Order, it states [20, Section 2]:

> Consistent with ... explicit statutory principles, ... we must maintain rates for basic residential service at affordable levels. We believe that rates for this service are generally at affordable levels today.

Thus, at present, the affordability issue seems to have been laid to rest. But as already remarked, and discussed further below, it remains to be seen how long it will stay there.

P5. Subsidies should be paid for by taxes that are as broadly based as possible.

Generally speaking, the more goods and services or individuals that are taxed to raise a subsidy, the less prices and incomes will be distorted from their free market values and the more prices will be able to do their job of steering resources to their highest-valued uses. But taxation policy is rarely based entirely on this principle. Rather, a subsidy is typically raised by taxing a specific group on the basis of some fairness principle – subsidize necessities by taxing luxuries, subsidize health care by taxing tobacco, etc. U.S. telecommunications is no exception. For example, from the beginning of telephony, business users have had to pay roughly twice what residential users pay for essentially the same service.

TA96 and the FCC's implementation of it continue this tradition. Instead of raising telecommunications subsidies broadly by taxing all goods and services in the economy, TA96 raises it by taxing only "telecommunications services". Moreover, the FCC has interpreted "telecommunications service" narrowly, exempting Internet service providers.

Further, in its access charge order, the FCC has raised the subscriber line charges on second residential lines and multiple business lines in order to raise a $2.25B subsidy to fund K-12 schools and libraries.

P6. Policy should distort the workings of the market as little as possible. It should avoid managed competition.

At present writing, the amount of economic distortions and managed competition that TA96 will create is not clear. Partly this is because pending important court cases have still to be decided and because State commissions still are struggling with how to deal with TA96 and the FCC's orders. However, perhaps most important is that the FCC has put off dealing with the affordability and geographic leveling provisions of TA96.

Specifically, in interpreting TA96, the FCC has drawn a distinction between its "shall" and "should" clauses. Referring to its task of eliminating implicit subsidies in favor of explicit ones, the FCC states in its access order (19, Section 9),

> Although Congress said in the Act that "support *should* be explicit" (emphasis added), it did not provide that support *shall* be explicit...Congress qualified its intention that "support mechanisms should be explicit, rather than implicit" with the phrase "[t]o the extent possible."

Thus, the FCC seems to be interpreting "shall" as meaning "is to be done immediately", and "should" as meaning "is to be done to the extent possible". This puts "should" clauses into a limbo of maybe partial implementation now, fuller implementation at some indeterminate future date.

It is significant that both the affordability and geographic leveling clauses are "should" clauses. Section 254(b)(1) (affordability) states:

> Quality services should be available at just, reasonable, and affordable rates.

Section 254(b)(3) (geographic leveling) states:

> Consumers in all regions of the Nation, including low-income consumers and those in rural, insular, and high cost areas, should have access to telecommunications and information services, including interexchange and advanced telecommunications services, that are reasonably comparable to those services provided in urban areas and that are available at rates that are reasonably comparable to rates charged for similar services in urban areas.

Taken literally, 254(b)(3) means that *any given* rural area should have rates that are "reasonably comparable" to *any given* urban area. This implies uniform nationwide pricing or geographic leveling. Further, taken literally, Sections 254(b)(1) and 254(b)(3) are a mandate for nationwide price controls *on all services, advanced as well as basic, telecom or information*, and, thus, managed competition with a vengeance.

Admittedly, this is an extreme interpretation of TA96, but nevertheless a possible one. Together, these clauses of TA96 plant what I expect will be the biggest time bomb of all. To date the FCC has dodged the implications. In its Universal Service Report and Order, it seems to have put off considering the combined implications of Sections 254(b)(1) and 254(b)(3) until the next century by stating, "as recommended by the Joint Board, we will convene a Federal-State Joint Board to review the definition of universal service on or before January 1, 2001" [20, Section 22].

P7. New policy should contain mechanisms that make its implementers accountable and that prevent them from creating ever-growing bureaucratic empires.

TA96 places a number of time limits on the economic actors that it affects. Furthermore, TA96 requires that every even-numbered year, beginning in 1998, the FCC is to hold sunset hearings. These are to review all regulations and "shall determine whether any such regulation is no longer necessary in the public interest as the result of meaningful competition between providers of such service" (Sec. 402).

To my knowledge, this is a novel sunset provision. At a minimum, it should allow interest groups favoring the removal of a particular regulation a hearing in which to plead their case. It remains to be seen how effective this provision will be in having regulation "wither away".

5. The Future – TA96's Time Bombs

TA96 has been much publicized as introducing a new, competitive era in telecommunications. Should its deregulatory promise not come to pass because of its built-in economic inefficiencies, we can expect the time bombs to start to go off. Indeed, at present writing, this has already happened; most LECs and many State commissions are challenging in the federal courts the FCC's self-assumed power under TA96 to micromanage telecommunications.

Even if this hurdle is quickly cleared, we are still left with other time bombs. As already stated, potentially the biggest time bomb is the combination of the "affordability" and geographic leveling provisions of TA96 (Sections 254(b)(1) and (3)). So far, the FCC and the Universal Service Joint Board have wisely kicked "affordability" back to the state level and declared that the rates it presently controls are "affordable". This means that a state which imposes price controls to maintain "affordability" is liable to lose out, in the interstate competition for telecom, to states that do not. Interstate competition thus has a chance to discipline politicians capitalizing on "affordability". However, it remains to be seen whether or not "affordability" will remain contained. My guess is that, long before January 1, 2001, consumer group political entrepreneurs will be working hard to make "affordability" and geographic leveling of prices, quality, and range of telecommunications and information service offerings into a national issue.

However, there are other time bombs as well. Ironically, when they go off, the cry may be for either less or more deregulation. Here are some of the obvious ones:

Waste – The media love to expose governmental waste; "the fleecing of America" is a standard feature of NBC's nightly news. TA96's social agenda could easily lead to such media features. Indeed, skeptical articles are already appearing in the press questioning the benefits of computerized instruction and educational use of the Internet [26,27]. If the educational benefits of the Internet should turn out to be marginal, the mandated subsidies for the rural, K-12, and libraries markets may backfire. We may have outcries that billions of dollars of taxpayers' money have been wasted on an ill-conceived and ill-thought-out project. As one teacher has written, "[t]he money would be better spent on books, equipment, teachers' aides, and tutors"[28]. The result might very well be reform legislation that eliminates the TA96-mandated K-12/libraries subsidy, or, more likely, legislation to redirect the subsidy to less wasteful uses.

Rural subsidies and the Internet subsidy (should it continue) are other time bombs.

Incentive deficiencies – These may show up in several ways:

Service disasters. A few service disasters like that in New York City in the early 1970s may expose inadequate incentives for LECs to continue to develop and maintain infrastructure. They will in turn give incentives for political entrepreneurs to cry out for more regulatory control.

Technological Change. Video-on-demand and 500 channel set top boxes have developed slower than expected; Netscape and other web browsers have pushed the Internet into broadband applications faster than expected. Technological change developing slower than expected may trigger a demand for a slowdown in deregulation; unanticipated technological developments might trigger a demand for its acceleration.

Foreign benchmarks. The movement to restructure the telecommunications industry is worldwide. New Zealand is thus far a benchmark; it seems to have successfully eliminated regulatory bodies like the FCC or State commissions in favor of relying almost completely on its antitrust laws [29]. Demonstration projects like New Zealand will put pressure on legislators and regulators to push TA96's free market agenda.

Managed competition. The FCC has angered congressmen and senators of both parties by aggressively converting its implementation duties under TA96 into mandates to micro-manage the U.S. telecom industry. House Telecom Subcommittee Chairman Billy Tauzin (R-La) has suggested eliminating the FCC and transferring whatever few essential duties it has to the Federal Trade Commission [30]. Foreign benchmarks like New Zealand can only feed such views, and may contribute to a significant demand for reform of TA96.

Inadequate accountability. TA96's mandate for the FCC to hold even-numbered-year hearings is a two-edged sword. It gives both those who wish to eliminate a regulation and those who wish to preserve it a soap box. If this results in a standoff, with the FCC doing little to reduce regulations, it would feed into the disenchantment with managed competition and a backlash.

6. Summary and Conclusions

TA96 is a complex "Christmas tree" bill, born of decades of negotiations among the major stakeholders in the U.S. telecommunications industry. It has been hailed as a major deregulatory initiative. In fact, TA96 is not very deregulatory. Instead, it marks the replacement of one regulatory regime with another. TA96's social agenda guarantees that the new regime will perpetuate significant government intervention in the telecommunications and information services markets.

However, TA96 may prove to be a bonanza for policy analysts. Its mandates for government intervention mean that it will spawn losers as well as winners, and will generate controversy. Most significantly, the speed of telecommunications technological change means that time scales will be compressed, and policy analysts will much more quickly be able to see whether or not their analyses are on target.

In this paper, I have tried my hand at analyzing TA96 and the FCC's actions to date, and to make predictions regarding time bombs that the Congress and the FCC have planted. I have based my analysis on a simple theory that partially follows from the results of laboratory experiments: legislation that creates institutions that are not firmly market-based invites future political entrepreneurs to mount fairness fights to change it. I have flushed out the theory by observing that legislation that does not conform to a benevolent despot economist's ideal will contain the potential for waste and economic distortions, and the greater the lack of conformity, the greater the possibilities for political entrepreneurs to mount fairness attacks.

I have then described seven different ways in which TA96 differs from the benevolent despot economist's ideal. This analysis has led me to identify time bombs that TA96 has planted in the way of opportunities for political entrepreneurs to exploit the waste and economic distortions that are inherent in TA96. Finally, I have pointed out that the combination of the "affordability" and geographic leveling clauses of TA96 (Sections 254(b)(1) and (3)) creates potentially the biggest time bomb of all.

References

[1] E.E. Zajac, An Analysis of U.S. and Arizona Telecommunications Policy. Arizona Issue Analysis #143, Goldwater Institute, Phoenix, Arizona, October, 1996.

[2] D.C. North, Institutions, Institutional Change and Economic Performance. Cambridge University Press, p. 84.

[3] G. Stigler, The Theory of Economic Regulation, *Bell Journal of Economics* **2**, (1971) 3-21.

[4] Peltzman, Toward a More General Theory of Regulation, *Journal of Law and Economics* **XIX** (1976) 211-40.

[5] V. P. Goldberg, Regulation and Administered Contracts, *Bell Journal of Economics* **7**, (1976) 426-48.

[6] M.P. Fiorina and R.G. Noll, Voters, Bureaucrats, and Legislators, *Journal of Public Economics* **9** (1978) 239-54.

[7] B.M. Owen and R. Braeutigam, The Regulatory Game: Strategic Use of the Administrative Process. Ballinger, 1978.

[8] E.E. Zajac, Political Economy of Fairness. MIT Press, 1995.

[9] R.M. Isaac, D. Mathieu, and E.E. Zajac, Institutional Framing and Perceptions of Fairness, *Constitutional Political Economy* **2** (1991) 329-370.

[10] J.Q. Wilson, editor, The Politics of Regulation. Basic Books, 1980, pp. 182-184.

[11] P.W. Huber, M.K. Kellogg, and J. Thorne, The Telecommunications Act of 1996. Special Report. Little Brown, 1996.

[12] FCC, Commission Adopts Rules to Implement Local Competition Provisions of the Telecommunications Act of 1996, NEWSReport No. DC 96-75, Aug. 1 (1996).

[13] FCC, Commission Reforms Interstate Access charge system (CC Docket No. 94-1)(CC Docket Nos. 96-262; 94-1; 91-213; 95-72), News Report No. CC 97-23, May 7 (1997).

[14] FCC, Commission Implements Telecom Act's Universal Service Provisions; Adopts Plan to Ensure Access to Affordable Telecommunications Services for All Americans (CC Docket No 96-45), Report No. CC 97-24, May 7 (1997).

[15] FCC, In the Matter of the Federal-State Joint Board on Universal Service, Notice of Proposed Rule Making and Order Establishing Joint Board, FCC 96-93, CC Docket No. 96-45, Mar. 8 (1996) Section 4, p. 5.

[16] A.G. Oettinger, Run Computer Run: The Mythology of Educational Innovation. Harvard University Press, 1969.

[17] D. Ackerman, Why is AT&T Afraid to Compete? *Wall Street Journal*, July 3, A10 (1997).

[18] AARP-CFA-CU, Reply Comments, In the Matter of Federal-State Board on Universal Service before the FCC, Docket No. 96-45, May 7 (1996).

[19] FCC, First Report & Order on Access Charge Reform, CC Docket Nos. 96-262; 94-1; 91-213; 95-72, May 7 (1997).

[20] FCC, Report & Order in the Matter of Federal-State Joint Board on Universal Service, CC Docket No. 96-45, May 7 (1997).

[21] J.H. Rohlfs, A Theory of Interdependent Demand for a Communications Service, *The Bell Journal of Economics and Management Science* **5** (1974) 16-37.

[22] S.C. Littlechild, Two-part Tariffs and Consumption Externalities, *Bell Journal of Economics* **6** (1975) 661-670.

[23] R.G. McLeod, FCC Bombarded With E-Mail on Phone Rate Proposal, *San Francisco Chronicle*, Feb. 14, A5 (1997).

[24] D.W. Pearce, The MIT Dictionary of Modern Economics, Fourth edition. MIT Press, 1992.

[25] FCC, In the Matter of the Federal-State Joint Board on Universal Service, Recommended Decision, Section V. Affordability, FCC 96J-3, CC Docket No. 96-45, Nov. 8 (1996).

[26] T. McNichol, Computers in Class: A Waste of $50 Billion? *Tucson Citizen, USA Weekend*, Feb. 14-16 (1997) 10-11.
[27] J. Kronholz, Getting Wired, U.S. Schools Face Steep Learning Curve, *Wall Street Journal*, March 19 (1997).
[28] S. Cooke, Should Classrooms be Wired? *Tucson Citizen, USA Weekend*, March 28-30 (1997) 11.
[29] New Zealand, Ministry of Commerce, The Treasury: Regulation of Access to Vertically-Integrated Natural Monopolies, A Discussion Paper, 1995.
[30] Washington Telecom Newswire: Tauzin to Focus on FCC Role, PCS Siting, Cellular Privacy. Warren Publishing, Feb. 25, 1997.

Telecommunications Transformation: Technology, Strategy and Policy
E. Bohlin and S.L. Levin (Eds.)
IOS Press, 1998

CHAPTER 21

Economic Analysis of Federal Communications Commission Decisions for Implementing Local Telephone Competition

John B. Meisel
Department of Economics, Southern Illinois University at Edwardsville, Edwardsville, Illinois 62026

Abstract. This paper presents an analysis of the rules developed by the Federal Communications Commission (FCC) to implement the local competition provisions of the Telecommunications Act of 1996. Positions of primary stakeholders, such as incumbent local exchange carriers and interexchange carriers, on crucial issues are analyzed from an economic efficiency perspective. Key issues analyzed include the division of power between federal and state regulators, the scope of unbundling and the ability to rebundle network elements, and the pricing methodologies for interconnection, unbundled network elements, collocation, and resale. This paper concludes that there is an appropriate role for the FCC to develop federal guidelines to govern state-arbitrated interconnection agreements, that an important FCC pricing methodology is flawed and can be viewed as anticompetitive, that the criteria for unbundling should include an essential facilities analysis, and that some of the complaints about FCC rules represent dissatisfaction with the federal legislation.

1. Introduction

In February, 1996, the Telecommunications Act of 1996 [14, the Act] became law. In part, the Act sets out an ambitious new direction for telecommunications regulation. Broadly, the Act opens up all telecommunications markets to competition and, specifically, contains provisions that open the local exchange and exchange access markets to competitive entry. The Act directs the Federal Communications Commission (FCC) and the state commissions to remove statutory, regulatory, economic, and operational barriers to local competition.

To implement the eight pages of federal legislation addressing interconnection (Section 251) and the procedures for negotiation, arbitration, and approval of agreements (Section 252), the FCC adopted a 650-page First Report and Order [8, the Order] six months after enactment of the Act as required by the legislation. Although many of the FCC's key decisions angered two groups of stakeholders, incumbent local exchange companies (ILECs) and 28 state commissions, including the National Association of Regulatory Utility Commissioners, these same decisions pleased other stakeholders, including interexchange carriers (IXCs), competitive access providers (CAPs), federal agencies, and consumer groups. In response to appeals by the ILECs and the state commissions, in October, 1996, the Court of Appeals for the Eighth Circuit suspended key decisions of the Order in *Iowa Utilities Bd. V.*

FCC [9]. In January, 1997, the Court heard oral arguments. On July 18, 1997, the Court overturned several parts of the Order.

The primary purpose of this paper is to analyze several of the disputed FCC decisions and to assess the criticisms of them from an economic perspective. Each FCC decision is analyzed to determine if it is consistent with the goal of promoting efficiency in the economy. The components of economic efficiency are (1) allocative efficiency, producing the correct level of output from society's perspective; (2) technical efficiency, producing the correct level of output at the lowest cost of society's scarce resources, and (3) dynamic efficiency, developing a market structure conducive to innovation.

The paper is organized as follows: Section B provides background material on Congress, the FCC, the challengers, and the status of appeals; Section C analyzes issues involving the division of power, unbundling, pricing methodologies, and other issues; and Section D offers a summary and conclusions.

2. Background

2.1. Congress

A major objective of the Act is to rely on competitive market forces to determine the future performance of telecommunications markets. To accomplish this objective, the Act gives potential entrants to the local exchange and exchange access markets three rights that parallel the different paths of entry into the markets. Corresponding to each right are special duties imposed by the Act on ILECs to facilitate entry.

First, the Act gives entrants a right to interconnection. This enables an entrant to connect its new network to the ILEC network so that the entrant's customers can exchange calls with the ILEC's customers. As a corollary to this right, all local exchange carriers have a duty to establish reciprocal compensation arrangements to terminate each other's traffic.

Second, the Act gives entrants a right to lease unbundled elements of an ILEC's network and to combine such elements as it sees fit. Unbundling the elements of the local exchange reduces the barriers to entry to local markets by enabling entrants to share in the economies of the existing local network. Mandatory unbundling permits an entrant to get inside the ILEC network and to use an ILEC's elements as if they were its own.

The Act dictates that the rates, terms, and conditions for interconnection and unbundled elements should be just, reasonable, and nondiscriminatory. Furthermore, rates should be based on cost and include a reasonable profit. However, the Act does not define any of these terms.

Third, the Act gives entrants a right to purchase an ILEC's retail services at wholesale rates and then resell those services to end users. The magnitude of the discount from the retail rate is determined by measuring the costs (such as marketing, billing, and collection) that will be avoided by the ILEC. Reselling services can be viewed as a less risky entry strategy for it involves no investment in facilities, but it does place the entrant at the mercy of the ILEC.

Congress granted these asymmetric rights to entrants (or what are referred to in the Act as additional obligations of incumbent local exchange carriers) in light of the unequal bargaining power ILECs possess relative to new entrants. Generally, ILECs have little incentive to open up their networks to entrants posed to compete for their customers. One carrot Congress provided to the Regional Bell Operating Companies to encourage them to open up their networks is the possibility of entering in-region interLATA markets as detailed in Section 271 of the Act. To support this observation on unequal bargaining power, Congress also developed elaborate procedures governing agreements reached by compulsory arbitration if voluntary negotiations failed. In such cases, state commissions ensure that these rights are protected.

2.2 FCC

Within six months of passage of the Act, the FCC's job was to "... complete all actions necessary to establish regulations to implement the requirements of this section" [Section 251(d)(1) of the Act]. Section 251 includes provisions concerning interconnection, unbundling, and resale. Most observers agreed that this required the FCC to address highly complex and contentious issues in a relatively short time. During the Commission's deliberations, the chairman of the FCC, Reed Hundt [7], identified three primary objectives:

(1) be pro-competition, not pro-competitor,
(2) begin the transition to a regulatory environment in which a minute is a minute is a minute, and
(3) efficiency.

To accomplish these objectives, the FCC issued its lengthy First Report and Order. Several themes characterize the Order.

First, the FCC argues that the Act created a new regulatory paradigm for the division of power between federal and state regulators. The old paradigm (a dual regulatory system established in the 1934 Act) divides responsibility based on the geography of a call. The new paradigm (a parallel regulatory system established in the 1996 Act) envisions a federal-state partnership that is independent of the jurisdictional nature of a service when addressing inter-carrier transactions and that is based on sharing power. The new paradigm views the FCC as developing general principles and guidelines (in part derived from state experiments with local competition) regarding the components of Section 251 arrangements and the states as expanding, if necessary, on these as well as applying the principles and guidelines to the specific circumstances of the respective states.

Second, the Order continues the process of a transition to competition in local telecommunications markets by converging the efforts of federal and state regulators. Pro-competition public policy efforts began in the late 1980s in interstate exchange access markets. In 1992, the FCC ordered the ILECs to submit tariffs governing interconnection with CAPs for the purpose of providing interstate special access and private line services. In 1993, the FCC ordered interconnection for CAPs for switched transport. In both cases, the ILECs were ordered to unbundle their tariffs so that a CAP would not have to pay for a part of an incumbent's service that it did not use to provide its competitive service. Concurrently, states eliminated statutory barriers to competition, authorized competitors for local intrastate services, and developed rules to facilitate local competition. States such as Washington, Illinois, Maryland, Michigan, and New York have been leaders in these efforts.

Third, the FCC views the ILECs as retaining the incentive and ability to impede competitive entry because of their control of an ubiquitous local network. This leads the FCC to develop rules and regulations that eliminate any real or imagined barrier to entry to these local markets. The zeal to make entry as easy as possible causes the FCC at times to develop economically inefficient rules, such as the federal criteria for determining the elements of the ILEC's network that must be unbundled. Related to this view of the world, the FCC places the burden of proof on the ILECs to address requests by entrants. For example, in developing rules to identify technically feasible points of interconnection within the local network, an entrant's proposed access point is assumed to be technically feasible unless an ILEC can demonstrate otherwise.

Fourth, the Order focuses on network facilities and elements and not the services produced by these inputs. Such a focus eliminates the need to distinguish between what is and what is not a local, state, or federal service.

Fifth, the FCC believes it was Congress' intent to make the transition to competition as rapid as possible. To support this contention, the FCC cites the short time frames contained in the Act (e.g., the FCC had only six months to develop its rules and the states have only 90 days to make an arbitration ruling once it has been requested). This interpretation caused the FCC to develop a critical pricing methodology that focuses on theoretical, static outcomes of a competitive market structure and that ignores the importance of the dynamic processes that competitive forces unleash in real world markets. As discussed later, this focus on outcomes and not process contributes to the adoption of a flawed pricing methodology for establishing prices for interconnection, unbundled network elements, and collocation. Moreover, and, in my opinion, overreaching its responsibility, the FCC develops a set of proxy prices based on its pricing methodologies to facilitate the state arbitration process and, thus, quicken the pace to competition.

2.3 Challengers

Major objections by challengers to the FCC decisions addressed in the paper include the following.

(1) The Order fails to respect the traditional jurisdictional division of authority by allowing federal regulators to usurp power from the states.
(2) The Order fails to respect Congress's intent " ... to accelerate rapidly private sector deployment of advanced telecommunications and information technologies ... " [13]. According to the ILECs, this goal means that Congress intended to promote facilities-based competition in conjunction with the preservation and enhancement of existing networks. However, according to the ILECs, these goals will not be realized because the federal pricing methodologies discourage facilities-based investment as well as continued investment in existing networks since (a) prices for interconnection and unbundled network elements are too low and (b) wholesale prices to resellers are too low.
(3) The Order fails to give proper consideration to statutory and economic factors in determining elements that must be unbundled.
(4) A combination of the Order's pricing and unbundling rules distorts the economic and statutory relationship between entry strategies based on leasing unbundled elements and reselling retail services.

(5) The Order fails to respect the Act's intent to rely on privately negotiated interconnection agreements.
(6) The Order imposes asymmetric duties on ILECs. For example, the Order lacks a reciprocal non-discrimination standard for entrants.

These objections will be discussed below. Some complaints are with the statute itself, reflecting policy decisions made by Congress, not the FCC.

2.4 Appeals, Appeals, Appeals

The process for making telecommunications policy provides alternative places to seek redress to FCC and state commission decisions, and the stakeholders in this process are availing themselves fully of these opportunities. As mentioned above, parts of the Order have been overturned by the Court of Appeals for the Eighth Circuit.

Generally, state arbitration decisions have been consistent with the rules and guidelines ordered by the FCC. [1] The ILECs have been especially frustrated with the states' general adherence to FCC policies and the use of proxies for pricing unbundled network elements and the magnitude of the resale discounts. The ILECs consider the federal pricing rules and proxies to favor non-facilities-based entrants at the expense of facilities-based entrants and incumbents. This frustration has manifested itself in numerous appeals of state arbitration decisions (utilizing Section 252 (e) (6) of the Act) to federal district courts. GTE and US West have led the parade of appeals. Not to be outdone, entrants have filed appeals to state arbitration decisions and have also filed complaints with the FCC.

2.5 Unfinished Business

The three major telecommunications goals of the Act are to (1) open the local exchange and exchange access markets to competition, (2) enhance the level of competition in other previously opened telecommunications markets, and (3) reform and preserve the system of universal service. The FCC addressed these goals in three related proceedings, referred to as the competition trilogy. The trilogy is comprised of Section 251, universal service reform, and access charge reform. There are common issues that cut across these three interrelated proceedings. Recently, the FCC issued orders to reform access charges and universal service. In the FCC's view, the regulatory framework to guide the transition to competition in local markets is complete. However, there are appeals of these orders also. It is evident that the regulatory framework to guide the transition to competition is more accurately described as a work-in-progress.

---

[1] A notable exception to this pattern concerns the price an entrant will pay if it uses only unbundled network elements which are then rebundled into a full local service package (what ILECs pejoratively call "sham unbundling"). The FCC failed to impose a local exchange facilities requirement to be able to purchase unbundled network elements. However, the Georgia and Louisiana commissions ruled that if AT&T provides local service in this manner it must pay the resellers' price rather than the lower prices for unbundled elements. See [4] and [10].

3. Issues

3.1 Division of Power

It makes economic sense for the FCC, a federal agency, to establish national guidelines to promote efficiency that a state public utility commission must follow when determining the rates, terms, and conditions for an arbitrated Section 251 agreement between an ILEC and an entrant to local exchange and exchange access markets. This conclusion is based primarily on the likely existence of important external effects arising from individual states pursuing their own objectives and is independent of legal analysis.

It is wellknown that under most circumstances decentralized decision-making in competitive markets leads to the attainment of allocative and technical efficiency. A significant qualification to this prediction arises, however, when third parties are affected in a substantial way by the actions of the directly affected parties in a market transaction. For example, a steel plant polluting the air of a surrounding community imposes costs on the residents of the community. Even if the plant were selling its output to its customers in a competitive market, this market will fail to achieve economic efficiency. Such external costs, or spillover costs, are a source of market failure, and corrective action is called for to internalize into the firm's decision-making process the external costs imposed on the residents.

As stated above, it is clearly the objective of the Act to stimulate efficiency in telecommunications markets, including local exchange and exchange access and long distance. If decentralized decision-making is relied upon exclusively to establish general guidelines, as well as to determine the specific rates, terms, and conditions for arbitrated section 251 agreements, it is possible that some states' rules and regulations would not be efficient, thus interfering with the attainment of a national goal. This means that it is at least possible and probably likely, based on the historical record of non-cost-based prices,[2] that public utility commission decisions in one state could adversely affect the economic welfare of residents in other states for which they do not have responsibility. In other words, there could exist significant spillover effects from decentralized decision-making.

For example, suppose a state commission decides not to base the price of unbundled elements on economic costs. It is likely that such a non-cost-based pricing methodology will result in allocatively inefficient prices. Furthermore, assume that a competitive local exchange carrier chooses to lease unbundled local loops to enter the local market in order to become a full service provider of telecommunications services. An artificially high price for local loops will negatively affect the ability of the competitor to compete locally, which one can argue is primarily a state concern. However, the effect of inefficiently high prices is not only felt in local markets. Since local services are complements in consumption with long-distance services, the allocative inefficiency in the former market is extended into the interstate market.[3] This means that the efficiency of interstate markets is adversely affected by a

---

[2] Even when a state commission would prefer to be pro-competition, state politics can lead to different objectives. The Texas Public Utility Act of 1995 contains provisions that make it difficult to lease Southwestern Bell's existing lines, that erect barriers to entry for large interexchange carriers who wish to offer local services, and that limit the magnitude of wholesale discounts to 5%. For details, see [15].

[3] Moreover, North Carolina Utilities Commissioner JoAnne Sanford indicates the interdependence of states' decisions when she laments the hostility of the parties involved in a state arbitration proceeding. The cause of the hostility is that, according to the Commissioner, "each side says the 'wrong' decision in any state will mean its ruin, because the decision will affect all other states where it operates." See [12] for details.

state's decision to pursue narrow state objectives. In this way, a state interferes with the attainment of the national goal to achieve efficiency in all telecommunications markets.

Similarly, application of the traditional dual regulatory paradigm can lead to technical inefficiency in local and interstate markets. Some of the facilities and equipment an entrant will use when it develops its network can be used to provide a variety of local and interstate services. Entrants are likely to lease unbundled elements of the ILEC's network and combine these with elements they supply themselves. Appropriate unbundling rules are needed to foster technical and dynamic efficiency in telecommunications. If a public utility commission orders too little unbundling it can cause a competitor to configure its network in a technically inefficient manner that will raise its costs of providing both local and interstate services. On the other hand, ordering too much unbundling can diminish the incentive of competitors to invest in new configurations for satisfying the demand for local and interstate services. Thus, the appropriate rule for determining elements that should be unbundled should be determined at the federal level to promote technical and dynamic efficiency.

Even though, based on economic analysis, there is a role for the FCC to play in developing national guidelines in the face of potential external effects from state decision-making, this does not mean that all decision-making authority should be exclusively at the federal level. Rather, the extent of federal rules should be limited to mirror the scope of the externality problem.

From an economic perspective, a new regulatory paradigm is a reasonable solution to this problem. The FCC should establish minimum, uniform national rules that promote efficiency and provide the boundaries for the states to decide rates, terms, and conditions in arbitrated Section 251 agreements.

Diagrammatically, the sharing of power is described in Figure 1:

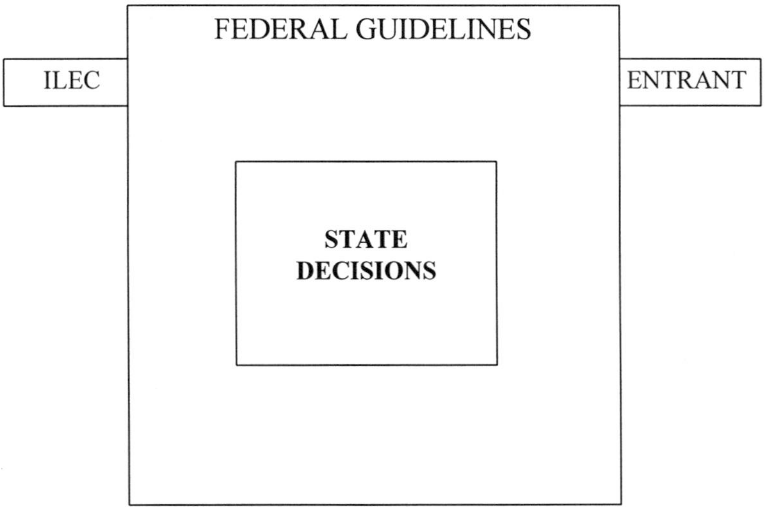

Figure 1. Sharing of Power Between Federal and State Regulators Governing an Arbitrated Interconnection Agreement

The box represents the universe of possible issues in an arbitrated Section 251 agreement. Federal guidelines reduce the universe of debatable issues (e.g., the criteria for mandatory unbundling, specification of the pricing methodology for setting rates for interconnection, unbundled network elements, and collocation). The area within the inner box represents decisions a state must make subject to their consistency with the guidelines. For example, specific prices for unbundled network elements should be included in the inner box.

The sharing of power depicted in Figure 1 is supported by stakeholders that do business or have rsponsibility *across* multiple jurisdictions. Supporters of federal guidelines include IXCs, CAPs, cable operators, federal agencies, and equipment manufacturers. Opponents to national rules, including the ILECs and most state commissions, tend to do business or have responsibility *within* a single jurisdiction. [4] It may be that the supporters of federal guidelines see them as consistent with their own self-interest. Thus, supporters and opponents are not taking a principled approach but rather a pragmatic one.

The primary consideration in favor of the new regulatory paradigm is based on the presence of external effects. A secondary consideration in favor of national rules concerns the ways federal rules can save society's scarce resources and, thus, promote economic efficiency. Cost savings may result from national guidelines for the following reasons.

(1) National rules permit regional and national entrants to take advantage of potential economies of scale that otherwise would be denied if they had to tailor their network configurations and marketing strategies to the differing specific rules in each state.
(2) National rules permit a reduction in transaction costs when negotiating interconnection agreements. National rules narrow the range of disputed issues and provide a uniform interpretation of the law.

In conclusion, it is economically efficient to develop a new regulatory paradigm that is based on sharing regulatory power and is independent of the jurisdictional nature of a service.

3.2 Unbundling

3.2.1 Scope of Mandatory Unbundling

As argued above, the FCC has an appropriate role to play in developing federal guidelines. One important guideline identifies the criteria for determining elements of an ILEC's network that should be unbundled. The criteria should be consistent with Chairman Hundt's efficiency objective and economic reasoning. Unfortunately, as developed below, the FCC's criteria are neither consistent with efficiency nor grounded in solid economic reasoning.

As outlined in the Order, the FCC's unbundling strategy attempts to achieve several objectives. First, unbundling should be accomplished in a fashion that will enable the ILEC to maintain operating efficiency. Second, unbundling should enable entrants to share the

---

[4] It would not be surprising to see that, after the Regional Bell Operating Companies receive permission to provide in-region interLATA long-distance service, they become more favorably disposed to federal rules.

economies of the existing network. The FCC asserts that this will reduce a significant barrier to entry to the local exchange, the sunk cost nature of the investment. Entrants are likely to begin facilities-based operations by providing their own switches, which in general are re-deployable, and lease loops, which are generally viewed as sunk. Loops may be an element in which, for certain locations and for some customers, facilities-based investment is not economically justifiable. That is, some network elements may retain the characteristics of a natural monopoly, and the Act's grant of a right to lease unbundled network elements creates a new market structure, a shared natural monopoly, for some elements of the local exchange.

Third, the FCC views the unbundling plan as the next logical step following its efforts to unbundle features and functions in the Open Network Architecture (ONA) proceeding [3]. The ONA plan unbundled basic serving arrangements (BSAs) from basic service elements (BSEs). By accepting the Regional Bell Operating Companies common model, a BSA still bundled access, switching, and transport elements while BSEs, defined as optional features and functions provided in general by the stored program component of a local switch, were available on an individual basis and, at least in the federal jurisdiction, priced on a cost-based basis.[5] The local competition Order extends features and functions unbundling to facilities unbundling, which, in the ONA Order, the FCC referred to as a radical reconfiguration of the local exchange network.[6]

The FCC decision process establishing the criteria for determining network elements[7] that must be unbundled is comprised of several sequential steps. The first step in the process considers all elements of the local exchange as candidates for unbundling for which it is technically feasible to unbundle without consideration given to the cost of physically unbundling an element. If it is not technically feasible to unbundle an element, then the element is excluded from further consideration. The remaining elements are divided into two categories, proprietary and nonproprietary, with a separate analysis for each.

The next step in the decision process requires the FCC to consider, at a minimum, two factors, a proprietary and an impairment standard, to determine the actual elements to unbundle. A nonproprietary element must be unbundled if denial of access to the element would decrease the quality or increase the cost of the entrant's service compared to providing the service by using other unbundled elements within the ILEC's network (this is called the impairment standard). On the other hand, if the entrant is not impaired by an ILEC's refusal to unbundle a specific nonproprietary element, then there is no public policy reason to require the element to be unbundled.

A higher standard is applied in considering whether to require a proprietary element to be unbundled. This makes sense given the negative effect on an ILEC's incentive to innovate if it is required to unbundle and, thus, make available to competitors a proprietary element. A proprietary element must be unbundled if denial of access would significantly impair an entrant (this is called the proprietary standard). If an entrant would not be significantly impaired, then no unbundling is required. If the entrant would be significantly im-

---

[5] ONA can be viewed as primarily an interconnection arrangement to accommodate enhanced service providers with some limited unbundling of tariffs.

[6] For details, see the ONA Order, paragraphs 70 and 72.

[7] Not surprisingly, the interpretation of the term "network element" is a source of contentious legal debate. The Court of Appeals supported a broad definition of the term "network element" proposed by the FCC.

paired without access to the proprietary element, it must be unbundled unless an entrant could substitute other nonproprietary elements within the ILEC's network to provide its service.

In its development of the criteria for mandatory unbundling, the FCC considered whether an element was available from a source other than an ILEC but rejected this as a relevant criterion. The rejection reflects the FCC's obsession with eliminating any supposed barrier to entry and speeding the transition to competition. Specifically, the FCC asserts that giving weight to such a consideration would require the entrant to duplicate an incumbent's network and delay entry.[8]

This reasoning is clearly faulty for it is not grounded in any economic principle. The appropriate economic analysis for determining the criterion for mandatory unbundling is based on the identification of essential facilities. According to Grieve and Levin [5, p.3], "unbundling is required where one local exchange carrier controls essential facilities – those which other competitors cannot economically duplicate in order to construct their own networks." Clearly, the phrase "unable to economically duplicate" should be interpreted to include situations in which other competitors cannot purchase these facilities from alternative suppliers and not just self-supply. Specifically, an essential facility (or element) exists when two conditions hold. First, the element must be monopoly-supplied by the ILEC and second, the element is necessary for the competitor to provide a telecommunications service.

The availability and number of alternative suppliers for a network element probably will vary with the size of market demand. Thus, it could be that the same network element is an essential facility in a small market but not in a large market. Such a determination should be made at the state level.

The FCC arguments for not considering the availability of alternative supplies are nonsensical. If actual alternative suppliers exist, then the entrant does not have to duplicate the ILEC's network, and entry will not be delayed. Assertions to the contrary lack common sense.

The FCC's criteria for determining network elements that must be unbundled are too broad and will lead to too much unbundling from an efficiency perspective. Thus, the criteria are not consistent with Chairman Hundt's stated objectives. The FCC should reconsider its criteria for determining the network elements to be unbundled. It should include in its criteria for unbundling network elements the availability of alternative suppliers, and it should force an ILEC to unbundle only an element (proprietary or nonproprietary) that meets the definition of an essential facility. Such an analysis would seem to be consistent with consideration of the statutory terms necessary and impair. The Court of Appeals found that the FCC's approach for determining the elements that must be unbundled is reasonable, but the Court's reasoning lacked any economic analysis.

3.2.2 Ability to Rebundle

The FCC did not require an entrant to own some of its own facilities in order to lease unbundled network elements. This decision seems consistent with the intent of Congress to provide a leasing option for entering the local exchange. The ILECs' objection to this deci-

[8] For details, see the First Report and Order, paragraphs 283 and 287.

sion reflects a private fear that entrants will rebundle existing elements to provide a service that is equivalent but priced lower than the service available by resale. The FCC's unbundling policy that effectively unbundles all elements requested by entrants contributes to this fear. According to the ILECs, rebundling not only could be cheaper than reselling service but, also, will circumvent the statutory prohibition on the joint marketing of resold local services and interLATA service prior to an ILEC's entry into interLATA markets. The extent to which rebundling is cheaper than reselling service depends on the relationship between the cost of local service and its price.

There is little evidence to date that entrants have pursued a rebundling strategy. Rather, many entrants (especially IXCs) have publicized plans to use resold local services to reach, in particular, residential customers. More generally, the FCC's argument that leasing is different than resale because of the greater risk to the entrant from the former strategy is compelling. In addition, leasing unbundled elements gives the entrant control over facilities where reselling does not, which suggests that leasing is more attractive even without a cost advantage. For example, leasing enables the entrant to collect access charges. The ILECs also recognize the difference between leasing and reselling, for they are likely to argue, in Section 271 applications for authorization to provide in-region interLATA service, that entrants who lease unbundled network elements meet the definition of a facilities-based competitor.

3.3 Pricing Methodologies

3.3.1 Rate Level Methodology

Perhaps the most important and controversial issue in the Order concerns the federal guidelines mandating the pricing methodology for establishing rates for interconnection, unbundled network elements, collocation, and local call termination. Much of the FCC's analysis is economically sound with one significant exception. The exception involves the decision to establish rates on the basis of theoretical forward-looking costs rather than actual forward-looking costs. The decision interferes with the dynamics of the competitive process and may lead to uneconomic decision-making.

The FCC's pricing methodology, referred to as total element long-run incremental cost (TELRIC), reflects an appreciation for sending proper economic signals to market participants by what it includes and what it excludes. According to the FCC, the price of an element should include two components:

(1) Forward-looking costs directly attributable to use of a specific element. These costs should include a reasonable return on the investment (to be determined by the state commission) needed to build the element.
(2) A reasonable share of forward-looking joint and common costs.

There are several noteworthy characteristics of the TELRIC method. First, the focus is on a network element, not on services, and, because multiple services can be produced by the same network element, this delineation should reduce the common costs associated with an element compared to those associated with a particular service. Second, the method looks to the future as economic costs do and excludes considerations of embedded costs which look to the past. Third, by including a contribution toward joint and common costs, the method includes a markup above incremental cost which is intended to recognize the unique obligations (not including universal service obligations which are addressed in a related proceeding) imposed by regulators on the ILECs. This means the pricing methodology departs from an exclusive focus on the goal of promoting economic efficiency. Given the complex issues involved in facilitating a transition from a regulated monopoly to a competitive market structure, this is a reasonable compromise.

Table 1 presents six options based on economic costs for determining the cost of a network element.

Table 1. Options for Determining the Cost of a Network Element

1.	TELRIC based on the current technology and the current network configuration
2.	TELRIC based on the most efficient technology and the current network configuration
3.	TELRIC based on the most efficient technology and the optimal network configuration
4.	LRIC based on the current technology and the current network configuration
5.	LRIC based on the most efficient technology and the current network configuration
6.	LRIC based on the most efficient technology and the optimal network configuration

Three critical questions in the analysis are:

(1) should a markup above long-run incremental cost (LRIC) be permitted?
(2) should the most efficient or the current technology be assumed?
(3) should the optimal network configuration or the current configuration be assumed?

As discussed above, the FCC decided to permit a markup above long-run incremental cost,[9] thus eliminating options 4-6.

The next step required the selection of either a theoretical or the actual technology and network configuration (i.e., the location of wire centers). The FCC decided, in what it considered a compromise, to select the most efficient technology but the existing network configuration (option 2 in Table 1). It reasoned that this would establish rates comparable to those reached in a long-run competitive equilibrium. This choice reflects the attempt to mandate the long-run results of a competitive process without allowing the competitive process to achieve these results. This pricing rule actually undermines the competitive process and thus can be described as anticompetitive, because it discourages competitors from entering the market by building their own facilities. The artificially low prices will discourage entrants from building their own facilities and, thus, uneconomically encourage them to lease network elements and/or resell local services. In a competitive market, firms base prices on their actual costs of production. A firm considering entering the market will com-

[9] Long-run incremental costs are the forward-looking costs that result from changing the output of a good or service when all inputs are varied. The FCC defines economic costs as those that result from adding a portion of forward-looking joint and common costs to the incremental costs.

pare what its costs will be to what the existing firms' actual costs are, as reflected in prices. If the entrants can produce at a lower cost, this will cause adjustments in the market toward these lower costs. The error the FCC made is attempting to define market long-run marginal costs before the competitive process begins when, in fact, the process will reveal what they are. Moreover, even if the FCC knew the long-run costs in a competitive market, according to Wenders [16, p.236], " ... they would be irrelevant for anything that the regulatory process might do now because that process cannot effectively leap to the long-run competitive state of affairs without going through the process of deregulating the industry and allowing the necessary forces and institutions of competition (to) develop."

There is preliminary evidence to support the ILECs' contention that the pricing methodology for network elements discourages competitors from making facilities-based investments. In spite of numerous interconnection agreements across the United States, plans by the IXCs to build networks have been delayed and a new emphasis has been placed on leasing lines and, especially, reselling local services. Since the passage of the Act, California is one of the few states to experience new competition in residential markets by major IXCs, and this is mainly resale competition. In addition, leading cable television companies, such as Tele-Communications, Inc., and Time Warner, are reassessing their plans to enter local service markets and are refocusing their attention on their core business.[10]

In sum, the pricing methodology that best meets the efficiency objective of the FCC is to use the TELRIC methodology but base rates on the actual forward-looking costs of the ILECs.[11] The Court of Appeals recent decision was silent on the merits of the TELRIC pricing methodology but instead overturned the FCC's pricing rules because the Court believed the FCC had overstepped its appropriate jurisdictional authority.

3.3.2 Rate Structure Methodology

Guidelines established by the FCC to govern the general structure of rates are economically efficient. The basic principle is that price differences should reflect cost differences. Historically, the ILECs' pricing structures have exhibited price discrimination in the form of differing rates for services having similar marginal costs (e.g., business vs. residential service) or similar rates for services whose marginal costs differ (e.g., urban vs. rural service).

The FCC guidelines attempt to structure prices on how costs are incurred. Specific guidelines include:

(1) prices for the use of dedicated facilities (e.g., unbundled loops, line ports, dedicated transport, interconnection, and collocation) should be flat-rated since the cost of using these types of facilities does not vary with use, and

[10] For representative articles, see [1] and [2]. The President of the National Cable Television Association attributes the reassessment to the " ... ongoing appeals court battle over the FCC's local competition rules, contentious interconnection negotiations between incumbent LECs and new entrants, and financial markets that appear to be "skittish about competition ... "; see [11].

[11] The FCC expressed concern about a price squeeze if the ILECs based rates on actual costs. This concern is correctly addressed by the proper imputation rule and is independent of the level of an input price.

(2) prices for the use of shared facilities (e.g., transmission links between tandem switches and end offices, tandem switching, signaling, and database services) should be usage-sensitive since the cost of using these types of facilities varies with use.

The FCC ordered geographic deaveraging for unbundled network elements and interconnection. The specific rule (requiring at least three zones) is an example of the agency overstepping its bounds by getting too detailed. It makes more sense for the FCC to establish the general guideline that when geographic costs vary significantly prices should reflect those differences, and leave implementation to the states.

3.3.3 Resale Discount Methodology

The FCC did not specify a particular pricing methodology for calculating the discount on retail prices called for the Act but rather said that states should use a methodology that is consistent with how it sets retail rates. According to the FCC, the discount should be based on the costs that are reasonably avoidable by providing wholesale services instead of retail services. The ILECs argue that the discount should be based on the costs that are actually avoided by providing wholesale services instead of retail services. The FCC is concerned that if the discount is based on actual costs avoided, then the ILECs would have an incentive to minimize the costs that are actually avoided in order to reduce the size of the discount given to the competitors. If the ILECs engaged in this type of strategic behavior, then they would be technically inefficient.

Resale is one issue where the potential entrants are not united in their opposition to the ILECs. CAPs and cable operators have joined the ILECs in opposition to significant discounts to facilitate the entry of resellers. This is an example of where facilities-based competitors oppose an artificial advantage given to non-facilities-based competitors.

The FCC overstepped its appropriate role when it went beyond guidelines and provided a specific range of discounts (17% to 25% off retail rates) based on reasonably avoidable costs to help the states in their arbitration decisions. The recent Appeals Court decision agrees with this observation. There is evidence that these default proxies altered the pattern of discounts established before the FCC Order. As of August 15, 1996, there were nine state commission decisions regarding the discount, and the average discount was 14.09%. Based on 22 representative state decisions made subsequent to the FCC Order and as of December 12, 1996, the average discount increased half as much again to 20.05%, within the FCC range. More fundamentally, it is not clear how a discount off retail rates promotes allocative efficiency in the economy. In many markets, efficiency is enhanced because discounted prices based on volume and term contracts result from market forces. However, the purpose of the resale discount mandated in the Act is to allow competitors to compete based on the margins between wholesale and retail rates. This purpose is unlikely to lead to an improvement in allocative efficiency in local markets, which would be to move the rates for residential exchange service closer to costs.

3.4 Other Issues

The ILECs are displeased with an FCC rule that allows new entrants to pick and choose among the terms and conditions of existing interconnection agreements. The ILECs believe that this rule undermines the voluntary negotiation process, a process they view as a hallmark of the Act. The FCC views the rule as protection against discriminatory behavior on the part of the dominant ILECs. Regardless of the relative merits of the arguments, it seems obvious based on a careful reading of Section 252 (i) of the Act that Congress' intention is consistent with the FCC implementation rule, and that the ILECs' problem is with the public policy determination made by Congress, not the FCC. However, the Court of Appeals recent decision supports the position of the ILECs.

A similar conclusion is reached with respect to the asymmetric obligations of the ILECs reflected in the FCC guidelines. The ILECs argue that it is unfair that they have obligations such as mandatory unbundling of network elements, access to operating support systems, and access to premises and rights-of-way while the competitors have no such obligations. The decision not to require reciprocal non-discrimination provisions reflects the strength of the incumbent position. The decision was made by Congress, not the FCC. The decision is mandated in the section of the Act that identifies the additional obligations of the incumbent local exchange carriers. There is no similar section for the entrants.

4. Summary and Conclusions

This paper analyzes key elements of the Order to implement the local competition provisions of the Telecommunications Act of 1996 from the perspective of economic efficiency. There is an appropriate role for federal guidelines in a new regulatory paradigm. However, the FCC has gone beyond developing rules based on efficiency considerations and has attempted to micromanage the transition to competitive markets. The FCC should reconsider immediately two key rulings. First, the FCC should reconsider the criteria for determining the network elements of the ILEC that must be unbundled. The FCC should incorporate an essential facility test into its analysis. Failure to do so imposes unnecessary costs on ILECs, including a dampening of the incentive to innovate, and reflects a misunderstanding of the role alternative suppliers play in determining the competitiveness of a market. Since the number of alternative suppliers will vary across network elements and across markets, the FCC should establish the general criteria for mandatory unbundling based on an essential facility test and leave the particular determinations of the network elements to be unbundled in each market to the state commissions.

Second, the FCC should reconsider the pricing methodology for interconnection, unbundled network elements, collocation, and local call termination. The FCC's analysis of the role of forward-looking costs is sound, but it erred in requiring rates to be based on the assumption of the most efficient technology. This assumption will create a level of rates that will undermine, not facilitate, competitive processes by discouraging construction of lower cost (relative to the ILECs' actual costs) facilities by entrants. There is preliminary evidence that the pricing rule is altering the entry strategies of competitors away from new facilities and toward leasing unbundled network elements and, especially, reselling local service. In short, the FCC should retain its TELRIC methodology but order the states to base rates on the existing level of technology and network design.

It is also time for the challengers to reconsider their positions on several issues. First, the challengers should accept the new regulatory paradigm for it will ensure that regulators in both jurisdictions work together in the pursuit of the efficiency goals of the Act for all telecommunications market.

Second, the challengers should accept or work to change policy decisions made by Congress, such as the pick-and-choose rule and the asymmetric obligations imposed on the ILECs, and not try to shift their dissatisfaction and undermine the rules by blaming the FCC for poor decision-making.

The Act held out the promise of altering fundamentally the structure of all telecommunications markets. Ironically, for several reasons, the transition to competitive markets may be slower as a result of the Act and the FCC's implementation of the legislation. Without the Act, it is highly likely that Ameritech's Customers First proposal would have been approved by the proper authorities and competition in local markets and in in-region long-distance markets would have been enhanced. Most likely other local companies would have followed Ameritech's lead and submitted similar proposals. Since passage of the Act, the FCC has not authorized any Regional Bell Operating Company to provide in-region inter-LATA service and the pace of local competition has slowed.

Second, several large states were developing the necessary rules to facilitate competition in local markets and the Act probably slowed down their work as these states had to divert their attention to following the new rules issued by the FCC. Moreover, the FCC rules have slowed the pace of facilities-based competition. In addition, the appeal of the rules and the FCC's likely appeal of the Court of Appeals decision further slow the transition to competition.

Finally, it appears that the IXCs, who were vocal supporters of the Act, are starting to see significant problems with the implementation of the Act. AT&T's discussions with SBC suggest that buying an ILEC may be the preferred method of entry into local exchange markets rather than utilizing the paths outlined in the Act.

References

[1] Big Carriers are Slow to Enter Local Markets, *Wall Street Journal*, January 28 (1997) B1.

[2] Consumers Have Yet to See Benefits of Telecom Reform, *St. Louis Post-Dispatch*, February 19 (1997) 5C.

[3] Filing and Review of Open Network Architecture Plans, 4 FCC Rcd 1 (1988).

[4] Georgia PSC Says AT&T Can't 'Rebundle' Elements, *TR Daily*, December 6 (1996) 4.

[5] Grieve, Willie A. and Stanford L. Levin, Of Mantras and Mischief, paper presented at International Telecommunications Society, Eleventh Biennial Conference, p. 3.

[6] Haring, John R. and Kathleen B. Levitz, The Law and Economics of Federalism in Telecommunications, *Federal Communications Law Journal* 41, no. 3, July (1989) 261-330.

[7] Hundt Eyes 'Radical' Overhaul of FCC's Regulatory Regimes, Telco Competition Report, July 18 (1996) 3-9.

[8] In the Matter of Implementation of the Local Competition Provisions in the Telecommunications Act of 1996. CC Docket No. 96-98, First Report and Order, August 1, 1996.

[9] Iowa Utilities Bd. v. FCC, Nos. 96-3321 et.al., Filed October 15, 1996.

[10] Louisiana Says 'Rebundlers' Must Pay Wholesale Rates, *TR Daily*, January 16 (1997) 5.

[11] NCTA's Anstrom Says LECs are 'Stonewalling.' *TR Daily*, January 30 (1997) 4.

[12] State Telephone Regulation Report. November 28, 1996, p. 5.

[13] Telecommunications Act of 1996 Conference Report, S. Rep. 104-230 at 113, February 1, 1996.

[14] Telecommunications Act of 1996, Public No. 104-104, 110 Stat. 56.

[15] Texas Defies Washington in Phone Deregulation, Protecting Its Local Bell Against Giant Rivals, *Wall Street Journal*, May 2 (1996) A16.

[16] Wenders, John, The Economics of Telecommunications, Theory and Practice. Ballinger Publishing Company, 1987, p. 236.

Telecommunications Transformation: Technology, Strategy and Policy
E. Bohlin and S.L. Levin (Eds.)
IOS Press, 1998

CHAPTER 22

Costing and Pricing for Local Exchange Competition: Experience under the U.S. Telecommunications Act

Timothy J. Tardiff

National Economic Research Associates, One Main Street, Cambridge, MA 02142 USA

Abstract. The U.S. Telecommunications Act of 1996 requires that incumbent local exchange (ILECs) provide unbundled elements and make retail services available for resale at a discount. This paper identifies the costing and pricing issues that have emerged to date in proceedings designed to implement these provisions. Alternative approaches for establishing prices carry very different views of competition. Negotiations tend to reflect a balance of interest among parties. In contrast, arbitrations often (1) presuppose imbalances in bargaining power, (2) assume time horizons of indefinite length, and (3) function on a premise that the ILEC maintains near monopoly control over all inputs to be sold to competitors. Major potential entrants have presented cost estimates that are typically about one-half of the ILEC's current (book) costs and approximately the same proportion of the incumbent's forward-looking cost estimates. These differences are explained by differences in (1) efficient network design and operation, (2) input prices, (3) depreciation and cost of capital, and (4) magnitude of common costs. The paper discusses the implications of particular price estimates for (1) the amount and type of competitive entry, and (2) the competitive prospects of incumbents and other competitors.

1. Introduction

The Telecommunications Act of 1997 and the FCC Interconnection Order were major developments that anticipate significant changes in the structure of communications industries in the United States. Among other things, the Act and Order establish rules for removing all barriers to entry into local exchange markets and, in fact, providing a jump-start to that competition. In particular, facilities and services provided by incumbent local exchange carriers (ILECs) are to be made available to local exchange entrants and certain rules must be followed in setting prices for these elements.[1]

[1] In October 1996, the Court of Appeals for the Eighth Circuit issued a stay of the pricing rules in the FCC's interconnection order. The Court overturned the FCC's pricing rules in July 1997. These actions did *not* affect (1) the requirement that services be made available for resale and that network elements be unbundled consistent with the FCC's rules and (2) the requirement that states must set prices for resold services and un-
(continued...)

Thus, US telecommunications law and regulation have decreed that wide-scale access to ILEC facilities is a prerequisite for competition. In effect, substantial parts of the ILEC's network are treated as essential facilities, which in legal and economic terms are those monopoly facilities that competitors need in order to compete with the ILEC. In fact, an extreme interpretation of the FCC's implementation of the Act is that the ILEC's network is still the natural monopoly it always has been under traditional regulation and that competition will be primarily at the retail (as opposed to facilities-based) level. Of course, others have argued that whatever natural monopoly the ILECs possessed in the past is rapidly eroding and that facilities-based competition is both inevitable and likely to be the dominant form in the future.

How the issues of the availability and price of ILEC services and facilities are set in ongoing regulatory proceedings will have a critical impact on the type of competition that emerges and how the firms in those markets will choose to compete. For example, setting prices in the belief that the ILECs are natural monopolies whose prices must be low in order to prevent competitive abuses will have the consequence (intended or otherwise) of promoting resale and unbundled element competition at the expense of facilities-entry. Similarly, the relative levels of the ILEC's retail, wholesale, and unbundled element prices will an important determinant of industry structure.

The remainder of this paper is organized into four sections. First, the various types of competitive entry are enumerated. Next, relevant features of the resale and unbundling provisions of the Act and Order are reviewed. The next section describes possible competitive strategies, given possible outcomes from the ongoing interconnection proceedings. The final section draws conclusions.

2. Local Exchange Entry

2.1 Types of Entry

There are three ways in which competing local exchange networks can enter and provide services. First, a *facilities-based* entrant builds its own facilities in order to compete with the ILEC. For example, prior to 1996, firms such as MFS and Teleport constructed high-capacity fiber facilities, primarily to service large business customers and interexchange carriers (IXCs) in urban areas. Another example is the use of existing cable television infrastructure, suitably upgraded, to provide telephone services. Time-Warner has been offering telephone service in Rochester since 1995 and other firms maintain plans to provide telephone service in the future.[2] Finally, it is interesting to note that in the early history of the telecommunications industry, multiple providers competed in particular local service areas. In fact, in order to reach all subscribers, a customer had to subscribe to all providers, a

---

(...continued)
bundled elements consistent with the schedule required by the Act. Further, states have the authority to base prices on rules similar to those enunciated by the FCC.

[2] Recently, TCI indicated that it was scaling back its ambitious telecommunications plans and returning its focus to its core market (*Wall Street Journal*, January 2, 1997). However, other firms such as Media One (formerly Continental Cablevision and now part of US West) continue to pursue telecommunications plans.

situation that the interconnection requirements of the Telecommunications Act are designed to avoid.

Resale entry occurs when a competitor purchases services from the ILEC, typically at a lower price than the ILEC's retail offering. The margin between the ILEC's retail and wholesale rates and/or additional value added by the new competitor are the basis for the competitor offering attractive services. Historically, resale competition was an important event in long-distance competition.[3] In fact, the existing facilities-based long-distance firms often built their customer bases by reselling other carriers' high-volume services and then extending their own networks to serve these customers. A large majority of the several hundred firms classified as long-distance carriers today are resellers. Indeed, the "big three" facilities-based carriers – AT&T, MCI, and Sprint – account for about 80 percent of the market.

Finally, firms can deploy a combination of resale and their own facilities. As mentioned above, the hybrid strategy is a means of expanding the customer base before ubiquitous facilities have been deployed. The hybrid approach dictated by the Act and Order differs from the historical example in that the unbundling requirements provide the new opportunity for an entrant to serve customers with a combination of their own facilities and unbundled elements; e.g. a long-distance company, such as AT&T, could connect customers to its existing network by obtaining unbundled loops from the ILEC.

Of course, these three entry strategies are not mutually exclusive, as the history of long distance competition illustrates. Indeed, on February 8, 1996, the day that the Telecommunications Act became law, Robert Allen, AT&T's Chairman, boldly proclaimed that his company intended to capture a large share of the local market by deploying a combination of the entry strategies enumerated above. AT&T's and MCI's vigorous participation in the regulatory cases that are establishing prices for resale and unbundled elements, at the same time they are pursuing strategies that will give them a facilities-based presence, is a further indication that firms will pursue mixed strategies.

2.2 Economics of Entry Approaches

Four fundamental economic principles should inform public policies that facilitate entry. *First*, when a regulated industry structure is no longer in the public interest, existing artificial barriers and impediments to entry should be removed. These include outright prohibitions, such as exclusive territories granted to regulated monopolies as well as policies that confer asymmetric advantages on particular providers. For example, until ILECs qualify to offer long-distance services in their own territories, the IXCs have the artificial advantage of being able to offer their customers "one-stop shopping" by packaging local and long-distance services. Such asymmetries should be removed at the earliest practicable date.

---

[3] An important difference between resale of long-distance services and the form of resale specified by the Act and Order is that historically resale has taken the form of an entrant purchasing an existing *retail* service (albeit from a high-volume tariff offering lower unit prices than the low-volume tariff). In contrast, for local resale, special wholesale offerings, with explicit discounts off of the retail price, must be developed and made available to entrants.

Second, the ILEC's essential facilities[4], to the extent they exist, should be provided on a non-discriminatory basis at prices that permit efficient competition between ILECs and entrants.[5] *Third*, unless there are "infant company" considerations, i.e., new competitors are unable to survive and thrive without special advantages conferred by the government or regulators, entry policies should neither favor nor disadvantage particular firms. The main reasons for this principle are that (1) artificial advantages (or disadvantages) distort competition, to the ultimate disadvantage of consumers, in general,[6] and (2) special treatment tends to establish vested interests that argue for maintaining the artificial advantages well beyond the time they are arguably helpful. Thus, policies designed to "jump-start" competition should be no stronger than needed and subject to elimination when competition has taken root.

Fourth, when access to the ILEC's facilities and resale of ILEC retail services are necessary for efficient competition, the prices and other terms and conditions should neither favor nor disadvantage the alternative entry approaches. Telecommunications markets are dynamic, with particular firms seeking special competitive advantages as they enter and perform in the market. Winners and losers should be determined by their competitive performances, unaided or unimpeded by artificial price advantages given to particular types of entry.[7]

3. Unbundling and Resale Requirements

The Telecommunications Act requires that ILECs make their retail service available for resale and provide elements of their network at prices that are cost-based. The particular form in which these requirements will be implemented is still being resolved as a result of agreements and arbitrations at the state level. Initial decisions are starting to roll in. However, even here, many of the decisions have established only interim rates, with permanent rates deferred to proceedings in which costs can be examined in greater detail.

3.1 Negotiation Versus Arbitration

The US Telecommunications Act specified two mechanisms for providing access to ILEC inputs necessary for competition. Even before the Act was passed, parties were negotiating

---

[4] An essential facility is (1) necessary for entrants to compete, (2) available from the incumbent on a monopoly basis, and (3) cannot be economically or technically duplicated. The same requirement of non-discriminatory access to essential facilities should also apply to new entrants if they possess such facilities. For example, when local networks compete, access to customers in order to terminate calls can be viewed as an essential facility. The mutual interconnection obligations under the Act can be interpreted in this context.

[5] In general, the price of an essential facility should contain no larger margin over cost than contained in the ILEC's competing retail service. See J. Hausman and T. Tardiff, Efficient Local Exchange Competition, *Antitrust Bulletin*, Volume 40, 1995, pp. 529-556.

[6] Customers of advantaged firms might, indeed, benefit. However, necessary cost increases for other customers more than offset such advantages.

[7] Inefficient prices that favor particular forms of entry (perhaps unintentionally) are similar in nature and consequences to "infant company" advantages. Companies that choose the favored approach would gain an artificial competitive advantage, even though other companies, naturally employing other strategies, may be more efficient.

unbundling and interconnection agreements. These negotiations tend to be based on mutual interests and of limited duration. Because different parties brought different interests to the table, the details of agreements reached by negotiation could differ. In short, the negotiated agreements can be viewed as a market-based process, albeit with regulatory oversight.

The Act also specifies arbitrations and the FCC's "pick and choose" requirement (subsequent interconnectors have the right to select the best features of previously negotiated agreements) has tended to steer the process in that direction, at least for the largest potential entrants, such as AT&T and MCI.[8] Arbitrations can take on much more of the flavor of traditional regulation than voluntary negotiations. In particular, the arbitration route carries the implicit assumptions that (1) the ILECs have an imbalance in bargaining power,[9] (2) they will maintain virtual monopoly control over the bulk of network elements indefinitely, and (3) as a result the prices established in arbitrations often assume long planning horizons, e.g., that it is possible to set a price for an unbundled local loop that can remain in effect for a long time.

3.2 Resale

The Telecommunications Act requires that ILECs provide a wholesale version of their retail services at a discount, based on costs that will be avoided. A major issue in establishing rules has been the definition and measurement of these avoided costs.

Differences in definition of avoided costs stem from different conceptions of the type of company that the reselling ILEC will be. The ILECs typically view their operation under resale obligations as one in which firms reselling their services function as new sales channels. Accordingly, in defining and measuring avoided costs, the ILECs ask how much cost is avoided if they no longer employ their own sales channels *for the volume of business that resellers can be expected to win*. In particular, costs that only go away if the ILEC got out of the retail business altogether are not considered to be avoided.

Entrants such as the IXCs (and the FCC's overturned pricing rules) view the ILEC as two separate companies – a wholesale-only company and a retail affiliate that "buys" wholesale services, just like any other reseller. Under this perspective, avoided costs include *all* costs that go away if the ILEC were to completely exit the retail business.[10] Because this perspective includes costs that will be avoided only under the unlikely scenario that the ILEC completely leaves the retail business, such costs are often called "avoidable". Since avoidable costs are generally larger than avoided costs on a per unit basis, i.e. costs

[8] The July 1997 decision of the Eighth Circuit Court also struck down the FCC's "pick and choose" requirement.

[9] Because the Act requires Regional Bell Operating Companies (RBOCs) to open their markets to competition as a prerequisite for gaining entry into interLATA long-distance, parties whose interest would be served by excluding the RBOCs from long-distance clearly have a countervailing bargaining chip.

[10] See for example, D.L. Kaserman and J.W. Mayo, An Efficient Avoided Cost Pricing Rule for Resale of Local Exchange Telephone Services, *Journal of Regulatory Economics*, Volume 11, 1997, pp. 91-107. Kaserman and Mayo's approach has the additional feature that margins over incremental cost in excess of "normal profits" would also be considered an avoidable cost and included in the calculation of the discount. As a result, entrants would receive deep discounts on high-margin services such as toll – services that have traditionally contributed to subsidized services, such as basic access. In the process, the "costs" that Kaserman and Mayo deem to be avoided would greatly exceed the costs the incumbent would actually avoid if it provided such high-margin services on a wholesale, rather than retail, basis.

avoided by losing less than the full retail business, the resulting discounts are larger than under the ILEC's perspective.

The following hypothetical example, depicted in Table 1, illustrates the different positions. Suppose the ILEC is currently selling 100 units of a retail service. Wholesale costs are $0.70 per unit. Retail costs are $20 of fixed costs, which go away only if the ILEC completely leaves the retail business, and $0.10 per unit. The ILEC's retail price is $1.00 per unit, which just recovers all retail and wholesale costs.

Table 1. Alternative Wholesale Discounts

Initial sales	100
Wholesale unit cost	$0.70
Retail fixed cost	$20.00
Retail unit cost	$0.10
Current cost	$100.00
Retail price	$1.00
Reseller sales	25
"Avoided" lost revenue	$25.00
"Avoided" cost	$2.50
"Avoided" cost discount	10%
"Avoidable" lost revenue	$100.00
"Avoidable" cost	$30.00
"Avoidable" cost discount	30%

Resellers are expected to win 25 percent of the market, or 25 units. Therefore, the ILEC would lose $25 dollars of revenue (25 x $1), but would also avoid the unit retail cost of these units, or $2.50 (25 x $0.10). Therefore, the avoided cost discount is 10% (2.5/25).

Under the avoidable cost perspective, the ILEC is assumed to lose all its retail sales, or $100 of revenue. Under this hypothetical assumption, the ILEC would shed both the fixed and unit retail costs, or $30 ($20 + 100 x $0.10). Therefore, the "avoidable" cost discount is 30% (30/100).

The impact of the alternative discounts on ILEC finances is also of interest. Under the avoided cost discount, the ILEC loses $25 in retail revenue. But this lost revenue is exactly offset by wholesale revenue of $22.50 (25 x $0.90) and $2.50 of avoided retail costs (25 x $0.10). Table 2 summarizes these calculations.

Table 2. Impact of Avoided Cost Discount on ILEC Finances

Reseller sales	25
Lost retail revenue	$25,00
Wholesale price – "avoided"	$0,90
Wholesale revenue – "avoided"	$22,50
Avoided retail cost	$2,50
Net	$0,00

Table 3 shows that under the 30 percent avoidable cost discount, the ILEC loses revenue. Because the wholesale discount contains a credit for a *pro rata* share of the fixed retail costs, which are not avoided when the ILEC does not leave the retail market, the wholesale price is not compensatory. In particular, wholesale revenue plus avoided costs add up to only $20, which is $5 short of the lost retail sales revenue.

Table 3. Impact of Avoidable Cost Discount on ILEC Finances

Reseller sales	25
Lost retail revenue	$25,00
Wholesale price – "avoidable"	$0,70
Wholesale revenue – "avoidable"	$17,50
Avoided retail cost	$2,50
Net	($5,00)

The hypothetical alternative discounts depicted in the example are quite close to what parties have been proposing in interconnection proceedings. The discounts selected by regulators as part of arbitration awards are typically somewhere between the parties' positions.

3.3 Unbundled Elements

The FCC's Order requires that certain network elements be provided on an unbundled basis. This part of the Order is unaffected by the Stay. Table 4 lists these elements. State regulators are free to order a greater amount of unbundling.

Table 4. FCC's Required Unbundled Elements

Loop
Local Network Interface Device
Switching Capability
- Local
- Tandem

Interoffice Transmission Facilities
Signaling Networks and Call-Related Databases
- Signaling Networks
- Call-Related Databases
- Service Management Systems

Operator Support Systems Functions
Operator Services and Directory Assistance

The list of unbundled elements covers the bulk of the ILEC's network, with loops and local switching accounting for most of the underlying costs.[11] Available cost models indicate that loops and local switching may account for well over 90 percent of network costs.

A great deal of controversy in interconnection arbitrations has involved the methods and models used to estimate network element costs. The FCC's overturned rules called for a total element long-run incremental cost (TELRIC) approach, which was its attempt to apply the economic concept of forward-looking costs plus an allocation of common costs. The FCC describes is methodology as follows:

> The forward-looking economic cost of an element equals the sum of: (1) the total element long-run incremental cost of the element ...; and (2) a reasonable allocation of forward-looking common costs.
>
> The total element long-run incremental cost of an element is the forward-looking cost over the long run of the total quantity of the facilities and functions that are directly attributable to, or reasonably identifiable as incremental to, such element, calculated taking as given the incumbent LEC's provision of other elements.[12]

The FCC further specifies that the TELRIC costs should be based on a reconstructed network, using existing LEC wire-center locations, and that cost should be based on forward-looking depreciation rates and cost-of-capital. A reasonable allocation of common costs is defined as one in which (1) the price of an element does not exceed its stand-alone cost and (2) the sum of the costs allocated to individual elements equals the total forward-looking common cost.

Two approaches have been advanced as means for setting unbundled element prices. AT&T, MCI, and other parties have developed the Hatfield model, a computer program that provides cost estimates for unbundled elements for individual ILECs. The ILECs themselves have typically produced their own forward-looking cost studies, based on cost models developed over the years for somewhat different purposes.

Not surprisingly, the cost estimates produced by the two approaches are not close. Hatfield model cost estimates tend to be about 50 to 60 percent of current (embedded) costs, while the LEC cost estimates are much closer to these historical figures. The major reasons for the differences include:

- Different prices for components such as switches, infrastructure items, etc.
- Different representations of efficient network design
- Different descriptions of how an efficient network will operate (e.g., how much spare capacity will be needed)

[11] Thus, no attempt to distinguish elements that are essential facilities from those that are not has been made. For example, while loops, particularly in low density areas, may qualify as essential facilities (perhaps only in the near term), other elements, e.g., interoffice transport, clearly are not, because competitive suppliers already exist.

[12] Federal Communications Commission, Implementation of the Local Competition Provisions of the Telecommunications Act of 1996 and Interconnection between Local Exchange Carriers and Commercial Mobile Radio Service Providers, CC Docket No. 96-98 and CC Docket No. 96-185, August 8, 1996, Rule 51.505.

- Different forecasts of the economic environment, represented by cost of capital and depreciation lives
- Different estimates of the magnitude of common costs

4. Competitors' Concerns

During the numerous arbitration proceedings, the concerns of competitors fall into two major areas: (1) price level and (2) relative prices.

4.1 Price Levels

As indicated above, parties interested in buying services from ILECs tend to want low prices and the ILECs tend to propose higher prices. In each case, the motivation is understandable. For example, competitors currently purchase carrier access services from the ILECs, an input that has accounted for 40 to 50 percent of the IXCs' production costs. Cost-based prices for access can be as low as one-tenth (or lower) of current rates. Similarly, as noted above, pricing all elements at the cost-based prices proposed by the Hatfield model would cause the ILECs to lose up to one-half of their current revenues.

4.2 Relative Prices

The comparative prices for (1) the ILECs' retail offerings, (2) wholesale services, (3) unbundled elements, and (4) the prices facilities-based entrants will be able to charge will be important factors in how local exchange competition emerges. Therefore, these relative levels may be an even more important concern than the absolute levels.

A major concern of potential entrants is that the margin between the ILECs' retail prices and what they buy from the ILECs is large enough to allow them to recover the costs of operations. The ILECs, of course, have the opposite concern over a margin that is too generous. In addition, particular ILECs have expressed concern that retail rates, wholesale rates, and unbundled element rates have a rough balance, consistent with underlying cost characteristics. For example, a well-designed wholesale discount would be undermined if competitors could build the same service from unbundled elements (which is permitted under the Act and Order).

Potential facilities-based entrants have a somewhat different concern. If wholesale prices and/or prices for unbundled elements are too low, they simply may not be able to compete. After all, the essence of unbundling and resale requirements is to provide potential entrants with a make/buy option when developing their own facilities. If the "buy" option is always more attractive as a result of excessively low prescribed prices, investment in alternative networks becomes problematic.[13]

[13] An executive of an ILEC recently characterized emerging resale requirements in the following colorful terms: "It's like requiring a local pizza parlor on Worcester Street in New Haven or on Franklin Avenue in Hartford to let a national chain like Pizza Hut use its brick oven whenever Pizza Hut wants..." "How Good a Deal is Telephony Resale for Cable?" *Cable World*, December 9, 1996.

4.3 Competitive Strategies

As the preceding sections suggest, what will emerge from the dictates of the Telecommunications Act and the FCC's implementation actions is still very much a work-in-progress. Therefore, definitive statements about competitive strategies are premature. Indeed, companies seem to be changing strategies by the day, as illustrated by TCI's apparent pull back from an all-out telephony strategy and the recent abandonment of the video joint venture by Pacific, Bell Atlantic and NYNEX. At present, there are more questions and possibilities than answers.

From the entrants' point of view, a key question is whether the prices ILECs eventually charge will be attractive enough and certain enough to make resale an attractive long-term proposition. A recent article in the cable television trade press suggests that the answer is not clear at present.[14] Various participants offered the following perspectives.

- Prices that are too generous would make CATV infrastructure less attractive to potential partners such as the IXCs
- Discounts emerging from regulatory deliberations do not provide enough margin
- Uncertainty regarding what regulators will finally do is inhibiting action
- There may be better prospects from reselling other services, such as long-distance and wireless.
- There is some question as to what brand, e.g., their own or an IXC's, provides the greatest competitive advantage.

There are also critical questions for the ILECs. For example, should the resale and unbundling requirements be viewed as a threat to the ILECs' existing business or an opportunity for expansion through the use of new sales channels? Of course, the absolute and relative price levels will be crucial, as was discussed above. If the prices are right, some ILECs may well determine that, at least for some services, the strength of a new sales channel can outweigh the erosion of the retail operation. For example, the *Cable World* article discussed above notes that: "NYNEX says it's looking at the resale market as another revenue stream and will hire 70 customer-service representative[s] to handle the market."[15]

A related, but even more fundamental question, is whether the fact that ILECs seem to view recently regulatory actions as constituting pervasive regulation for the foreseeable future might lead to voluntary restructuring as a means of mitigating the most stringent aspects of that regulation. A few years ago, Rochester Telephone (now Frontier) proposed a separation of wholesale and retail offerings into different subsidiaries. More recently, Southern New England proposed and in June 1997 the Connecticut Department of Public Utility Control, its regulator, approved a plan to separate the wholesale and retail operations of the company. Such restructuring (either by voluntary separations or combining through alliances) may be as important in the emerging competitive landscape as particular competitive strategies under current corporate structures.

---

[14] *Cable World*, December 9, 1996, *op. cit.*

[15] The article also suggests that the requirement that the local exchange markets must satisfy certain requirements, including the availability of resale, before ILECs are allowed to compete in long-distance markets could be an incentive for ILECs to promote resale.

5. Conclusions

Important decisions on the nature and prices of the services incumbent local exchange companies will be required to offer to competitors are being made throughout the country. Although there is a great deal of uncertainty regarding the outcome of these decisions and the shape of the local exchange industry that emerges in response to these prices, there is no doubt that the structure of the industry will change in fundamental ways. In their attempts to provide a "level playing field," regulators run the risk of tilting that field to the advantage of particular players.

The experience in efforts to establish prices through regulatory processes instead of through market outcomes and/or voluntary negotiations raises an interesting issue. In the guise of promoting competition, regulators have arguably increased the scope and level of regulatory activity. Indeed, the unbundling and resale requirements extend regulation into every facet of the ILECs' operations. Further, the models used to establish prices typically assume that the elements are to be priced as if they were truly monopoly services for the indefinite future. The fundamental question is whether this level of regulatory activity is superior to less regulation and/or voluntary negotiation, and whether it can be sustained for very long into the future.

Telecommunications Transformation: Technology, Strategy and Policy
E. Bohlin and S.L. Levin (Eds.)
IOS Press, 1998

CHAPTER 23

Designing Regulation to Achieve Universal Service Goals: Unilateral or Bilateral Rules

Barbara A. Cherry

Northwestern University, Department of Communications Studies
1881 Sheridan Road, Harris Hall, Evanston, IL 60208-2236, USA

Abstract. Government may use various mechanisms to achieve social goals that would not otherwise be fulfilled in an unregulated marketplace environment. For example, in the United States the new federal Telecommunications Act of 1996 (TA96) embraces competition in the telecommunications industry but also creates requirements to fulfill social goals, such as universal service. Applying the typology of unilateral-bilateral rules and associated principles of sustainable regulation developed by Steve Wildman and Barbara Cherry, this chapter analyzes the ability to achieve universal service goals in a competitive environment under the recent FCC Universal Service Order. Conclusions include that, although some of the FCC rules – such as those implementing the federal levy on carriers to fund universal support – appear on their face to be symmetrically imposed unilateral rules, they are problematic in application due to their asymmetry in effect among providers and their incompatibility with other state and federal rules. Other rules, such as Lifeline and Linkup programs for low income support, are properly designed and sustainable as bilateral rules. Finally, but perhaps most importantly, although the FCC appears to have corrected the Federal-State Joint Board Recommendations by imposing carrier of last resort obligations as part of a bilateral rule, the sustainability of such obligations is threatened by Congress' asymmetric imposition under TA96 of certain unbundling and resale requirements on only incumbent local exchange carriers.

1. Introduction

Historically, telecommunications services have been provided through heavy governmental intervention. This intervention has consisted of some form of monopoly status conferred on the telecommunications provider. In its earliest stage, monopoly status was granted by virtue of telephone patents. After expiration of these patents, and an ensuing period of competition, monopoly status was again conferred in a different manner. In the United States, beginning in the late nineteenth and early twentieth centuries, state regulatory commissions were created over time by their respective state legislatures, with the authority to grant monopoly franchise rights to local exchange providers in geographic serving areas. Internationally, most countries also adopted a monopoly structure by providing telecommunications services through government-owned entities.

This monopoly structure has been breaking down in recent decades. In the United States, it began with a series of orders by the Federal Communications Commission (FCC) to open the interstate long-distance telecommunications market to competitive providers [1]. Since the 1980s, many states have also opened the intrastate, intraLATA toll market to competition [2].

This trend towards competition is now extending to the local exchange market. Although a few states in the United States had permitted local exchange entry in recent years, Congress has now mandated in the Telecommunications Act of 1996 (TA96) that all states permit local exchange competition. Internationally, this same development is occurring through privatization of government-owned telecommunications entities and liberalization of telecommunications markets, driven by actions such as directives of the European Union and the recent agreement of the World Trade Organization.

A shift from monopoly to competition in the provision of telecommunications service, particularly as to the provision of local exchange service, poses new challenges for governmental bodies. This is because, although pressing for competitive provision of telecommunications services, governments still seek to pursue other social goals that would not necessarily be fulfilled in an unregulated marketplace. Such social goals include the provision of universal service, that is, the notion that all citizens should have access to the telecommunications network at affordable prices.

In the past year, significant developments have occurred concerning universal service. In the United States, a Federal-State Joint Board was established pursuant to section 254(a)(1) of TA96 and issued recommendations on November 8, 1996, for implementing the universal service provisions of section 254 [3]. Considering these recommendations, the FCC then released an order [hereinafter referred to as the "FCC Universal Service Order"], pursuant to section 254(a)(2), on May 8, 1997, in Docket No. 96-45 [4].

It has been widely acknowledged by industry members, regulators, and economists that the traditional regulatory mechanisms for fulfilling universal service goals are not sustainable in a competitive environment. In this context, sustainability means that the underlying universal service goals are in fact achievable over time through the coexistence of the regulatory requirements and competition. Because fulfillment of universal service goals depends, in part, upon the financial viability of firms bearing universal service obligations, sustainability also includes the ability of firms to remain financially viable given the governmental obligations imposed upon them. The question for policy makers is how to design governmental interventions for universal service that are sustainable with competition.

The author and Steve Wildman have developed a typology of economic regulation to provide an analytical framework for describing and designing regulatory mechanisms. This typology – based on a distinction between unilateral and bilateral rules – maps social goals concerning marketplace activities with regulatory interventions that are necessary to accomplish those goals. This typology has been used for purposes of understanding and designing sustainable government interventions for universal service goals in a competitive telecommunications environment [5].

This chapter describes the unilateral-bilateral rule typology and associated principles for designing sustainable regulation. It then applies the typology and principles to various rules embodied in the FCC Universal Service Order. It is shown that the underlying universal service goals are likely to be sustainable under some of the rules but not others. First, some rules, such as a universal service levy, can be sustainable when imposed as unilateral requirements. However, the FCC rules in this regard are problematic because they impose some asymmetric effects among providers and are inappropriately combined with other rules. Second, other rules must assume a bilateral form in that the telecommunications providers must receive some form of compensation or consideration from government to better ensure fulfillment of the associated governmentally imposed requirements. In this regard, the FCC has properly designed Lifeline and Linkup programs for low income support. However, appropriate treatment of carrier of last resort obligations is complicated by the asymmetric imposition under TA96 of unbundling and resale requirements on incumbent local exchange companies. Finally, some rules, such as federally funded discount rates for certain educational institutions, may be unnecessarily burdensome in scope given the unilateral promises – that is, the legally unenforceable activities – of industry players. Specific modifications, which are required to make various of the FCC rules sustainable, are also discussed.

2. Unilateral/Bilateral Rules and Unilateral Promises

Social goals, and economic activities for accomplishing them, may be provided to society in various ways – through unregulated or regulated activities. Regulated activities themselves may take on many forms. Government may supply the product or service itself, or it may regulate privately owned suppliers. This typology focuses on the characteristics of regulatory mechanisms which are created to induce certain behavior by privately owned companies. These characteristics are reflected by the classifications of unilateral or bilateral rules.

Unilateral rules are performance requirements imposed by government on firms as a condition for providing services without any assurance by government that the affected firms will be able to generate revenues sufficient to cover the associated costs. Such rules may also be viewed as granting a benefit or privilege, such as a tax credit, but this paper is concerned with unilateral rules that impose performance requirements. Examples of unilateral performance requirements are taxes or levies, minimum wage laws, workplace safety requirements, and product safety standards.

Bilateral rules differ from unilateral rules in that the affected firms receive some form of compensation or special consideration in exchange for meeting the governmentally specified performance obligations. With a bilateral rule, the government and regulated firm acknowledge mutual and specific obligations toward each other, as in legal contracts.

Within the category of bilateral rules, there are two types. *Bilateral agreements* are performance requirements imposed by government on firms which are coupled with financial compensation to cover some of the costs associated with the requirements and to enable their performance. *Bilateral commitments* are performance requirements accepted by firms in exchange for which government accepts some degree of responsibility and provides some form of assurance for the financial health of the firms taking on the obligations. This includes the provision of safeguards against the threat of regulatory expropriation of the investments required to provide service.

Lifeline and Linkup programs in the U.S., which provide funding to local exchange companies for the provision of services to low-income customers, are examples of bilateral agreements. In this case, the government provides explicit funding to the firms but assumes no responsibility for their overall financial health.

Traditional monopoly franchises granted to public utilities in the U.S. are a form of bilateral commitment. This is because the regulated private firm agrees to provide service at a certain price in exchange for a monopoly franchise granted by government. The monopoly franchise, in turn, gives the firm a reasonable expectation of providing service at compensatory rates for a sufficiently long period to realize a fair return on its sunk investments. As so eloquently expressed by Goldberg, this arrangement developed to solve economic problems requiring a long-term contract, and is often referred to as a regulatory contract [6]. Patent laws are another example of a bilateral commitment, where the prospect of earning a return on investments in innovations is given through exclusive use – that is, a patent – of the innovation over a substantial period of time.

In any event, all unilateral and bilateral rules created by units of government are subject to certain limitations. For example, in the United States, all rules by any unit of government are subject to the provisions of the U.S. Constitution. Furthermore, all state and local units of government are subject to the provisions of the respective state constitutions. In Europe, the rules enacted in each member State are subject to the directives of the European Union as well as the provisions of the respective Member State's constitution. This typology of economic regulation is depicted by a flowchart in Figure 1.

Given that unilateral and bilateral rules are forms of governmental interventions, they are distinguishable from the voluntary activities, or *unilateral promises*, of private entities. In this context, unilateral promises refer to non-enforceable pledges or contributions by private parties, whether in terms of money, grants, or provision of in-kind goods or services. Examples of such non-enforceable pledges or contributions include corporate foundation grants, donations of equipment, and sponsorships of art performances or exhibitions. Unilateral promises have played a significant role to date in providing educational institutions with computers and access to the public telephone switched network in the U.S. Policy makers should always keep in mind that the impact of unilateral promises in the marketplace may be sufficiently significant to mitigate or even obviate the need for direct government intervention.

3. Principles for Sustainable Regulation

In developing this typology, several principles for designing sustainable, governmental interventions for achieving universal service in a competitive environment were identified [5]. As previously stated, sustainability in this context refers to the achievability of policy goals, here universal service goals, given the full breadth of regulatory requirements imposed on telecommunications carriers. This concept includes the ability of carriers to remain financially viable entities due to the nature and burden of the governmental obligations imposed on them.

First, there must be compatibility among policy goals and governmental interventions. In this regard, two forms of compatibility are required. The first form, referred to here as single goal-rule compatibility, requires compatibility between a given goal and the associated governmental obligation. In other words, there must be a "cause and effect" relationship between the governmental requirement and goal so that an entity's fulfillment of the governmental requirement will likely lead to achievement of the goal for which the requirement was implemented. For example, subsidized prices for *local* rates will not increase telephone subscribership among households which do not subscribe to service due to high *toll* bills.

The second form of compatibility, referred to here as combinatorial compatibility, requires compatibility among any combination of goals and their associated obligations. This requires a different form of cause-and-effect relationship among goals and rules so that combinations of goals and rules do not work at cross-purposes, rendering certain goals necessarily unfulfillable. Thus, even though each goal-rule pair may satisfy the principle of single goal-rule compatibility, a combination of goals and/or requirements may not. For example, fiber-optic cable to the home and low-cost local service are examples of goals that cannot be achieved simultaneously, at least using current technology.

Second, governmental interventions must be scrutinized for transitional sustainability problems which arise from the effects of government action on the preexisting circumstances and investments of private parties. In other words, some parties may have made reasonable investments under a given regulatory regime which would be adversely affected by a proposed *change* in regulatory conditions. Such adverse effects may include rendering any commercial use of private real property impossible or a firm financially unviable. Transitional sustainability problems are most commonly known as confiscation problems. In the U.S., confiscation (or trapping of costs) claims are typically raised under the Takings or Due Process Clauses of the U.S. Constitution. However, adverse effects of government action on preexisting investments have also been found under the Supremacy and Commerce Clauses. Remedies for claims based on adverse consequences to preexisting investments of private parties include government payment of compensation for damages to the private party or even invalidation of the governmental regulation in question.

Third, government interventions must also be analyzed for long-term sustainability problems, which arise independent of preexisting investments. Long-term sustainability problems typically arise when unilateral rules are applied asymmetrically among otherwise similarly situated parties. For example, the viability of a firm may be affected by the asymmetric application of a tax or fee to one type of firm in an industry but not to other firms within that industry providing competing services. In the U.S. such long-term sustainability problems are addressed by the Equal Protection Clause of the U.S. Constitution. In this regard, various state tax laws have already been held unconstitutional in the telecommunications industry because they were applied differently between resellers and facilities-based providers [7, 8].

Fourth, to fulfill the preceding principles, unilateral and bilateral rules must be applied in a competitively neutral manner. As to unilateral rules, this means that obligations imposed on private parties must be competitively neutral, not only on their face, but in their effects on the parties. Some rules, although imposed on all parties as a matter of law, may have differential effects among parties because some parties may be in a better position to evade enforcement of the rule. This concern, for example, has already been raised in the telecommunications industry with regard to common carrier obligations. Given that incumbent local exchange carriers currently have facilities deployed throughout their service area whereas new local exchange entrants do not, new entrants can more easily avoid serving less profitable customers and engage in cream-skimming.

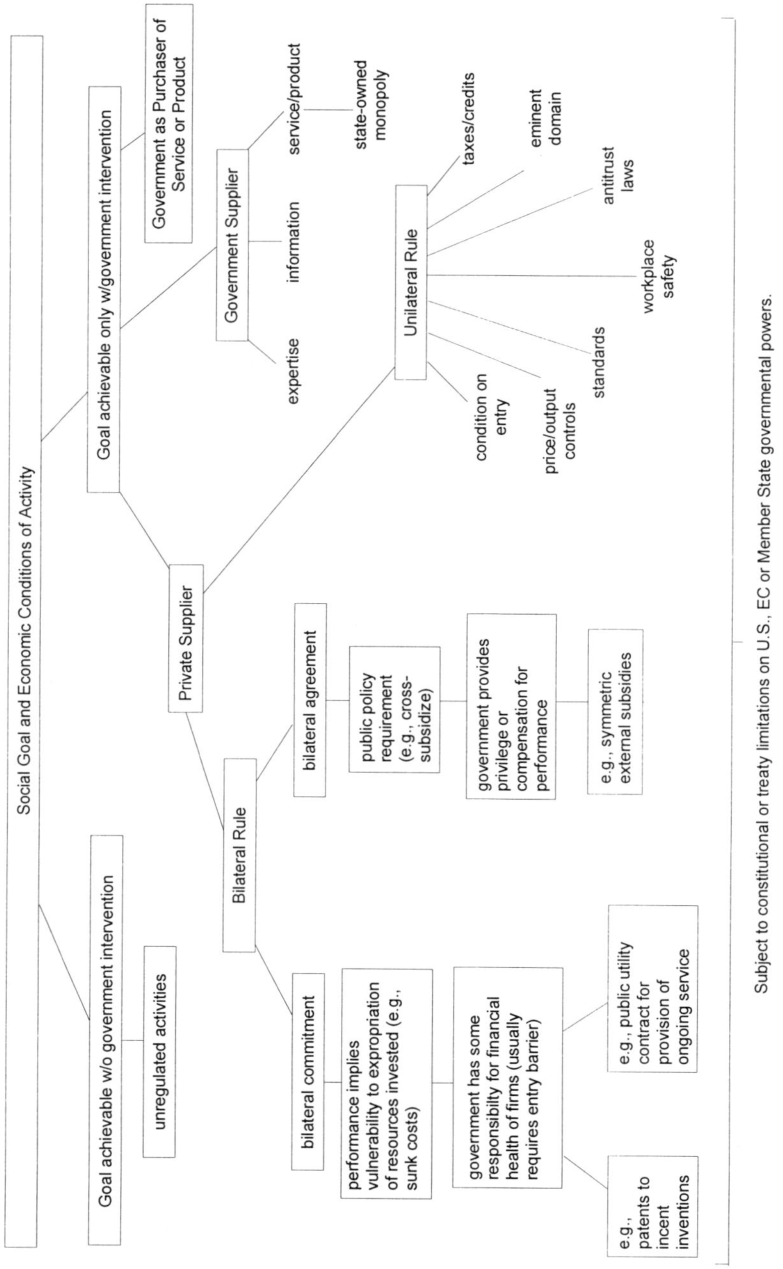

Figure 1. Typology of Economic Regulation based on Government Actions to Accomplish Social Goals
Source: [5]. Reprinted with permission.

As for bilateral rules, competitive neutrality means that there is appropriate matching of compensation and performance requirements so that the amount of consideration is adequate and provided only to those parties who assume the requirements. To the extent that entities receive compensation without assuming the corresponding obligations – a possibility which may exist with regard to carrier of last resort obligations discussed in section 4.5 – such entities receive a financial windfall which confers a competitive advantage relative to other competing entities.

Given these principles, some universal service goals can be achieved through imposition of unilateral rules, symmetrically applied, on telecommunications firms. Other goals, however, require the use of bilateral rules – that is, government must provide some form of compensation or privilege in exchange for performance by the private provider. Furthermore, different types of bilateral rules may be required, depending upon the private parties' vulnerabilities to expropriation of investments, such as with carrier of last resort obligations.

Thus far, governmental bodies have not fully appreciated the necessity of these four principles, nor the importance of this distinction between unilateral and bilateral rules, in designing sustainable regulatory requirements for universal service in a competitive environment. This is understandable in that the industry was traditionally regulated as a monopoly, which, by definition, rendered competitive sustainability concerns irrelevant. With competition, however, the requirements for sustainable regulation – and the role that unilateral and bilateral rules play in satisfying them – can no longer be ignored.

As the following discussion shows, in its Universal Service Order the FCC has recommended some rules that satisfy the above principles but others that do not. By analyzing what makes some of the proposed rules appropriate or inappropriate, important insights for designing sustainable government interventions for universal service goals can be gained. These insights are applicable for designing sustainable government interventions in any nation, not just the U.S.

4. Sustainability of the Rules in the FCC Universal Service Order

4.1 The Collection Mechanism

In section 254(b)(4), Congress states that the collection mechanism for universal service funds should consist of equitable and nondiscriminatory contributions from all providers of telecommunications services. Authority is granted to the FCC to require contributions from providers of interstate telecommunications services (sec. 254(d)), and to the state commissions to require contributions from carriers that provide intrastate telecommunications services (sec. 254(f)).

These collection mechanisms, analogous to taxes, are unilateral rules. To implement the federal collection mechanism, or levy, in its Universal Service Order, the FCC implemented a number of new rules. All of these rules, in turn, are unilateral rules, which can be categorized in terms of: how they identify the entities to be subject to the levy, the methodology for calculating the relevant revenue base upon which the levy is imposed, and the extent to which the levy can be recovered in service prices to customers. Each of these rules affects the sustainability of the overall unilateral rule – or collection mechanism – of which they are a part. Some of them may prove to be problematic due to their disparate impact, either standing alone or in combination with other rules, among providers. Modifications are then required to restore competitive neutrality and long-term sustainability of universal service.

4.1.1 Entities Subject to the Levy

In applying section 254(d), the FCC decreed that the definition of a carrier that provides interstate telecommunications services be construed as broadly as possible (pars. 779-791).[1] Thus, such carriers are to include both wireless and wireline providers (par. 780). By broadly defining the carriers subject to the levy, the FCC's unilateral rule is symmetrically applied to all carriers of interstate services. This is likely to be a sustainable rule, at least among all entities considered to be telecommunications carriers.

However, to the extent that there are substitutable services available from entities not considered to be telecommunications carriers under TA96, the long-term sustainability and competitive neutrality principles are violated. This is because, as frequently found by courts in the U.S. applying the Equal Protection Clause of the U.S. Constitution, the levy would have a disparate impact among competing providers, here carriers and non-carriers [7] [8]. This scenario is likely to occur with respect to Internet and enhanced service providers, because they are currently considered not to be telecommunications carriers under federal law and the FCC has declined to apply the levy to them (par. 788). To correct this potential sustainability problem, the entities subject to the levy would need to be expanded to include all the relevant providers of closely substitutable services.

4.1.2 Revenue Base

There are two aspects of the revenue base for the federal universal service fund. First, there is the jurisdictional aspect, that is, whether the levy applies to interstate and/or intrastate revenues of the telecommunications providers. In this regard, the FCC rules impose a levy on both the interstate and intrastate revenues of interstate providers to support the federal fund for educational institutions and health care providers (par. 818). As with other taxes and levies, this is a unilateral rule, but it is applied asymmetrically to providers of interstate services as opposed to providers of intrastate-only services because the intrastate revenues of only the former class of providers are subject to the rule. Such an asymmetric rule may pose long-term sustainability and competitive neutrality problems because of the disparate impact of placing a federal levy on intrastate revenues of interstate providers but not on the intrastate revenues of competing intrastate-only providers. The severity of the impact will depend, of course, on whether there are significant intrastate revenues earned by providers of only intrastate services.

It is unlikely that this asymmetry can be corrected by simply extending the federal levy to the intrastate revenues of intrastate-only providers, because to do so would be beyond Congress' authority under the Commerce Clause of the U.S. Constitution. For this reason, other remedies must be sought in order to avoid any category of provider – providers of interstate-only, interstate and intrastate, and intrastate-only service – bearing a different tax burden on its services that compete with those of another provider.

[1] The paragraph numbers refer to paragraphs in the FCC Universal Service Order [4].

The best available remedy is to: (a) permit the federal levy to be applied only to the interstate revenues of interstate carriers; and (b) require the state levies, if they exist, to be applied only to the intrastate revenues of intrastate service providers.[2]

Analogously, for purposes of low income and high cost support, the FCC assesses the federal levy on the interstate revenues of interstate providers (pars. 824, 831), where interstate revenues include the international revenues billed to domestic end users (par. 779). In this case, an asymmetry is created by assessing a levy on the international revenues of domestic, but not of non-domestic, carriers. Elimination of the asymmetry requires either removal of the levy from any international revenues or its imposition on the international revenues of non-domestic carriers as well.[3]

The second aspect of the revenue base is the definition of the relevant revenues upon which the levy is calculated. In this respect, the FCC Universal Service Order applies the federal levy to end-user revenues of telecommunications services (par. 844), including revenues from the subscriber line charge (par. 844) but excluding revenues from the universal service support fund itself (par. 857). The same rule applies to all telecommunications carriers and providers subject to the levy, and, as such, is a unilateral rule symmetrically applied to all providers.

At first glance, it would appear that this methodology is competitively neutral in that, by its terms, it does not place a double tax on any service, as would happen if the levy were placed simply on gross revenues. However, as described in the next section, the levy on end user revenues is not competitively neutral because of a combinatorial compatibility problem with other rules.

4.1.3 Levy on Net Revenues and Differential Abilities to Pass Through the Levy

In applying the federal levy to end-user revenues, the FCC expressly declined to adopt the Joint Board recommendation that the levy be assessed against net telecommunications revenues. This is because:

> Although the two assessment methods are theoretically equivalent, we conclude that, in practice, the net telecommunications revenues approach is likely to cause distortions that could be avoided by using the end-user telecommunications revenues approach. For example, *the theoretical equivalence of the two methods assumes that all carriers will be able to recover fully their contributions from their customers.* Some carriers, however, particularly those with long-term contracts, may be unable to recover fully those costs....Basing contributions on end-user telecommunications revenues[,]...[c]ontributors will not have more of an incentive to build their own facilities or purchase services for resale in order to reduce their

[2] Even this remedy, however, would not eliminate all arbitrage problems if the burdens of the federal and state levies on interstate and intrastate services differ. For example, if the intrastate levy is higher, there would still be the incentive to reroute intrastate calls in order to make them interstate ones and subject to the lower levy.

[3] In this situation, arbitrage between domestic interstate revenues and international revenues would occur if the levy applies only to domestic interstate revenues. However, it is unclear whether the FCC has the authority to extend the levy to the international revenues of foreign carriers and thereby to remove providers' incentives to reroute interstate calls to international ones in order to avoid the levy.

> contribution because, regardless of how the services are provided, their contributions will be assessed only on revenues derived from end users. (par. 850; emphasis added)

However, the intended superiority of the end-user revenue approach does not exist because of the coexistence of other FCC and state rules. As a result, as described below, a levy on end-user revenues is not competitively neutral *in effect* because all carriers will not have the same ability to pass the levy through to customers due to the existence of other regulatory rules.

There are two sources of other rules that prohibit the pass-through of levies, but they apply only to incumbent local exchange carriers (ILECs). The first source is the FCC Universal Service Order itself, which prohibits the ILECs from adjusting their rates for unbundled network elements to account for the amount of the levy (par. 851). Since competitive local exchange providers are not required to sell unbundled network elements, and the prohibition by its terms refers only to ILECs, this prohibition is a unilateral rule asymmetrically applied to ILECs. Given that LEC's are permitted to pass through their federal levy only onto interstate – primarily access – services, the prohibition of a pass-through to unbundled elements will effectively make interstate access services relatively more expensive. In this way, competitors may be induced to purchase ILEC-unbundled elements rather than access services. Furthermore, competitors will be induced to build their own facilities, even when economically inefficient to do so, in order to avoid paying for the pass-through to access charges. Thus, the prohibition in par. 851 creates the economic distortions supposedly avoided by a levy on end-user revenues described in par. 850. This is because the levy is not recovered in a competitively neutral manner across different forms of access to end users, creating greater difficulty for the ILECs to pass through *and collect* the levy from customers rather than their shareholders. To correct this sustainability problem, the prohibition must be eliminated.

The second source of rules consists of requirements in some state jurisdictions. In particular, many states have price caps or rate freezes, which may have the effect of prohibiting the pass-through of ILECs contributions to their customers. Yet, competitive local exchange providers are not subject to such requirements. This creates a higher levy burden on the shareholders of ILECs compared to the shareholders of their competitors. For this reason, such state requirements also undermine the long-term sustainability of the levy structure.

To *ensure* that carriers have the same ability to pass through the levy for universal service support, the FCC should provide an explicit mechanism whereby the carriers can collect the levy from customers. Such a mechanism should preempt all state requirements to the extent that compliance with them would prohibit pass-through of the levy by ILECs to the same extent permitted by competitive providers. This mechanism could also consist of requiring carriers to add a separate line item to customers' bills to recover the levy.

However, the FCC Universal Service Order expressly fails to provide an explicit mechanism whereby carriers can pass through the levies assessed to them into customers' rates. In particular, the FCC declines to mandate an end-user surcharge (par. 853). Furthermore, if a carrier seeks to pass any portion of the contributions in customer bills, carriers may not characterize its contribution as a surcharge (par. 855). Instead, the FCC merely permits, but does not require, carriers to recover contributions to the federal fund through rates for interstate services only (pars. 825, 829, 838). In reality, however, carriers will pass through the levies to customers, to the extent that the marketplace will allow, rather than pay the levies from retained earnings. Since ILECs will, due to the previously described rules, have greater difficulties in passing through the levies to customers, ILEC shareholders will bear a greater financial burden than the shareholders of competitive providers.

Thus, a combinatorial compatibility problem exists among the federal levy for universal service support and other federal and state regulatory rules. This incompatibility creates competitive non-neutrality, in effect, among providers. If the disparate financial impact to ILECs is sufficiently severe, their financial viability – at least for providing service for certain geographic areas, customers, and services – will undermine the long-term sustainability of ubiquitous service at affordable rates.

4.2 Rules for Low-income Customers

Section 254(b)(3) of TA96 provides that low-income customers are to have access to telecommunications and information services that are reasonably available, and at rates that are reasonably comparable to those provided in urban areas. With regard to this section, the FCC implemented several rules. Three of these rules will be analyzed below.

One method by which the FCC Universal Service Order provides for the benefit of low-income customers consists of revisions to existing Lifeline and Linkup programs (pars. 346-382). The key feature of these programs is that carriers would ultimately receive compensation from an explicit fund in exchange for providing lower rates under Lifeline and lower installation charges under Linkup. As such, these programs are bilateral agreements, applied symmetrically to all providers of Lifeline and Linkup services. So long as the amount of compensation paid to carriers (either under this rule or in combination with other rules) is sufficient to cover the amount by which rates or installation charges fall below costs for these customers, this bilateral agreement satisfies the competitive neutrality principle and is sustainable in the long run.

Another means of providing universal service to low-income customers is the requirement that carriers provide toll limitation services without charge to Lifeline customers (par. 385). This is also structured as a bilateral agreement in that carriers will be compensated from universal service support mechanisms for the incremental cost of providing such toll limitation services (par. 385). Given its symmetric application to all carriers providing Lifeline service, this rule should also be sustainable in the long run.

The FCC Universal Service Order also imposes some unilateral rules. For example, it prohibits carriers providing Lifeline services from disconnecting such service for non-payment of toll charges (par. 390). This rule is applied symmetrically to all providers of Lifeline services.

This rule would pose a sustainability problem if the costs to carriers of not being able to disconnect service, such as uncollectibles or bad debt, are large enough to threaten the financial ability of the carriers to continue providing universal service to customers. If so, there are various ways to remedy this problem. One is simply to remove the prohibition, but that may not result in sufficient numbers of low-income customers remaining on the network. A second remedy is to allow carriers to require customers to select toll blocking in order to prevent disconnection. This would limit the scope of possibly new, perhaps uncollectible, debt by such customers for non-toll services, which is usually much smaller than for toll services. A third remedy is to compensate providers for the uncollectibles they incur from such customers from the universal fund. The second remedy would effectively create a revised unilateral rule, whereas the third remedy would convert the unilateral rule to a bilateral agreement.

4.3 Rules for Services to Educational Institutions and Libraries

Section 254(h)(1)(B) of TA96 requires that telecommunications carriers provide telecommunications services, considered to be within the definition of universal service, at a discount to certain educational institutions and libraries. Sections 254(h)(1)(B)(i) and (ii) also provide that telecommunications carriers are to be compensated in an amount equal to the amount of the discount, either by an offset to contributions they would otherwise owe to universal service support mechanisms or by direct reimbursement from such support mechanisms. Thus, by law, Congress requires discounts by carriers as part of a bilateral agreement.

The FCC Universal Service Order imposes several rules to implement this bilateral agreement. Several aspects of these rules will be reviewed for their sustainability properties.

First, the FCC Order interprets section 254(h)(1)(B) to mean that carriers must be permitted to choose either reimbursement of money or an offset against the federal levy for purposes of being compensated in exchange for providing discounts (par. 586). So long as the methodology for calculating the amount of compensation (whether for reimbursement or offset) is sufficient to recover the true difference in rates charged to educational institutions and libraries as opposed to similarly situated parties for similar services, this bilateral agreement should satisfy the competitive neutrality principle with universal service sustainable in the long run. This is particularly true since carriers have the reimbursement option. Problems with providing only an offset option will be discussed in the next section where rules related to health care providers are analyzed.

Second, the FCC Universal Service Order states that discounts for educational institutions and libraries be provided not only for all telecommunications services (pars. 431-435), but also for Internet access (pars. 436-449) and internal connections (pars. 450-463). In this regard, Internet access includes information services and electronic mail (par. 444). Internal connections include the installation and maintenance of inside wire (pars. 451-456), as well as routers, hubs, network file servers, and wireless local area networks, but not personal computers (pars. 459-463). Third, by including discounts for Internet access and internal connections, the FCC Order establishes the amount of funding for educational institutions at $2.25B per year (par. 425).

In addition to asserting the illegality of including Internet access and internal connections, many parties question the wisdom of including these items and thereby creating such a large fund. Many businesses and community volunteers are already contributing to the connection of schools and libraries to the public switched telecommunications network. Telecommunications carriers have provided millions of dollars in donated computer equipment as well as grants to educational institutions. Communities have been holding "Net Days", whereby volunteers install wiring in schools and libraries, using materials and expertise donated by business. In 1996, over 30 states held Net Days; and on March 9, 1996, California's Net Day alone brought 20,000 volunteers into 2,000 California schools. Such activity has been encouraged through the Clinton Administration's political support for improving use of communications and computer technology in educational institutions.

With such support from businesses and communities, inclusion of internal connections in the universal service fund may be unnecessary and even wasteful. This is because, as many parties point out, use of the technology in schools will occur only if there are commensurate investments in training and integration of technology into course curricula. Therefore, more attention and dollars may need to be targeted to those purposes rather than just to more technology.

For these reasons, the voluntary activities – or unilateral promises – occurring in the private sector should reduce, if not eliminate, the amount of universal service funding dedicated to telecommunications infrastructure and equipment to educational institutions. The need to encourage, but not mandate, private investment in the technology infrastructure for education has already been recognized by Congress in TA96. In section 708, Congress expressly recognizes the existence of the National Education Technology Funding Corporation – a private, nonprofit corporation established in the District of Columbia – for the purpose of providing authority for federal departments and agencies to provide assistance to the Corporation in leveraging resources and stimulating *private investment* in education technology infrastructure.

Furthermore, in section 706 of TA96, Congress directs the FCC to take action, if necessary, "to accelerate deployment of [advanced telecommunications] capability [to all Americans, including, in particular, elementary and secondary schools and classrooms] by removing barriers to infrastructure investment and by promoting competition in the telecommunications market." Such action may utilize, for example, price cap regulation or regulatory forbearance, but does not encompass universal service support funds.

4.4 Rules for Services to Health Care Providers

In section 254(h)(1)(A), Congress requires telecommunications carriers to provide telecommunications services to any public or nonprofit health care provider (HCP) that provides health care services to persons residing in rural areas at rates that are reasonably comparable to rates charged for similar services in urban areas in that State. This section also provides that the difference between the urban and rural rates shall be treated as a universal service obligation under the support mechanisms established under the Act. Thus, Congress has created a bilateral rule for "subsidized" service to HCPs.

As to this bilateral rule, the Federal-State Joint Board recommended that compensation to carriers be provided only in the form of an offset towards their universal service support obligations, not direct reimbursements from universal service funds (pars. 734-737). Permitting offsets but no reimbursements creates an asymmetric impact between carriers that are annual net payors and those that are annual net recipients of federal universal service funds. Net recipients would effectively receive their compensation, if ever, at a later time because offsets would need to be carried over to subsequent years. This disparate impact would create a long-run sustainability problem if there are carriers whose financial viabilities are threatened because they are perpetually net recipients – which is likely, for example, for smaller local exchange providers. However, the FCC corrected this potential sustainability problem by permitting reimbursements to carriers to the extent that the amount of universal service support owed a carrier exceeds its universal service obligation (par. 734).

4.5 Rules for Carrier of Last Resort Obligations and Eligible Carriers

As with low-income consumers, under section 254(b)(3), Congress also establishes as a universal service principle that consumers in rural, insular, and high-cost areas should have access to telecommunications and information services that are reasonably available, and at rates that are reasonably comparable to, those provided in urban areas. Under section 214(e), providers are also eligible for distributions from universal service funds for providing service to rural and high-cost areas. Thus, TA96 creates a bilateral rule with respect to high cost support.

The imposition of carrier of last resort (COLR) obligations – customarily known as the requirement to expand capacity to serve customers and a barrier to exit – is a tool which has been used historically to ensure that all customers receive telecommunications services. The obligations have traditionally been placed on monopoly providers as part of a bilateral commitment, and have been particularly critical for ensuring provision of service in rural, insular, and high-cost areas.

However, the Federal-State Joint Board recommended that, for purposes of distributing high cost support, the COLR obligations should not be part of the bilateral rules. In particular, it recommended that the definition of an eligible carrier under section 214(e) be interpreted so that carriers are *not* required to take on COLR obligations in order to be eligible for universal service support. Instead, the Joint Board recommended that only ILECs be required to bear COLR obligations, and that all eligible carriers should receive the same amount of compensation for high cost support.

By imposing COLR obligations only on ILECs and not including them as part of the obligations of an eligible carrier (found in pars. 156-157 of the Joint Board recommendations), the Joint Board essentially proposes that COLR obligations be removed from a bilateral rule and be fulfilled through asymmetric imposition of a unilateral rule. Such a unilateral rule is unsustainable, not only because it is asymmetrically applied (i.e. not competitively neutral) but because COLR obligations are not sustainable as unilateral rules in a competitive environment.

Cherry and Wildman explain why COLR obligations are not sustainable as unilateral rules but require bilateral rules, whether in the short or long run [5]. First, significant cross-subsidies are unlikely to be enforceable as unilateral rules because of the fundamentally non-remunerative nature of the cross-subsidy requirement. In addition, symmetric enforcement of cross-subsidies would be extremely difficult, with asymmetries favoring those best positioned to avoid the cross-subsidy requirements. Since the very need to enforce COLR obligations implies that service must be provided at non-compensatory rates, such obligations also pose the threat of burdensome cross-subsidy and the associated problems of enforcement as unilateral rules. Furthermore, since new entrants can assure themselves lower burdens than incumbents through selective entry, unless symmetry in geographic coverage is required, asymmetric incidence of COLR obligations (i.e. non-competitive neutrality in effect) among ILECs and CLECs is virtually assured.

Second, the type of bilateral rule that is required depends upon the degree of vulnerability to expropriation of investment that is created by fulfilling the COLR obligations. To the extent that incremental sunk costs necessary to meet COLR obligations are large, then bilateral commitments – not simply bilateral agreements – are required. This is because large sunk costs pose substantial vulnerability to expropriation of investment, which is the economic problem that regulatory contracts are developed to address. However, it may be possible to use a bilateral agreement, instead of a bilateral commitment, if the vulnerability to expropriation of investment can be sufficiently lowered by allowing a different form of technology to be utilized by a carrier in fulfilling its COLR obligations. This might be achieved, for example, by permitting carriers to fulfill their COLR obligations on a wireless rather than a wireline basis. If a bilateral agreement can be used, then it also needs to be structured properly. At a minimum, compensation must be provided to carriers for the additional costs incurred in complying with their COLR obligations

The FCC Universal Service Order appears to mitigate the sustainability problems associated with the Joint Board's position on COLR obligations by restoring symmetry in application of the COLR obligations among providers and making COLR obligations part of the bilateral rule for high cost support. For example, some commenters maintained that all eligible carriers should bear the same COLR obligations. In reference to that aspect of COLR obligations which consists of exit barriers, the FCC found it unnecessary to impose additional exit barriers on competitive local exchange carriers (CLECs) comparable to those imposed on ILECs "because section 214(e)(4) already imposes exit barriers similar to the protections imposed by traditional state COLR regulation" (par. 143). In addition, in response to a request to permit carriers to serve something less than the entire service area (defined by the state), the FCC found that "the terms of section 214(e) do not allow us to alter an eligible carrier's duty to serve an entire service area" (par. 141). Thus, the FCC Universal Service Order incorporates COLR obligations as part of a symmetrically applied bilateral rule for high cost support eligibility by requiring all eligible carriers to stand ready to serve the entire service area and to bear similar exit barriers.

However, combinatorial compatibility problems may still jeopardize fulfillment of the COLR-type obligations contained in the high cost support eligibility rule. For example, a carrier may be an eligible carrier even if it intends to rely in whole or in part on unbundled network elements to provide service in a high-cost area (par. 154-156). Furthermore, a carrier may be eligible if it relies in part, but not solely, on resale (par. 178). Yet, TA96 imposes asymmetric requirements for unbundling and resale between ILECs and CLECs. Section 151(c) requires only ILECs to unbundle network elements at any technically feasible point and to offer any retail telecommunications service to other carriers at wholesale rates.

Such asymmetric unbundling and resale obligations means that a CLEC's acceptance of designation as an eligible carrier for high cost support, if based on some portion of resold service or unbundled network elements, effectively binds the underlying ILEC provider(s) to the obligations of an eligible carrier as well. Yet an ILEC's acceptance of an eligible carrier designation does *not* similarly impose any eligible carrier obligations on any CLEC. As a result, ILECs will, in actuality, bear greater COLR obligations than CLECs, and with fewer options (e.g. use of unbundled elements from other carriers) than CLECs as to how to fulfill such obligations.

Essentially, a CLEC will have the power to bind the underlying ILEC provider to a long-term supplier contract and to do so through strategic behavior. The sustainability of the preceding combination of rules will depend in part on how these long-term supplier commitments of the ILEC are addressed. Will CLECs be required to commit to minimum and/or maximum levels of unbundled elements over "contract periods" in order to enable ILECs to forecast demand for their facilities and to build adequate but not excessive capacity? Will

CLECs likewise be required to provide ILECs with forecasts of resale demand, enabling ILECs to plan for capacity requirements, and to pay for such default capacity even if it remains unused? Will the unbundling and resale obligations of ILECs remain asymmetric even after CLEC's build some of their own facilities? Until combinatorial compatibility problems – such as those arising from the treatment of unbundled access and resale of service – are adequately addressed, the sustainability of COLR-type obligations as a means of ensuring ubiquitous provision of service is seriously in question.

5. Summary and Conclusion

New technologies and increasing competition are rapidly dismantling the monopoly provision of telecommunications services in the global economy. To manage the transition to a more competitive regime, governments need to recognize the important distinction between unilateral and bilateral rules, and the circumstances under which it is appropriate to use these different types of rules. In some situations, it is appropriate for government simply to impose requirements, or unilateral rules. In others, the requirement must be accompanied by some form of compensation or consideration from government to the private party, that is, government must use bilateral rules. Furthermore, in some instances, government intervention may be mitigated or even eliminated given the scope of voluntary activities, or unilateral promises, of entities in the private sector.

Government interventions must also satisfy certain sustainability principles. These principles include single goal-rule compatibility as well as combinatorial compatibility among goals and rules. They also include the need for transitional and long-term sustainability, where the former derives from the effects of change in regulation on preexisting investments but the latter from the prospective application of regulation independent of preexisting investments. Finally, to satisfy the preceding principles, government intervention must be competitively neutral, both on its fact and in effect. This requires symmetric application of unilateral rules and a proper matching of obligations and compensation under bilateral rules.

Using implementation of the universal service provisions of the U.S. Federal Telecommunications Act of 1996 as an example, it has been shown that, in some cases, the FCC has appropriately structured unilateral rules and bilateral agreements. Imposition of a levy on *all* telecommunications carriers, broadly defined, to fund universal service support is an example of an appropriate, symmetrically applied unilateral rule. Generally, such a levy satisfies the principles of competitive neutrality. Unfortunately, further aspects of implementing this levy pose combinatorial compatibility problems. In particular, the coexistence of other federal and state rules effectively creates differential abilities among carriers to pass through the levy to customers, posing sustainability problems and requiring modifications to these rules.

The basic compensation arrangements to carriers for Lifeline and Linkup programs and for discounts provided to schools and libraries are examples of appropriate bilateral agreements. They satisfy the competitive neutrality principles with an appropriate matching of compensation to the performance requirement under the bilateral rule. However, other aspects of the FCC rules for low income support may require modifications to ensure sustainability and competitive neutrality. This is because some of the rules – such as prohibitions on disconnection of local service for non-payment of toll and use of service deposits – are imposed as unilateral rules but may require compensation, or a bilateral rule, due to the costs incurred by the providers.

In some cases, no government rule may be required. For example, some of the proposed funding in technology for educational institutions may be inappropriate given the level of voluntary support, or unilateral promises, already coming from the private sector.

Perhaps the greatest potential problem is the treatment of carrier of last resort obligations in a competitive environment. In this regard, the FCC appears to have corrected a serious problem with the Joint Board recommendations. The Joint Board had recommended imposition of COLR obligations as a unilateral rule. Instead, the FCC considered COLR obligations to be part of a bilateral rule, by clarifying that all eligible carriers for high cost support must provide service throughout the entire serving area and bear similar exit barriers. However, combinatorial compatibility problems – arising from the asymmetric opportunity of CLECs to use unbundled network elements and resale (at wholesale rates) to satisfy eligible carrier obligations, and the asymmetric obligations of ILECs to provide CLECs with unbundled elements and resale at wholesale rates – effectively create greater COLR obligations for ILECs. Such asymmetry (in effect) threatens the sustainability of ubiquitous service to high-cost areas through reliance on COLR-type obligations. Consequently, sustainability of service to all high-cost customers requires satisfactory resolution of such combinatorial compatibility problems.

References

[1] R. Horwitz, The Irony of Regulatory Reform. Oxford University Press, New York, 1989, pp. 126-138, 221-244.

[2] T.W. Bonnett, Telewars in the States. Council of Governors' Policy Advisors, Washington, D.C., 1996, pp. 57-74.

[3] Federal-State Joint Board on Universal Service, Recommended Decision, CC Docket No. 96-45, 12 FCC Rcd 87 (1996).

[4] In the Matter of Federal-State Joint Board on Universal Service, FCC Report and Order, CC Docket No. 96-45, May 8 (1997).

[5] B.A. Cherry, S.S. Wildman, A Typology of Unilateral and Bilateral Rules for Managing the Transition to Competition While Meeting Universal Service Goals, forthcoming in Universal Service in Context: A Multidisciplinary Perspective (Cherry, B., Hammond, A., & Wildman, S., editors). Muhwah, NJ, Lawrence Erlbaum & Associates.

[6] V. Goldberg, Relational Exchange: Economics and Complex Contracts, *American Behavioral Scientist* **23** (1980) 337-352.

[7] *MCI Telecommunications Corp. v. Limbach*, 68 Ohio St. 3d 195 (1994).

[8] *GTE Sprint Communications Corp. v. Wisconsin Bell*, 155 Wis 2d 184 (1990).

Telecommunications Transformation: Technology, Strategy and Policy
E. Bohlin and S.L. Levin (Eds.)
IOS Press, 1998

CHAPTER 24

The U.S. Telecommunications Act of 1996, The Internet, and Access Charges

A Case Study

William F. Maher, Jr. and Jeffrey Magenau

Halprin, Temple, Goodman & Sugrue
2808 Valley Dr. Alexandria, VA 22302, USA

Abstract. This paper presents a case study of certain pricing policy issues linking the provision of traditional telecommunications services and Internet service in the United States. It recounts the historical rationale of the U.S. access charge regime, surveys recent calls for change, and analyzes the potential application of access charges, in some form, to Internet service providers. The relevant policy underpinnings of the U.S. Telecommunications Act of 1996 are also examined. The paper describes the Federal Communications Commission's implementation of its "competition trilogy" in its effort to develop fully competitive markets for telecommunications and information services in the United States. The study concludes that, given the difficulties of distinguishing among the various users of telecommunications networks, an access charge regime that allows greater flexibility for access providers to recover the costs incurred by access customers is both fair and efficient.

1. Introduction

U.S. implementation of the Telecommunications Act of 1996 (the "1996 Act")[1] may be viewed by other administrations as an ongoing regulatory experiment. This paper describes the policy underpinnings of the 1996 Act and their effect on two important U.S. information industries – specifically, the "traditional" telecommunications industry and the Internet industry. It analyzes an ongoing dilemma that implementation of the 1996 Act has escalated: the uneasy relationship between Internet services and "access charges" – the U.S. system of charges for interstate access to the networks of local exchange carriers, codified in a complex set of rules of the U.S. Federal Communications Commission ("FCC").[2] These charges, which were first adopted in 1983, are a major component of the compensation ar-

[1] 1996 Act, Pub. L. No. 104-104, 110 Stat. 56. The 1996 Act amends the Communications Act of 1934, 47 U.S.C. § 151 et seq. Hereinafter, all citations to the 1996 Act will be to the relevant sections of the United States Code unless otherwise noted.
[2] See 47 C.F.R. § 69.1 *et seq.*

rangements in the U.S. telecommunications system; for example, more than 25% of local carriers' total regulated revenues come from interstate access services.[3]

Spurred by passage of the 1996 Act, the FCC undertook a broad review of its access charge rules. After reviewing public comments from a wide range of industry participants and others, the review culminated in the release of a rulemaking known as the "Access Charge Order".[4] Because of the explosive growth and high visibility of Internet usage in the United States, a major issue in that review concerned the type of access charges, if any, that should be imposed on Internet service providers ("ISPs"). Currently, ISPs are treated as "end users" under the FCC's access charge rules, and as such are able to obtain flat-rated "business line" services under state regulation to receive Internet traffic. In contrast, long-distance providers such as AT&T, MCI, and Sprint pay usage-based rates for FCC-regulated access to local carriers' networks, based on their classification as "interexchange carriers" under the FCC's rules. In the new environment spawned by the 1996 Act, the access charge treatment of ISPs has profound implications for the development of U.S. communications.

2. The Telecommunications Act of 1996

The 1996 Act relies on substantial governmental intervention to establish competitive telecommunications markets through which social goals are to be met. The preamble of the 1996 Act states that its ultimate goals are to "secure lower prices and higher quality services for American [*i.e.*, U.S.] telecommunications consumers and encourage the rapid deployment of new telecommunications technologies".[5]

Although parts of the 1996 Act speak of deregulation, it is more accurately described as both highly regulatory and heavily market-oriented. The 1996 Act embraces a market philosophy by permitting entry into markets, particularly long-distance and local telephony and cable television, that previously were subject to governmental barriers to entry. However, it also contains detailed regulatory requirements that nominally seek to ensure the development of competition in such markets. For example, the FCC has listed 38 separate issues raised in the 1996 Act for it to address in implementation proceedings.[6] The 1996 Act thus generally regulates for competition, rather than deregulates. As discussed below, such statutory ambivalence affects different industries in different ways.

---

[3] See Access Charge Reform, Price Cap Performance Review for Local Exchange Carriers, Transport Rate Structure and Pricing, Usage of the Public Switched Network by Information Service and Internet Access Providers, Notice of Proposed Rulemaking, CC Docket Nos. 96-262, 94-1, 91-213, 96-263, FCC 96-488 (rel. Dec. 24, 1996) ("Access Charge Notice") at para. 29.

[4] Access Charge Reform, First Report and Order, CC Docket Nos. 96-262, 94-1, 91-213, 95-72 (rel. May 16, 1997) ("Access Charge Order").

[5] S. Conf. Rep. No. 104-230, 104th Cong., 2d Sess. 1 (1996).

[6] See Draft FCC Implementation Schedule For The Telecommunications Act of 1996, <http://www.fcc.gov/Reports/implsched.html> (rev. Mar. 24, 1997), on file with authors.

3. Traditional Telecommunications Industry

For "traditional" U.S. telecommunications industries – particularly the long-distance and local telephone businesses and users – the 1996 Act contemplates potentially earth-shaking changes, but only after completion of administrative processes based on those associated with traditional regulation. The Act seeks to introduce competition into U.S. long distance and local markets where it previously had been limited by regulation or judicial decree. However, consistent with the 1996 Act's emphasis on the development of new forms of regulation designed to promote competition, market-opening efforts to date primarily have been contested before regulatory agencies and the U.S. courts rather than in the markets themselves. For example, numerous individual proceedings are ongoing in which state regulatory commissions arbitrate differences in interconnection arrangements proposed by incumbent local exchange carriers and potential new entrants.

Most prominently, to implement the 1996 Act, the FCC undertook three major rule-making proceedings that it refers to as the "competition trilogy".[7] The component of the trilogy of most interest for our purposes is the FCC's access charge reform proceeding. The Access Charge Order was preceded by a notice of proposed rulemaking ("NPRM"), which reviewed broadly the overall interstate access charge structure.[8] A motivation for that review was the FCC's recognition that the existing access rules were not designed to address the potential effects of competition in U.S. markets for local exchange and exchange access services.

The proceeding also includes a more preliminary notice of inquiry ("NOI") into the use of the U.S. public switched network by Internet access and information service providers.[9] An NPRM differs from an NOI in that an NPRM can result in the adoption of final rules on the issues it raises. An NOI, by contrast, is a tool for gathering information that may be the basis for a future NPRM. The NOI sought to "identify what FCC policies would best facilitate the development of the high-bandwidth data networks of the future, while preserving efficient incentives for investment and innovation in the underlying voice network".[10]

Another part of the FCC's trilogy, the interconnection proceeding, sought to implement the interconnection, dialing parity, and nondiscriminatory access requirements imposed on so-called incumbent local exchange carriers.[11] These rules are designed to permit new entrants into local telephony to use incumbent local exchange carriers' facilities to compete effectively. Several portions of the rules adopted in that proceeding, including the pricing rules, were vacated by the U.S. Court of Appeals for the Eighth Circuit.[12]

The third part of the trilogy sought to implement the provisions of the 1996 Act that are designed to preserve and advance "universal service" in the United States. These provisions are based on principles that address quality and rates, access to advanced services, access in rural and high-cost areas, equitable and nondiscriminatory contributions to universal service funds, specific and predictable funding mechanisms, and access to advanced telecommu-

---

[7] See Implementation of the Local Competition Provisions of the Telecommunications Act of 1996, CC Docket No. 96-98, First Report and Order, 11 FCC Rcd 15499 (1996) ("Interconnection Order") at 15507.

[8] See Access Charge Notice at paras. 21-299.

[9] See *id.* at paras. 311-318.

[10] *Id.* at para. 311.

[11] See Interconnection Order, *supra* note 7; Order on Reconsideration, 11 FCC Rcd 13042 (1996); vacated in part, Iowa Utilities Board et al. v. FCC, No. 96-3321 and consolidated cases (8th Cir., July 18, 1997).

[12] See Iowa Utilities Board et al. v. FCC, supra n. 11.

nications services for schools, libraries and rural health care providers.[13] Historically, support for U.S. universal service policies has been funded through a complex mix of implicit and explicit subsidies sanctioned by the FCC and state regulators. Pursuant to the 1996 Act, the FCC formed a "Joint Board" of FCC and state regulatory commissioners and a state-appointed consumer advocate to address universal service issues. In the fall of 1996, the Joint Board issued its Universal Service Recommended Decision, for consideration by the FCC.

On May 8, 1997, the FCC released a Report & Order ("Universal Service Order") adopting most – but not all – of the recommendations of the Joint Board. Of most interest for our purposes, the Universal Service Order extends to ISPs eligibility to receive universal service funds to permit them to offer discounted "conduit" services for Internet access to schools and libraries. Under the 1996 Act, schools and libraries are entitled to discounts for certain telecommunications services pursuant to gaining access to the Internet. Some parties to the proceeding argued that the statutory reference to supporting "access" to information services referred only to the basic telecommunications service linking a school or library to an ISP. The FCC determined, however, that "we are authorized to provide discounts on the data links and associated services necessary to provide classrooms with access to [Internet content]".[14]

While acknowledging that the services of ISPs are properly classified as information services, and, as such, arguably are not within the definition of services to be supported by universal service funds, the FCC determined that certain "information services are essential for effective transmission service, *i.e.*, 'conduit' service [to the Internet],"[15] and thus are to be supported by universal service funding. Such information services include navigational systems, introductory information, and e-mail. Although it is far from clear that the U.S. Congress intended to mandate discounts for such services – in addition to the basic telecommunications service needed to access the Internet – the FCC nonetheless determined that discounts must apply to such services offered by ISPs to eligible schools and libraries.[16] The Universal Service Order also concludes that the FCC is "authorize[d]... to permit schools and libraries to receive [at discounted rates] the telecommunications and information services provided by telecommunications *carriers* needed to use the Internet".[17] Because the FCC has stated on the record that ISPs are not telecommunications carriers, it is uncertain that the FCC has properly extended universal service support to the services of ISPs.

The Joint Board had determined that the record before it was insufficient to justify recommending support for Internet access to rural health care providers. In the Universal Service Order, the FCC determined that the record developed in response to the Joint Board's Recommended Decision indicated that "rural health care providers often incur large telecommunications toll charges and that these charges are a major deterrent to full use of the Internet for health-related telecommunications services".[18] Thus, the Universal Service

[13] See 47 U.S.C. § 254(b); see also Federal-State Joint Board on Universal Service, Recommended Decision, CC Docket No. 96-45, FCC 96J-3 (rel. Nov. 8, 1996) ("Universal Service Recommended Decision").
[14] Federal-State Joint Board on Universal Service, Report and Order, CC Docket No. 96-45 (rel. May 8, 1997) ("Universal Service Order") at para. 441.
[15] *Id.*
[16] See *id.* at paras. 591-599.
[17] *Id.* at para. 436.
[18] *Id.* at para. 631.

Order determines that "limited toll charges incurred by health care providers that cannot obtain toll-free access to an Internet service provider" should be supported by universal service funds.[19] Because the 1996 Act explicitly limits support for health care providers to the provision of telecommunications services, however, the Universal Service Order limits the support to telecommunications toll charges, rather than extending support to the underlying Internet access as it does for schools and libraries.

Because they are not telecommunications carriers, ISPs are not obligated to contribute to universal service funding. Thus, while the 1996 Act requires all interstate telecommunications carriers to contribute to universal service, ISPs benefit from universal service subsidies without paying into universal service funds. An additional legal argument thus exists that the FCC exceeded the authority of the 1996 Act by including ISPs among those eligible for universal service support.

4. Internet Service Providers

The 1996 Act does not directly address the economic structure of the newest of the U.S. information industries: ISPs. Although a portion of the 1996 Act sought to control so-called "indecency" on the Internet, most of that portion was held unconstitutional by the U.S. Supreme Court.[20] More generally, a policy of the 1996 Act is to "preserve the vibrant and competitive free market that presently exists for the Internet and other interactive computer services, unfettered by Federal or State regulation".[21] Perhaps in line with this policy, little else in the statute specifically addresses pricing structure, entry conditions, or availability (universal service) as they apply to the Internet, even though the statute addresses such issues comprehensively for the traditional telecommunications industry. As a practical matter, the FCC can determine the extent of government regulation of the Internet by interpreting the applicability of the 1996 Act to ISPs.

As part of the "competition trilogy" described above, the FCC's massive Interconnection Order mentions the Internet only twice. In the first instance, the FCC acknowledges that Internet use may affect periods of peak usage of local telephone lines.[22] In the second, it notes that some parties believe that access to the Internet serves the interest of the public and schools in accessing information.[23] The FCC declines, however, to adopt any specific rules in response to such parties.[24] The main thrust of the Interconnection Order – the rules and regulations for competitive access to network elements – is limited to traditional telecommunications services, *e.g.*, voice and data telephony.

The Universal Service Joint Board concluded that "provision of Internet service does not meet the statutory definition of a 'telecommunications service.'"[25] Although the Universal Service Order does not explicitly recognize the Joint Board's conclusion, it does find that "the information service component of Internet access cannot be supported under section

---

[19] See *id.*

[20] See ACLU v. Reno, slip op. 521 U.S. (June 26, 1997)(No. 96-511).

[21] 47 U.S.C. § 230(b)(2).

[22] See Interconnection Order, *supra* note 7, at para. 756.

[23] See *id.* at para. 1267.

[24] See *id.* at para. 1268.

[25] Universal Service Recommended Decision, *supra* note 13, at para. 69.

254(c)(1), which describes universal service as 'an evolving level of *telecommunications services.*'"[26] Thus, the Universal Service Order implicitly concludes that Internet service is not a telecommunications service, and thus its provision to residential consumers is not eligible for universal service support. In addition, the Universal Service Order adopts the Joint Board's finding that "access to the Internet... is provided though voice-grade access to the public switched [telephone] network,"[27] which is already covered by universal service rules. However, as discussed earlier, the FCC found independent grounds in the 1996 Act for extending universal service support to the "conduit" aspects of Internet service provision to schools and libraries.

5. Interstate Access Reform and Internet Service Providers

In the Access Charge proceeding, the FCC specifically considered whether to permit local exchange carriers to assess interstate access charges on information service providers, including ISPs. The FCC tentatively concluded that "information service providers should not be required to pay interstate access charges as currently constituted".[28]

In the Access Charge Order, the FCC adopted its tentative conclusion, finding that "the existing pricing structure for ISPs should remain in place". Thus, incumbent LECs are prohibited from assessing access charges for ISP access to the LECs' networks.[29]

As noted above, because ISPs are treated as end users under the FCC's access charge rules, they may obtain flat-rated "business line" services for receiving calls, rather than the usage-sensitive access services that must be taken by U.S. long-distance carriers. The business line rates paid by ISPs are significantly lower than the interstate access charges that long-distance carriers must pay. As the FCC recognizes, this in part reflects the fact that although business line rates often include usage-sensitive charges for outgoing calls, ISPs generally receive calls from their subscribers. In the Access Charge Order, the FCC states that "[w]e think it possible that had access rates applied to ISPs over the last 14 years, the pace of development of the Internet and other services may not have been so rapid".[30] The FCC does not offer any empirical or anecdotal evidence supporting this claim.

The basis for this treatment of ISPs is the FCC's distinction in the access charge rules between "carriers" and "end users".[31] In 1983, the FCC declined to treat as "carriers" a class of firms known as "enhanced service providers," treating them instead as end users.[32] In doing so, the FCC highlighted its regulatory distinction, in place since 1976, between enhanced and basic services. Enhanced services are defined in the FCC's rules as services offered over common carrier facilities that employ computer processing applications that act on the format, content, code, protocol, or similar aspects of the subscriber's transmitted in-

---

[26] Universal Service Order, *supra* note 14, at para. 83.
[27] Universal Service Recommended Decision, *supra* note 13, at para 69.
[28] Access Charge Notice, *supra* note 3, at para. 288.
[29] See Access Charge Order, *supra* note 4, at para. 344.
[30] *Id.*
[31] See MTS and WATS Market Structure, Third Report and Order, CC Docket No. 78-72, Phase I, 93 FCC 2d 241, 245 (1983).
[32] See MTS and WATS Market Structure, Memorandum Opinion and Order, Docket No. 78-72, 97 FCC 2d 682 (1983); See also Amendments of Part 69 of the Commission's Rules Relating to Enhanced Service Providers, CC Docket No. 87-215, Order, 3 FCC Rcd 2631 (1988).

formation; provide the subscriber additional, different, or restructured information; or involve subscriber interaction with stored information.[33]

In contrast, the FCC's concept of "basic service" is typified by an ordinary voice telephone call – transmitted over the common carrier facilities of long-distance and local telephone companies – the content of which is neither controlled, interpreted, nor modified by the carrier.[34] Historically, enhanced services have not been regulated by the FCC, while basic services have been subject to the FCC's full regulatory jurisdiction. When the FCC decided to treat enhanced service providers as end users for purposes of its access charge rules, it did so out of concern that, among other things, such charges could cause "rate shock" that would stunt the growth of the enhanced service industry.[35]

The 1996 Act does not contain a definition of "enhanced service," but does define "information service" in a very similar way. The FCC recently has used these terms somewhat interchangeably, and includes ISPs within the definition of enhanced service providers.[36]

In the Access Charge proceeding, the FCC noted that it has "long sought to avoid unnecessary regulation of information services".[37] The FCC also observed that as Internet usage continues to grow, such use may have a significant impact on the networks of the local exchange carriers – the public switched telephone network or "PSTN". As a result, the NOI asks how the FCC can provide incentives for investment and innovation in the public networks that would lead to optimum efficiency and reliability in managing Internet traffic in addition to traditional telecommunications services.

As noted above, the Access Charge Order concludes that ISPs should not be subject to interstate access charges. The Order finds that the access charge system contains non-cost-based rates and inefficient rate structures. The Order claims to begin the process of removing such inefficiencies and moving towards eventually recovering access costs through a flat-rated mechanism. Moreover, because the Order determines that "it is not clear that ISPs use the [PSTN] in a manner analogous to IXCs,"[38] it concludes that it would not be reasonable to extend to ISPs a regime designed to recover the costs of providing access to IXC end users.

The Access Charge Order also concludes that "the nonassessment of access charges [does not] result in ISPs imposing uncompensated costs on incumbent LECs".[39] The Order finds summarily that incumbent LECs are adequately compensated from: ISPs paying for their connections under state tariffs; incremental revenue from second lines; usage of dedicated data lines by ISPs; and subscriptions to incumbent LEC Internet access services.[40] The Access Charge Order does not attempt to match the costs incurred by ISPs with the costs to be recovered from them. In addition, the Order concludes that to the extent that

---

[33] See 47 C.F.R. §64.702(a).

[34] See Amendment of Section 64.702 of the Commission's Rules and Regulations, Docket No. 20828, Final Decision, 77 FCC 2d 384 (1980).

[35] See MTS and WATS Market Structure, Memorandum Opinion and Order, *supra* note 32, 97 FCC 2d at 715.

[36] See Access Charge Notice *supra* note 3, at para. 284 ("The category of enhanced services, which includes access to the Internet...").

[37] *Id.* at para. 282.

[38] Access Charge Order, *supra*, note 4, at para. 345.

[39] *Id.* at para. 346.

[40] *Id.*

intrastate tariffs do not adequately recover costs incurred through provision of service to customers with high volumes of incoming calls, such matters may be addressed to state regulators.[41]

In responding to incumbent LECs' allegations that ISPs threaten the reliability of the PSTN, the Access Charge Order finds that "the extent to which [ISP] usage creates congestion [in the PSTN] depends on the ways in which incumbent LECs provision their networks, and ISPs use those networks".[42] Thus, the Access Charge Order determines that "[t]he public interest would best be served by policies that foster... technological innovation in the network".[43] Even when stripped of its current inefficiencies, the Access Charge Order concludes, "[the access charge system] may not be the most appropriate pricing structure for Internet access and other information services".[44] The FCC does not say, however, what the "most appropriate" pricing structure is.

While the pronouncement of the Access Charge Order regarding ISPs and access charges is definitive for the moment, presumably it does not represent a permanent moratorium on U.S. policy development. In light of the fact that the fundamental structure of U.S. access charges was not changed in the Access Charge Order, many basic questions remain regarding the application of that structure to ISPs. For example, the end user/carrier distinction of the prior rules has been maintained, and the complex system of flat- and usage-based charges has been modified but not simplified. Moreover, notwithstanding the FCC's conclusion that access charges should not be imposed on ISPs, the argument still remains that if the distinct usage-sensitive costs caused by ISP use of the LECs' networks can be determined, some form of pricing that contains usage-based elements would be a more rational method for recovering those costs.

In the NOI on these matters that is still pending, the Commission observed that because its rules have been designed for traditional circuit-switched voice networks, they may hinder the development of packet-switched networks such as those used for much Internet communication.[45] In the Access Charge Order, however, the FCC limits the scope of the NOI by determining that when it considers the record filed in response to the NOI, "[it] will consider solutions to network congestion arguments other than the incumbent LECs' recommendation that [the FCC] apply access charges to ISPs".[46]

Among the questions posed in the NOI are: (1) whether existing state or federal rules might prevent or inhibit the deployment of alternative network access arrangements for Internet users; (2) what the actual costs are to LECs of ESPs' use of the PSTN, and the revenues generated by such services; (3) whether the Commission should distinguish between different categories of enhanced services, on the basis of, *e.g.*, different network usage characteristics; and (4) whether and how issues associated with the Interconnection and Universal Service proceedings affect ISPs.[47] In addition, the NOI requests comment on jurisdictional, metering, and billing issues in light of the inherent difficulties of applying

[41] *Id.*
[42] *Id.* at para. 347.
[43] *Id.*
[44] *Id.*
[45] See *id.* at para. 311.
[46] *Id.* at para. 348.
[47] See *id.* at paras. 311-18.

jurisdictional separations and time-sensitive rates to packet-switched networks such as the Internet.

Those who have supported change in the FCC's rules regarding access charges and ISPs base their arguments on both economic efficiency and fairness grounds. The rationale for change on the basis of economic efficiency is that prices charged should be based on costs incurred. Network users, like ISPs, incur both fixed and usage-sensitive costs. Thus, the contention is that ISPs should pay rates with both fixed and usage-based components. "Economic efficiency" here is shorthand for the goal of devising access charge rules that capture the benefits of a competitive market for access services.

As noted above, U.S. local exchange carriers argue that ISPs cause congestion in the PSTN that threatens the availability and reliability of basic telecommunications services. Studies performed by some local exchange carriers indicate that because Internet users stay on the line longer on average (per day and per call) than voice callers, ISPs place a burden on the PSTN which is not recovered in the flat rates for each line, which are derived from voice call usage patterns.[48] In addition, proponents of change claim that the access rules' current treatment of ISPs provides a disincentive for ISPs to manage their facilities or make subscriber pricing decisions in a way that would result in local exchange carriers efficiently recovering the costs of carrying ISP traffic.[49]

The fairness rationale for change is that, to the extent ISPs are not paying for the usage-sensitive costs they incur, others pay those costs. Proponents of this argument point out the similarities, from a network usage perspective, between long-distance carriers and ISPs. However, those who oppose changes to the current treatment of ISPs compare ISPs to other end users that do not pay access charges. "Fairness," as used here, is shorthand for a goal of equitably distributing the burden imposed by regulation.

Those who oppose changes to the current treatment of ISPs argue from an efficiency perspective that the best way to encourage innovation in local exchange carriers' telecommunications facilities is to place the costs of such innovation on the carriers themselves.[50] Some contend that the local exchange carriers are best able to control their own network costs, and that allowing local exchange carriers to pass along to ISPs costs associated with ISPs' data traffic will actually provide a disincentive to move Internet traffic off existing circuit-switched networks and onto more efficient packet-switched networks.[51]

Further, ISPs and others argue that the local exchange companies' revenue from additional telephone lines – so-called "second lines", ordered by some telephone subscribers for dedicated Internet use covers or exceeds the cost of network upgrades to accommodate the added Internet traffic.[52] Of course, to the extent that local exchange carriers rely on fixed revenue sources to recover usage-sensitive costs, inefficiencies and fairness issues will persist. Moreover, the purchase of second lines could compound rather than alleviate these problems, particularly because customers who use their second lines for Internet service

[48] See, e.g., GTE Comments to FCC on Notice of Inquiry in CC Docket No. 96-263 (filed Mar. 24, 1997) at 11-13; U S West Comments at 17. All references to "Comments" in the following notes refer to those filed in CC Docket No. 96-263 on or about March 24, 1997.

[49] See, e.g., U S West Comments at 6; GTE Comments at 25-26.

[50] See, e.g., Internet User Coalition Comments at 11.

[51] See, e.g., *id;* Internet Access Coalition Comments at 25.

[52] See, e.g., Internet User Coalition Comments at 32-34; Internet Access Coalition Comments at 7-9; Information Industry Association Comments at 3.

generally do not order additional revenue-generating services such as call waiting and caller identification.[53]

Critics of the existing access charge regime argue that the "rate shock" rationale that the FCC used in the past for exempting enhanced service providers from access charges is no longer justified and that ISPs should now pay for the costs they cause. ISPs counter that the market for Internet service provision is so highly competitive that most ISPs will neither be able to absorb the additional costs posed by access charges, nor pass those costs on without losing their customers.[54] In addition, smaller ISPs argue that such pricing would provide ISPs owned by local telephone companies with a competitive advantage.

From a fairness perspective, Internet user groups, and others promoting Internet use as a means of enhancing participatory democracy, argue that current ISP pricing – free from interstate access charges and, thus, typically flat-rated – has allowed growing numbers of people to explore cyberspace. They point to surges in ISP subscribership as a result of the now industry-wide standard of U.S. $19.95 per month for unlimited Internet access. It is clear, however, that subscribership to an ISP does not translate to effective access to the Internet if subscribers are unable to sign on due to congestion. Well-publicized congestion problems in ISP facilities have resulted from increased subscribership related to flat-rated pricing. These problems have apparently limited Internet access in some cases.[55]

Many members of Internet-related industries contend that any increase in the cost of access to LEC facilities will inhibit the growth of the Internet and thus the emerging industries that depend on it.[56] In particular, some argue that the full realization of electronic commerce depends on mass market participation in the sale of goods and services over the Internet and other information services. Thus, they argue, because any increase in the price of access to LEC facilities may inhibit continued growth in Internet use, the many industries turning to the Internet to expand their markets may be adversely affected as well.

The ubiquitous nature of the Internet has been used, somewhat paradoxically, as a reason to maintain the current regime of various state-regulated rates for ISP access. Some parties claim that "it is neither possible nor useful to determine the geographic location of Internet destinations".[57] As a result, they claim, it would be unreasonable for the FCC to claim federal jurisdiction over the rates for ISP access, since it is difficult to discern whether ISP traffic is international, interstate, or intrastate for such purposes. Others point out that settled U.S. law establishes that if a service has a significant interstate use or character, but cannot readily be broken down into distinct interstate and intrastate components, the service can be treated as subject in its entirety to the FCC's jurisdiction.[58]

6. Summary and Conclusions

The issue of the FCC's treatment of access charges for ISPs is not the issue of whether to regulate the operations of ISPs. Putting aside the fundamental question of regulating con-

[53] See, e.g., GTE Comments at 24; United States Telephone Association ("USTA") Comments at 13-14.
[54] See, e.g., Internet User Coalition Comments at 15.
[55] See, e.g., USTA Comments at 9-13.
[56] See, e.g., Internet User Coalition Comments at 14.
[57] *Id.*
[58] See, e.g., *Louisiana Pub. Serv. Comm'n v. FCC*, 476 U.S. 355, 375-79 (1986).

tent, there is no need to impose such regulation, as FCC staffers themselves have recognized.[59] The issue of determining the charges that these unregulated ISPs should pay local exchange carriers for network access depends on the FCC's vision of the telecommunications competition intended by the 1996 Act. More narrowly, answers to such questions depend on the FCC's ability to avoid continued reliance on traditional concepts of regulating access charges. In 1987-1988, when the FCC last considered modifying its access charge treatment of enhanced service providers – and declined to do so – the legal monopoly status of U.S. local telephone companies was long-established and the Internet in its present form did not exist. In that relatively more stable environment, the regulatory classifications of "carriers" and "end users," and "enhanced" and "basic" services, were a basis for sustaining different prices for similar access services under the FCC's access charge rules.

Prompted by the 1996 Act, however, long-standing basic assumptions regarding the U.S. telecommunications landscape have changed. In particular, the advent of competition in local U.S. telecommunications markets will render untenable the price differences supported by the FCC's current access rules. Because of the sophistication of present-day communications services, the long-standing regulatory distinctions mentioned above are both more difficult for regulators to apply reasonably and more enticing for market participants to exploit for at least short-term advantage. Thus, it is increasingly difficult to distinguish among many ISPs, long-distance carriers, and other users of local networks. In light of these complexities, an access charge regime that provides greater flexibility for access providers to recover the costs incurred by access customers is both efficient and fair.

To the extent that alternative suppliers of local telecommunications services ultimately offer ISPs some form of access to Internet users, the operation of the marketplace will best match the demand for access with its supply, while providing proper economic signals for investment. This, of course, assumes that government regulatory intervention does not disrupt such competition. This assumption is by no means a certainty, in light of the political and interest-group pressures on the FCC and state regulators.

As Internet use grows internationally, the relationship between the Internet and the established telecommunications networks of many countries will have to be defined or, perhaps, redefined. The U.S. experience with ISPs and access charges is one point of comparison and information for such purposes.

---

[59] See Kevin Werbach, Digital Tornado: The Internet and Telecommunications Policy, FCC Office of Plans and Policy Working Paper No. 29 (Mar. 1997) at 47 ("The initial assumption ought to be that new Internet-based services should not be subject to the regulatory constraints of traditional services.").

Telecommunications Transformation: Technology, Strategy and Policy
E. Bohlin and S.L. Levin (Eds.)
IOS Press, 1998

CHAPTER 25

Winning and Losing the Telecommunications Sweepstakes of 1996

Robert M. Frieden

The Pennsylvania State University
201-D Carnegie Building, University Park, Pennsylvania 16802, USA

Abstract. In 1996 Congress finally legislated a broad-sweeping overhaul of the Communications Act. The Telecommunications Act of 1996 orders the abandonment of most market entry restrictions and creates ample opportunities for vertical and horizontal market integration. In reviewing the expected impact of the law the press and most stakeholders heralded a new age of robust resale and facilities-based competition in both local and long-distance markets. In 1997 the FCC became fettered with dozens of regulatory proceedings and litigation blocked or delayed implementation of several key provisions of the '96 Act.

This paper will identify the winners and losers in the shifting legal, regulatory and marketplace opportunities created by the '96 Act. The paper also will consider whether and how the Telecommunications Act of 1996 creates opportunities and impediments for both incumbent and newcomers in local and interexchange telephony, cable television, Internet access and other market segments. After identifying which operators have significant current market shares, the paper will provide educated guesses about who will maintain or acquire significant market share in the near- and mid-term. The paper will specify who wins and who loses primarily through a qualitative assessment of what synergies, cost-saving and market integration opportunities the legislation provides.

1. Introduction

When Congress enacted the Telecommunications Act of 1996 ("'96 Act")[1] most interested parties agreed that the law would generate almost immediate opportunities for robust competition and enhanced consumer welfare.[2] The '96 Act supersedes prior laws, court deci-

[1] Telecommunications Act of 1996, P.L. 104-104, 110 Stat. 56, *codified at* 47 U.S.C. §§ 151 et seq. (1996) [hereinafter cited as '96 Act].

[2] "This comprehensive bill strikes a balance between competition and regulation. New markets will be opened, competitors will begin to offer services, and consumers will be better served by having choices among providers of services. I urge my colleagues to adopt this bill. I myself would go further in several areas covered by the legislation, and not as far in other areas. But I have seen that, unless we adopt a comprehensive approach to legislation, any one sector of the telecommunications industry can stop this bill and checkmate the others. Telecommunications reform is too important to let this opportunity go by. This conference agreement is an equitable approach to most of the areas covered by the bill, and I urge my colleagues to
(continued)

sions, consent decrees and regulations that collectively foreclosed or conditioned facilities-based or resale competition in many market sectors including local exchange and interexchange, "long-distance" telecommunications as well as equipment manufacturing. On the other hand, the law establishes significantly more regulatory burdens on both the Federal Communications Commission ("FCC") and state public utility commissions, at least for an indeterminate transition period leading to full competition. The '96 Act creates more, not less, regulatory requirements of telecommunication carriers by creating a presumption that telecommunications carriers will operate as common carriers and by providing greater specificity on the duties which this status entails.[3]

Perhaps the promise of resale and facilities-based competition will take years, rather than months to evolve.[4] Certainly litigation over the FCC's interpretation of the '96 Act has resulted in implementation delays, a strategy for the Bell Operating Companies ("BOCs") that paradoxically leads to both victory (forestalled local exchange competition) and defeat (forestalled entry into long-distance services).[5] During the transition period, competition will occur primarily in the hearing room as advocates contest Congressional intent and seek marketplace advantages. For example, the '96 Act provides greater specificity on the duties of common carriers and creates a presumption that telecommunication service providers will operate as common carriers,[6] but also authorizes the FCC to abandon almost all traditional common carrier requirements[7] if doing so would serve the public interest.[8]

(continued)
support it." Congressional Record – Senate, Proceedings and Debates of the 104th Congress, Second Session, S 687, Unanimous Consent Agreement, Senator Ernest Hollings, 142 Cong. Rec. S687, S688 (February 1, 1996).

[3] "A telecommunications carrier shall be treated as a common carrier under this Act only to the extent that it is engaged in providing telecommunications services..." "96 Act, Section 3(44), *codified at* 47 U.S.C. 153(44).

[4] One skeptic suggests less rosy prospects for robust competition: "There is no guarantee, however, that true competition will flourish, and it is certainly possible that unregulated fiefdoms will soon dot the electronic landscape. The 1996 Act is an experiment, as, one would have to admit, all telecommunications regulation is an experiment." Michael I. Meyerson, Ideas of the Marketplace: A Guide to the 1996 Telecommunications Act, 49 Fed. Comm. L.J. 251,253 (Feb. 1997).

[5] See, e.g., Mark Landler, Stymied, SBC Seeks to Void Portion of Law, *The New York Times* (July 3, 1997), available at http://www.nytimes.com/yr/mo/day/early/070397 sbc-lawsuit.html (reporting that SBC Corporation filed an appeal in federal District Court challenging the constitutionality of provision in the '96 Act conditioning BOC entry into interexchange markets); Iowa Utilities Board et al., v. Federal Communications Commission, File Nos. 96-3321 et al., slip op. (8th Cir. rel. July 18, 1997) available at: http://ls.wustl.edu/8th.cir/ fcc.html (finding the FCC exceeded its jurisdiction in establishing pricing rules for local exchange access, including resale discounts and unbundling of network elements).

[6] "A telecommunications carrier shall be treated as a common carrier under this Act only to the extent that it is engaged in providing telecommunications services..." '96 Act, 47 U.S.C. §153(44).

[7] For an examination of the benefits and liabilities of the common carrier and private carrier classification see Rob Frieden, Schizophrenia Among Carriers: How Common and Private Carriers Trade Places, 3 *Mich.Telecom & Tech. L. Rev.* No. 2 (1997) available at http://www.law.umich.edu/mttlr/volthree/frieden.html; see also Robert M. Frieden, Contamination of the Common Carrier Concept in Telecommunications, 19 *Telecommunications Policy* No. 9, 685-697 (December, 1995).

[8] The Act authorizes the FCC to forbear from enforcing any regulation or provision of the Act if the Commission determines that such enforcement is unnecessary to guard against discrimination, to ensure just and reasonable services, to safeguard consumers and serve the public interest. '96 Act, Title IV – Regulatory Reform, Sec. 401(a)(1)-(3), *codified at* 47 U.S.C. § 160, (a)(1)-(3).

The '96 Act generates a number of paradoxical outcomes and creates more, rather than less, certainty as to regulatory classification. Some common carriers, i.e., incumbent local exchange carriers ("ILECs"), will incur expanded interconnection obligations, but on the other hand any common carrier can qualify for virtual deregulation.[9] The '96 Act appears to have intended for the FCC to establish a national model for establishing interconnection and resale terms and conditions. Yet the law also clearly intended for the states and state regulatory commission to arbitrate and approve what competing LECs negotiate.

This article will examine the '96 Act with an eye toward predicting which stakeholders win and which lose in the implementation phase. Even though the FCC has not yet completed all implementation proceedings, and litigation may run for several years, one can begin to identify how the legislation comparatively supports or handicaps new market access opportunities. The article concludes that immediate opportunities exist for incumbents to accrue greater economies of scale and scope through mergers and acquisitions. Incumbents, to the detriment of consumers, also generate short-term benefits by forestalling implementation of the '96 Act's requirements on interconnection, resale and disaggregation of services into separately available network elements. In the longer term, resale and facilities-based competition should evolve, creating the belated onset of enhanced consumer welfare.

2. Summary of '96 Act Provisions Affecting Telephony

The '96 Act overhauls the Communications Act of 1934 to establish a law that expressly favors competition and envisions a "network of networks"[10] with localities served by a multiplicity of operators each interconnected with all others.[11] The law and FCC's implementation of it strive to stimulate more resale and facilities-based competition, while at the same time fostering a level competitive playing field among incumbents and market entrants, and while promoting a more robust concept of what constitutes universal service in telecommunications.

The '96 Act can be characterized as removing previous legislative and judicially imposed barriers to market entry. For example, it eliminates cross-ownership restrictions on tele-

---

[9] Soon after the '96 Act became law, the FCC proposed to eliminate the tariff filing requirement of non-dominant interexchange carriers, something the Commission had unsuccessfully attempted prior to enactment of the law. See Policy and Rules Concerning the Interstate Interexchange Marketplace Implementation of Sec. 254(g) of the Communications Act of 1934, as amended, CC Docket No, 96-61, Report and Order, 11 FCC Rcd. 9564 (1996). The FCC also proposed to eliminate the requirement that carriers seek approval from the Commission before expanding phone services to consumers within any new geographic area that they are otherwise eligible to serve. See Elimination of Section 214 Applications for Extension of Lines, CC Docket No. 97-11, Notice of Proposed Rulemaking, FCC 97-6 (rel. Jan. 9, 1997). The Commission also proposed to forbear from applying any remaining Section 214 authority to carriers who may be subject to price-cap regulation, average schedule carriers, and domestic, non-dominant carriers offering local or long-distance services.

[10] See Eli M. Noam, Beyond Liberalization-From the Network of Networks to the System of Systems, 18 *Telecommunications Policy* No. 4, pp. 286-294 (1994).

[11] For a more comprehensive summary of the '96 Act provisions, see Christopher H. Sterling, Understanding the Telecommunications Act of 1996, book reviews, 49 Fed. Com. L. J. No. 2 509 (Feb. 1997); Carl B. Kress, The 1996 Telekommunikationsgesetz and the Telecommunications Act of 1996: Toward More Competitive Markets in Telecommunications in Germany and the United States, 49 Fed. Com. L. J. No. 3, 551 (April, 1997).

phone company provision of cable television service and vice versa. Likewise, the '96 Act eliminates the remaining two lines of business restrictions imposed on the Regional Bell Operating Companies ("RBOCs") by the Modification of Final Judgment[12], viz., inter-Local Access and Transport Area ("LATA")[13] long-distance telephone service and telecommunications equipment manufacturing, after the companies meet a fourteen-point competitive checklist evidencing full and fair interconnection and facilities access by other carriers.[14]

Congress leaves to the FCC and state public utility commissions the task of ensuring full and fair interconnection as well as a host of other issues requiring complex economic analysis. For example, the FCC and state public utility commissions must determine appropriate discounts for resold LEC services, and the actual cost of providing switching and routing of traffic generated by interconnecting with competing LECs, interexchange carriers ("IXCs") and end-users. Billions of dollars in carrier revenues and cross-subsidies to promote universal service are at stake as the FCC attempts to define a national framework for rational cost-based pricing and to move from implicit to explicit universal service subsidies. For example, IXCs pay LECs an access charge of about 3 cents a minute for the use of LEC facilities to originate and terminate long-distance toll calls. IXCs estimate that the actual cost of such access lies in a range from 0.4-1.2 cents, meaning that current access charges exceed cost by 250-700 percent and provide a large sources of funds for either profits or universal service subsidies.[15] The FCC and state regulatory agencies have the unenviable task of sorting out complex pricing issues and determining competitively neutral facilities, interconnection and access issues for local and long-distance carriers, as well as for end-users.

2.1 Local Competition

The '96 Act orders the removal of statutory, regulatory, and operational barriers to local telephone services competition. The local competition provisions of the legislation added new sections 251, 252, and 253 to the Communications Act of 1934. Section 251 estab-

---

[12] United States v. Western Elec. Co., 552 F. Supp. 131 (D.D.C. 1982), *aff'd sub nom.* Maryland v. United States, 460 U.S. 1001 (1983).

[13] Under the '96 Act a "local access and transport area" (LATA) is "a contiguous geographic area (A) established before the date of enactment of the ['96 Act] by a [BOC] such that no exchange area includes points within more than 1 metropolitan statistical area, consolidated metropolitan statistical area, or State, except as expressly permitted under the AT&T Consent Decree; or (B) established or modified by a [BOC] after such date of enactment and approved by the Commission." 47 U.S.C. § 153(25). LATAs were created as part of the Modification of Final Judgment's (MFJ) "plan of reorganization" under which the BOCs were divested from AT&T. United States v. Western Elec. Co., 552 F. Supp. 131 (D.D.C. 1982), *aff'd sub nom.* Maryland v. United States, 460 U.S. 1001 (1983); United States v. Western Elec. Co., 569 F. Supp. 1057 (D.D.C. 1983) (Plan of Reorganization), *aff'd sub nom.* California v. United States, 464 U.S. 1013 (1983); see also United States v. Western Elec. Co., No. 82-0192 (D.D.C. Apr. 11, 1996) (vacating the MFJ). Pursuant to the MFJ, "all BOC territory in the continental United States [was] divided into LATAs, generally centering upon a city or other identifiable community of interest." United States v. Western Elec. Co., 569 F. Supp. 990, 993 (D.D.C. 1983).

[14] See 47 U.S.C. §§ 271(c)(2)(B)(i)-(xiv).

[15] "AT&T asserts, for instance, that the current average per-minute access rates of the BOCs are nearly seven times the forward-looking economic cost of providing that service, and that total interstate access charges collected today from interexchange carriers exceed forward-looking economic cost by $11 billion, or 70 percent of the total." Access Charge Reform Price Cap Performance Review for Local Exchange Carriers, CC Docket No. 96-262, Notice of Proposed Rulemaking, Third Report and Order, and Notice of Inquiry, FCC 96-488, 1996 WL 733469 (F.C.C.) at ¶ 11 (rel. Dec. 12, 1996).

lishes general interconnection obligations for all telecommunications carriers,[16] delineates further obligations for LECs[17] and prescribes additional requirements for ILECs.[18] Section 251 requires an ILEC to provide interconnection at any technically feasible point, with equivalent quality as that provided to itself and affiliates and at just, reasonable and nondiscriminatory rates.[19] Section 252 establishes procedures that state commissions, ILECs, and new entrants must follow to implement the requirements of section 251, including examination and approval of carrier-to-carrier interconnection arrangements. Finally, Section 253 bars state and local regulations that prohibit or have the effect of prohibiting entities from offering telecommunications services.[20]

The '96 Act contemplates ILECs and market entrants negotiating terms and conditions for interconnection and resold services.[21] If an ILEC and requesting carrier are unable to reach a negotiated agreement, either party may ask a state regulatory agency to arbitrate the disputed issues. As required by the '96 Act, ILECs must provide interconnection and nondiscriminatory access to network elements on an unbundled basis.[22] In implementing the legislation, the FCC identified the following minimum set of network elements that ILECs must provide to requesting telecommunications carriers, many of which are analogous to interstate access rate elements: network interface devices; local loops; local and tandem switches (including all software features provided by such switches); interoffice transmission facilities; signalling and call-related database facilities; operations support systems and information; and operator and directory assistance facilities.[23] States may require unbundling of additional elements.

Section 251(a) of the '96 Act imposes general common carrier interconnection obligations on all telecommunications carriers. Section 251(b) imposes on all LECs additional requirements, including the obligation to lease services to other telecommunication carriers for resale to end-users, to provide access to rights-of-way, and to establish reciprocal compensation arrangements for transport and termination of traffic, including that generated by Commercial Mobile Radio Service providers, e.g., mobile cellular radio and personal communication service operators. In addition to the above requirements, Section 251(c) of the '96 Act requires ILECs to make available to new entrants interconnection and access to unbundled network elements, and to offer LEC retail services for resale to telecommunications carriers at wholesale rates.[24] Access to unbundled elements and resale opportunities

---

[16] 47 U.S.C. § 251(a)(1)-(2).

[17] 47 U.S.C. § 251(b)(1)-(5).

[18] 47 U.S.C. § 251(c)(1)-(6).

[19] See especially 47 U.S.C. §§ 251(c)(2) and (c)(3).

[20] 47 U.S.C. § 253(a). Section 253 also authorizes the Commission to preempt any law or regulation that is violative of this section. 47 U.S.C. § 253(d).

[21] 47 U.S.C. § 252(a)(1).

[22] 47 U.S.C. § 251(c)(3).

[23] Implementation of the Local Competition Provisions in the Telecommunications Act of 1996, CC Docket No. 96-98, Notice of Proposed Rulemaking, FCC 96-182, at ¶ 366, 61 Fed. Reg. 18311 (Apr. 25, 1996).

[24] Section 251(f)(1) of the '96 Act provides for exemption of the requirements in section 251(c) for rural telephone companies (as defined by the 1996 Act) under certain circumstances. Section 251(f)(2) permits LECs with fewer than 2 percent of the nation's subscriber lines to petition for suspension or modification of the requirements in sections 251(b) or (c).

States are primarily responsible for interpreting the provisions of section 251(f) through rulemaking and adjudicative proceedings, and are responsible for determining whether a LEC in a particular instance is entitled to exemption, suspension, or modification of section 251 requirements.

makes it possible for new ventures to enter local exchange markets without incurring the time and expense in duplicating the existing facilities operated by incumbent carriers.

In implementing the '96 Act, the FCC sought to promote full facilities-based entry, resale of ILEC services, or a hybrid of the two whereby market entrants purchase only those unbundled network elements they need to erect a complete service. To achieve such facilities-based and resale competition, Section 251(c)(2) requires ILECs to provide interconnection to any requesting telecommunications carrier at any technically feasible point. The interconnection must be at least equal in quality to that provided by the ILEC to itself or its affiliates, and must be provided on rates, terms, and conditions that are just, reasonable, and nondiscriminatory.

Additionally, Section 251(c)(6) requires ILECs to provide physical collocation of equipment necessary for interconnection or access to unbundled network elements at the ILEC's premises, except that the ILEC may provide virtual collocation if it demonstrates to the appropriate state regulatory commission that physical collocation is not practical for technical reasons, or because of space limitations. ILECs are required to provide any technically feasible method of interconnection or access requested by a telecommunications carrier, including physical collocation, virtual collocation, and interconnection at meet points, a location where one carrier's geographical service coverage ends and another carrier's begins, or starts to overlap.

The '96 Act also requires the FCC to establish procedures for nondiscriminatory access by cable television systems and telecommunications carriers to poles, ducts, conduits, and rights-of-way owned by utilities or LECs.[25] The Commission's Local Competition First Report and Order includes several specific rules as well as a number of more general guidelines designed to facilitate the negotiation and mutual performance of fair, pro-competitive access agreements without the need for regulatory intervention.[26] Additionally, an expedited dispute resolution is provided when good faith negotiations fail, as are requirements concerning modifications to poles, ducts, conduits, and rights-of-way and the allocation of the costs of such modifications.

---

[25] See Amendment of Rules and Policies Governing Pole Attachments, CS Docket No. 97-98, Notice of Proposed Rule Making, FCC 97-94, 1997 WL 119618 (F.C.C.) (rel. March 14, 1997).

[26] Implementation of the Local Competition Provisions in the Telecommunications Act of 1996, CC Docket No. 96-98, First Report and Order, 11 FCC Rcd. 15499, 15612 (1996) [hereinafter cited as Local Competition First Report and Order], Order on Reconsideration, CC Docket No. 96-98, 11 FCC Rcd. 13042 (1996), Second Order on Reconsideration, FCC 96-476 (rel. Dec. 16, 1996), *petition for review pending and partial stay granted, sub nom.* Iowa Utilities Board v. FCC, No. 96-3321 and consolidated cases (8th Cir., Oct. 15, 1996), *partial stay lifted in part*, Iowa Utilities Board v. FCC, No. 96-3321 and consolidated cases, WL 589204 (8th Cir. Nov. 1, 1996), *partially vacated* slip op. July 18, 1997. Implementation of the Local Competition Provisions in the Telecommunications Act of 1996, CC Docket No. 96-98, Second Report and Order, and Memorandum Opinion and Order, FCC 96-333 (rel. Aug. 8, 1996) (Second Interconnection Order); *appeal docketed sun nom.* Bell Atlantic Telephone Companies v. FCC, No. 90-567 (D.C. Cir. Sept. 16, 1996), People of the State of California v FCC, No. 96-3519 (8th Cir. Sept.23, 1996), SBC Communications Inc, v. FCC, No. 96-1414 (D.C. Cir. Nov. 1, 1996).

2.2 Number Portability, Dialing Parity and Other Numbering Issues

Section 251(b) requires all LECs to offer, "to the extent technically feasible, number portability in accordance with requirements prescribed by the Commission."[27] The '96 Act defines number portability as "the ability of users of telecommunications services to retain, at the same location, existing telecommunications numbers without impairment of quality, reliability, or convenience when switching from one telecommunications carrier to another."[28] In implementing this requirement the FCC first required all LECs to implement a long-term number portability method in the 100 largest Metropolitan Statistical Areas ("MSAs") according to a phased deployment schedule concluding on December 31, 1998.[29] Thereafter, for areas outside the 100 largest MSAs, each LEC must make long-term number portability available within six months after a specific request by another telecommunications carrier. The First Report and Order also required all cellular, broadband personal communications services and many Specialized Mobile Radio providers to be able to deliver calls from their networks to dialed numbers by December 31, 1998 and also required these wireless carriers to offer number portability throughout their networks, including service to temporary, "roaming" users by June 30, 1999.

Rather than choosing a particular technology for the provision of number portability, the FCC established performance criteria that any long-term number portability method must satisfy. The Commission required the maintenance of regional numbering databases by one or more independent administrators selected by the North American Numbering Council empaneled to address telephone numbering issues.[30]

Section 251(b)(3) requires all LECs to provide dialing parity to providers of telephone exchange and toll services. This means that customers can choose different carriers for different services without having to dial extra digits. To achieve this objective, the Commission established minimum federal standards for interstate, intrastate, local and toll calls.[31] The Commission also concluded that customers should have the option of selecting different carriers for routing intraLATA and interLATA toll calls. It imposed a deadline of no later than February 8, 1999 for LECs to make it possible for customers the option of pre-

---

[27] 47 U.S.C. §251(b)(2).

[28] 47 U.S.C. §153(30).

[29] See Telephone Number Portability, CC Docket No. 95-116, First Report and Order and Further Notice of Proposed Rulemaking, 11 FCC Rcd. 8352 (1996); First Memorandum Opinion and Order on Reconsideration, FCC 97-74, 1997 WL 106479 (F.C.C.) (rel. March 11, 1997).

[30] "[T]he North American Numbering Council will make recommendations to the Commission, develop policy, initially resolve disputes and guide the North American Numbering Plan Administrator. The North American Numbering Plan Administrator will process number assignment applications and maintain administrative number databases." Administration of the North American Numbering Plan, CC Docket No. 92-237, Report and Order, 11 FCC Rcd. 2588 (1995). See also Administration of the North American Numbering Plan, Report and Order, 11 FCC Rcd 2588, 2593-94 (1995).

[31] Implementation of the Local Competition Provisions of the Telecommunications Act of 1996, CC Docket No. 96-98, Notice of Proposed Rulemaking, FCC 96-182, ¶¶ 202-219 (rel. Apr. 19, 1996), 61 Fed. Reg. 18311 (Apr. 25, 1996); Second Report and Order, FCC 96-333 (rel. Aug. 8, 1996) (implementing the dialing parity, non-discriminatory access, network disclosure and numbering administration requirements of the 1996 Act).

selecting two different "Primary Interexchange Carriers" for direct dial (0+ and 1+) access.[32]

Section 251(e)(1) vests the FCC with exclusive jurisdiction over "those portions of the North American Numbering Plan[33] that pertain to the United States . . . [with authority to delegate to] State commissions or other entities all or any portion of such jurisdiction."[34] On the matter of abbreviated dialing codes, i.e., the use of fewer than seven digits to complete a call as is the case with 911 access to police and emergency services, the FCC authorized the continuing use of 311 for a non-emergency service, and 611 and 811 for access to LEC repair and business service offices.[35] The Commission prohibited LECs from offering enhanced services via any three-digit code unless it offers access to the code on a reasonable, nondiscriminatory basis to competing enhanced service providers operating in the local service area.

2.3 New Structural and Accounting Safeguards and Methodologies

To ensure a level competitive playing field between incumbent and recent market entrants, the FCC had to erect significant structural[36] and accounting safeguards,[37] particularly in view of new market opportunities for the Bell Operating Companies ("BOCs"), viz., interLATA long-distance services and manufacturing, and the need for new LECs to access the facilities and services of ILECs. Also the FCC and state public utility commissions needed a basis for assessing the what constitutes fair compensation when a market competitor requires access to an ILEC's installed network, or simply desires discounted charges when reselling local retail services of the ILEC.

Section 271 and 272 condition a BOC's entry into interLATA long-distance telephone services and manufacturing. Congress required the BOCs to establish separate manufacturing and long-distance service subsidiaries to ensure that in the absence of full competition in the local exchange marketplace, the BOC would not engage in anticompetitive, discriminatory or cost-shifting practices designed to favor its new ventures which would enter competitive markets. Under traditional rate-of-return regulation and even under incentive regu-

---

[32] The Commission applied dialing parity requirements to pay telephones. See Implementation of the Pay Telephone Reclassification and Compensation Provisions of the Telecommunications Act of 1996, CC Docket No. 96-128, Report and Order, FCC 96-388,
1996 WL 547458 (F.C.C.) (rel. Sep. 20, 1996).

[33] The North American Numbering Plan is the basic numbering scheme, e.g., area codes, for the telecommunications networks located in Anguilla, Antigua, Bahamas, Barbados, Bermuda, British Virgin Islands, Canada, Cayman Islands, Dominica, Dominican Republic, Grenada, Jamaica, Montserrat, St. Kitts & Nevis, St. Lucia, St.Vincent, Turks & Caicos Islands, Trinidad & Tobago, and the United States (including Puerto Rico, the U.S. Virgin Islands, Guam and the Commonwealth of the Northern Mariana Islands).

[34] 47 U.S.C. § 251(e)(1).

[35] Use of N11 Codes and Other Abbreviated Dialing Arrangements, CC Docket No. 92-105 First Report and Order and Further Notice of Proposed Rulemaking, FCC 95-51, 1997 WL 68052 (rel. Feb. 19, 1997).

[36] Implementation of the Non-Accounting Safeguards of Section 271 and 272 of the Communications Act of 1934, as Amended, CC Docket No. 96-149, First Report and Order and Further Notice of Proposed Rulemaking FCC 96-489, 1996 WL 734160 (F.C.C.)(rel. Dec. 24, 1996)[hereinafter cited as Non-Accounting Safeguards].

[37] Implementation of the Telecommunications Act of 1996: Accounting Safeguards under the Telecommunications Act of 1996, CC Docket No. 96-150, Report and Order, FCC 96-490, 1996 WL 734166 (F.C.C.)(rel. Dec. 24, 1996).

lation involving price caps, a BOC might "allocate improperly to its regulated core business costs that would be properly attributable to its competitive ventures."[38]

The FCC has developed a cost allocation methodology for use by state regulators when establishing actual rates for interconnection and the purchase of unbundled elements. The Commission used a controversial cost-based pricing methodology based on "forward-looking" economic costs, an estimate of the costs an efficient LEC would incur to provide such interconnection rather than an estimate of the carrier's actual or historical costs. The Commission pricing methodology is based on a staff study of the "Total Service Long-Run Incremental Cost" ("TSLRIC") incurred by LECs to provide network elements needed for interconnection and access to unbundled elements. Because the TSLRIC studies pertain to network elements, the FCC renamed them "Total Element Long Run Incremental Cost" ("TELRIC") studies. The Commission ordered that prices be set at TELRIC plus a reasonable share of forward-looking joint and common costs. The Commission contemplated that most states regulatory would use its proposed methodology when assessing the merits of a LEC access proposal, or when resolving a request for arbitration between ILECs and their competitors. For states not yet adopting this methodology, the Commission also established default proxies, i.e., actual rate ranges of 0.2-0.4 cents per minute for switching, and a ceiling of 0.15 cents per minute for tandem switching, that a state commission may use to resolve arbitrations even before completion of a TELRIC study. Language in the '96 Act also directs the FCC to determine rates on an "avoided cost standard," i.e., reducing charges to carriers on the basis of costs the incumbent carrier would not have to incur in providing such access or services for resale, e.g., sales and marketing expenses.

2.4 The FCC's Order Fails to Pass Full Muster With a Reviewing Court

Despite its best efforts the FCC could not implement the '96 Act without delay caused by litigation, a predictable outcome given the stakes and the ability of litigation to tilt the competitive playing field in favor of one category of operator over others. Soon after the FCC released its Local Competition First Report and Order in August, 1996 many ILECs and state public utility commissions filed motions for partial or complete stays on grounds that the Commission exceeded its jurisdiction in establishing pricing policies that address local intrastate telecommunication services. The petitioners convinced the Eighth Circuit Court of Appeals that it should temporarily stay pricing provisions and the "pick and choose"[39] rule[40] of the Local Competition First Report and Order.

In deciding the merits of numerous appeals, the court identified the following defects in the FCC's Local Competition First Report and Order:

---

[38] Non-Accounting Safeguards at ¶11. "For example, a BOC may have an incentive to degrade services and facilities furnished to its affiliate's rivals, in order to deprive those rivals of efficiencies that its affiliate enjoys. Moreover, to the extent carriers offer both local and interLATA services as a bundled offering, a BOC that discriminates against the rivals of its affiliates could entrench its position in local markets by making these rivals' offerings less attractive. With respect to BOC manufacturing activities, a BOC may have an incentive to purchase only equipment manufactured by its section 272 affiliate, even if such equipment is more expensive or of lower quality than that available from other manufacturers." *Id.*

[39] The FCC's "pick and choose" rule would have allowed requesting LECs the opportunity to incorporate any individual provision of any interconnection agreement that an ILEC previously had negotiated with one or more other carriers.

[40] Iowa Utilities Bd. v. FCC, 109 F.3d 418 (8th Cir.), *motion to vacate stay den.*, 117 S.Ct. 429 (1996).

- establishing pricing rules based on the forward-looking TELRIC methodology;
- allowing requesting LECs the "pick and choose" option when negotiating an interconnection agreement;[41]
- specifying the terms and conditions exempting rural and small LECs from the ILEC requirements;[42]
- establishing the FCC, instead of federal district courts, as the forum for reviewing and hearing complaints about ILEC agreements approved by state commissions;[43] and
- requiring state commission approval of interconnection agreements negotiated prior to enactment of the '96 Act.[44]

2.5 Universal Service

Section 254 of the '96 Act establishes an explicit mandate for the FCC to promote universal access to telecommunication services.[45] The Act directs the Commission to commence a proceeding to implement sections 254 and 214(e) and to refer such proceeding to a Federal-State Joint Board. The Joint Board was given nine months to make recommendations to the Commission, including a definition of the services to be supported by federal universal service support mechanisms and a timetable for the implementation of such recommendations. The FCC initiated the Joint Board proceeding in March 1996,[46] and the Joint Board issued its *Recommended Decision* in November 1996.[47]

The '96 Act established several requirements for federal universal service support mechanisms. It directed the Joint Board and the FCC to base the preservation and advancement of universal service on six general principles, but stipulated that additional appropriate principles might be considered. These six principles are:

- Quality services should be available at just, reasonable, and affordable rates.
- Access to advanced services should be available in all regions of the nation.

---

[41] The Commission based its "pick and choose" rule, codified at 47 C.F.R. § 51.809(1996), on subsection 252(i) of the '96 Act that provides: "A local exchange carrier shall make available any interconnection, service, or network element provided under an agreement approved under this section to which it is a party to any other requesting telecommunications carrier upon the same terms and conditions as those provided in the agreement." 47 U.S.C. § 252(i)(1996).

[42] Local Competition Decision at 156-158.

[43] "We believe that state commissions retain the primary authority to enforce the substantive terms of the [interconnection] agreements made pursuant to sections 251 and 252." *Id.* at 160.

[44] *Id.* at 162-163.

[45] 47 U.S.C. § 254.

[46] Federal-State Joint Board on Universal Service, Notice of Proposed Rulemaking and Order Establishing Joint Board, CC Docket No. 96-45, FCC 96-93 (rel. Mar. 8, 1996)[hereinafter cited as Universal Service NPRM].

[47] Federal-State Joint Board on Universal Service, CC Docket No. 96-45, Recommended Decision, FCC 96J-3, 1996 WL 656113 (F.C.C.) (rel. Nov. 8, 1996) [hereinafter cited as Joint Board Recommended Decision].

- Access to basic and advanced services should be available to customers in rural and high-cost areas and to low-income consumers at rates comparable to those in urban areas.
- Equitable and nondiscriminatory contributions should be made by all telecommunications providers to the preservation and advancement of universal service.
- Specific and predictable support mechanisms should exist at both the federal and state levels.
- Schools, health care facilities, and libraries should have access to advanced telecommunications services.

The Commission, after receiving the recommendations of the Joint Board, had to identify specific services for federal universal service support,[48] and target subsidies for the provision, maintenance and upgrading of the specifically identified facilities and services.[49] Such support is to be available to all telecommunications carriers eligible to receive universal service subsidies.[50] The legislation requires explicit universal service funding[51] and mandates equitable and non-discriminatory sharing of the financial burden among all telecommunications carriers providing interstate telecommunications services.[52]

In its *Recommended Decision*, the Federal-State Joint Board concluded that several universal service mechanisms currently implemented through the jurisdictional separations and access charge structures must be replaced or modified to meet the amended Communication Act's requirements that support mechanisms be explicit, specific, predictable and sufficient to preserve and advance universal service. Accordingly, the Joint Board recommended that changes be made to the high cost assistance fund,[53] and that the Dial Equipment Minutes ("DEM") weighting program[54] and Long Term Support ("LTS")[55] be phased out and replaced by a new explicit universal service mechanism.[56] The Joint Board's recommendations will obligate the FCC to revise its access charge rules to eliminate any implicit universal service subsidization, to prevent ILECs from recovering the same costs twice, and to provide the same subsidies to non-ILECs as are provided to ILECs for serving high-cost or low-income subscribers.

---

[48] 47 U.S.C. § 254(c).

[49] 47 U.S.C. § 254(e), (k).

[50] 47 U.S.C. §§ 254(e), 214(e); see also Joint Board Recommended Decision at ¶¶ 155-62; Joint Statement of Managers, S. Conf. Rep. No. 104-230, 104th Cong. 2d Sess. at 131 (1996) ("The conferees intend that only eligible telecommunications carriers should receive support from specific Federal universal service support mechanisms . . . .")[hereinafter cited as Joint Explanatory Statement].

[51] 47 U.S.C. 254(e); see also Joint Explanatory Statement at 131 ("In keeping with the conferees' intent that universal service support should be clearly identified, [section 254(e)] states that such support should be made explicit . . .").

[52] 47 U.S.C. § 254(d).

[53] 47 C.F.R. §§ 36.601 et seq.

[54] 47 C.F.R. § 36.125(b).

[55] 47 C.F.R. §§ 69.105, 69.502, 69.603(e), 69.612.

[56] Joint Board Recommended Decision at ¶¶ 268-82.

2.6 Access Charge Reform

The third major prong of the FCC's implementation of the '96 Act involves reforming the charges imposed on interexchange carriers ("IXCs") accessing local exchange carrier ("LEC") facilities to originate and terminate long-distance telecommunications. In view of the cost reductions imposed by the '96 Act on interconnection charges between LECs and the impending entry by the BOCs into interexchange services, the FCC also had to consider the even greater disparity in rates imposed by LECs on IXCs.[57] The technological nature of interconnection is quite similar for access among LECs and between an IXC and a LEC. Accordingly, absent compelling public interest justifications, like an explicit requirement that IXCs primarily underwrite universal service funding, the charges incurred for local and interexchange carriers should be similar.

In late 1996 the FCC contemporaneously initiated two proceedings to consider reforms to the access charge regime: 1) a Notice of Proposed Rulemaking seeking to reform the access charge system in view of the '96 Act and with an eye toward fostering efficiency; and 2) a Notice of Inquiry on Internet and interstate information services seeking to identify what policies would best facilitate the development of the high-bandwidth data networks of the future, while preserving efficient incentives for investment and innovation in the underlying voice network.[58]

The FCC commenced a review of its Part 69 interstate access charge rules, together with its Part 61 price cap rules:

> to determine the extent to which we must revise these rules to take account of the local competition and Bell entry provisions of the '96 Act and state actions to open local networks to competition; to reflect the effects of potential and actual competition on incumbent LECs' pricing for interstate access; to implement the '96 Act's direction to end implicit universal service subsidies in favor of a system of explicit

---

[57] "Current access charges distort competition in the markets for local exchange access. Our access charge rules create incentives for IXCs to bypass the LEC switched access network for reasons that have nothing to do with the economics of operating an access network. This uneconomic bypass may occur for a variety of reasons; rates may be too high, or our access charge rules may require rates for a LEC access service to be too high in relation to the rates for an alternative LEC service or for a comparable service offered by an alternative supplier. Inefficient entry may occur if the price for a package of jointly-provided services is above economic cost, even if the LEC would actually be the most efficient provider of the service. Conversely, if a package of jointly-provided services, including access, is priced too low because of regulatory requirements, efficient entry by an otherwise efficient provider may be precluded. In either case, the total cost of telecommunications service will not be as low as it could be if all services were priced at economic levels, thereby providing accurate price signals to all market participants. High access charges may also keep long-distance rates higher than they would otherwise be, which restricts demand for service and harms long-distance consumers. We describe more fully some of the causes of uneconomic bypass below." Access Charge Reform, Price Cap Performance Review, Transport Rate Structure and Pricing, Usage of the Public Switched Network by Information Service and Internet Access Providers, CC Docket Nos. 96-262, 94-1, 91-213, 96- 263, Notice of Proposed Rulemaking, Third Report and Order, and Notice of Inquiry, FCC 96-488 at ¶ 42, 1996 WL 733469 (F.C.C.) (released Dec. 24, 1996).

[58] Access Charge Reform, Price Cap Performance Review, Transport Rate Structure and Pricing, Usage of the Public Switched Network by Information Service and Internet Access Providers, CC Docket Nos. 96-262, 94-1, 91-213, 96- 263, Notice of Proposed Rulemaking, Third Report and Order, and Notice of Inquiry, FCC 96-488, 1996 WL 733469 (F.C.C.) (released Dec. 24, 1996)[hereinafter cited as Access Charge Reform NPRM].

> subsidies; and to establish fair rules of competition for both the local exchange and interexchange markets, especially as carriers begin to offer service packages that bundle local and interexchange offerings.[59]

The Commission noted that the current access charge system imposes traffic-sensitive, per minute carrier common line ("CCL") fees, which compensate LECs for the shortfall resulting from less than fully compensatory flat-rated Subscriber Line Charges,[60] viz., monthly access payments by end-users, even though local loop costs do not vary with the amount of traffic carried by the loop.[61] Likewise the Commission's access charge rules required IXCs to pay per-minute charges for local switching, even though a portion of such costs correlated to the number of lines connected to the switch, rather than the number of minutes of traffic routed by the switch. The Commission's access charge rules also imposed a non-facilities-based, per-minute transport interconnection charge ("TIC") on all switched access customers regardless of whether they used the incumbent LEC's transport facilities.

The Commission concluded that "[r]ather than fostering efficient pricing and competition, these mandatory rate structures inflate usage charges and reduce charges for connection to the network, in essence overcharging high-volume end-users in order to reduce rates for low-volume end-users."[62] The current access charge rules place:

> the incumbent LEC at a regulatory agency-imposed disadvantage in competing for high-volume end-users, and jeopardizes the source of revenue that permits the incumbent LEC to cover its costs of providing service to low-volume end-users. At the same time, these inefficient rate structures and implicit support flows also create artificial impediments to any new entrants that might seek to serve the subsidized end-users, because they must attempt to do so without the benefit of a subsidy. As a result, these access rate structures may inhibit the development of competition for service to low-volume end-users."[63]

In the sections that comprise the Third Report and Order in the Price Cap Performance Review for Local Exchange Carriers, the FCC eliminated the lower service band indices,

---

[59] Access Charge Reform NPRM at ¶ 5.

[60] "[I]n the original Access Charge Order, [93 FCC 2d at 279] the Commission found that recovering NTS [non-traffic-sensitive] costs through flat monthly charges imposed on end users by incumbent LECs would promote optimal utilization of telecommunications facilities. The Commission decided at that time, however, to place a limit on the SLC [Subscriber Line Charge], and, consequently, required incumbent LECs to recover the remainder of their common line costs through per-minute CCL rates. The current CCL charge has been uniformly criticized by both incumbent LECs and IXCs because it discourages efficient use of the network and encourages uneconomic bypass." Access Charge Reform NPRM at 58.

[61] "We tentatively conclude that several provisions in Part 69 of our rules compel incumbent LECs to impose charges for access services in a manner that does not accurately reflect the way those LECs incur the costs of providing those services. For example, generally the costs associated with the local loop are non-traffic-sensitive (NTS), but our rules require incumbent LECs to recover a portion of those costs through per-minute CCL charges. Similarly, at least some portion of the costs of local switching is NTS, but our rules require incumbent LECs to recover all local switching costs through per-minute charges. Access Charge Reform NPRM at ¶55.

[62] *Id.* at ¶ 7.

[63] *Id.* at ¶8.

which it considered unnecessarily restricting the ability of LECs subject to price cap regulation to lower their access prices.[64]

2.7 NOI on Internet Access and Service Providers

On the subject of Internet Service Providers and other "enhanced service providers" the Commission noted that previously it exempted such ventures from paying access charges in addition to their ordinary line rental fees. In 1983, the FCC classified enhanced service providers as "end-users" rather than "carriers" for purposes of the access charge rules.[65] The Commission tentatively concluded that ISPs should not be subject to access charges as currently constituted,[66] i.e., that enhanced and Internet service providers should only have to pay "business line rates and the appropriate subscriber line charge, rather than interstate access rates."[67] However, the Commission did note that "usage continues to grow, [and that] such services may have an increasingly significant effect on the public switched network."[68]

For the time being, the Commission chose to consider the issue broadly in terms of how to "provide incentives for investment and innovation in the underlying networks that support the Internet and other information services"[69] rather than narrowly a matter of whether enhanced and Internet service providers should pay access charges.[70] With that perspective in mind, the Commission noted:

> the development of the Internet and other information services raise many critical questions that go beyond the interstate access charge system that is the subject of this proceeding. Ultimately, these questions concern no less than the future of the public switched telephone network in a world of digitalization and growing im-

---

64 "Under our existing rules a price cap LEC must specifically justify a proposal to lower its access charges below the pricing floors established by the indices. Thus, our rules currently discourage price cap LECs from lowering prices even when it would be economically efficient to do so. These rules also hamper a price cap LEC in responding to lower-priced access service offerings by competing access service providers. To encourage the development and prompt deployment of new switched access services, we also streamline the process for price cap LECs to offer such services." *Id.* at ¶ 12.

65 "In 1983 we adopted a comprehensive 'access charge' plan for the recovery by local exchange carriers (LECs) of the costs associated with the origination and termination of interstate calls [citing MTS and WATS Market Structure, Memorandum Opinion and Order, 97 FCC 2d 682 (1983)]. At that time, we concluded that the immediate application of this plan to certain providers of interstate services might unduly burden their operations and cause disruptions in provision of service to the public. Therefore, we granted temporary exemptions from payment of access charges to certain classes of exchange access users, including enhanced service providers." Matter of Amendments of Part 69 of the Commission's Rules Relating to Enhanced Service Providers, CC Docket No. 87-215, Notice of Proposed Rulemaking, 2 FCC Rcd. 4305 (1987)(proposing to impose access charges on enhanced service lines), *terminated*, 3 FCC Rcd. 2631(1988)(proposal abandoned on ground that despite the apparent discrimination in charges "a period of change and uncertainty" besetting the enhanced services industry justified ongoing exemption from access charge payments). Currently the FCC requires users of ISDN services to pay only one Subscriber Line Charge, an access payment, despite the fact that ISDN circuits can derive more than one voice-grade equivalent channel.

66 Access Charge Reform NPRM at ¶283.

67 *Id.* at ¶285.

68 *Id.* at ¶282.

69 *Id.* at ¶ 283.

70 "The mere fact that providers of information services use incumbent LEC networks to receive calls from their customers does not mean that such providers should be subject to an interstate regulatory system designed for circuit-switched interexchange voice telephony." *Id.* at ¶288.

> portance of data technologies. Our existing rules have been designed for traditional circuit-switched voice networks, and thus may hinder the development of emerging packet-switched data networks. To avoid this result, we must identify what FCC policies would best facilitate the development of the high-bandwidth data networks of the future, while preserving efficient incentives for investment and innovation in the underlying voice network. In particular, better empirical data are needed before we can make informed judgments in this area.[71]

The FCC initiated a Notice of Inquiry seeking comment generally on the implications of information services such as Internet access for the telephone network and what it should do to encourage development of packet switching hardware able to route data traffic around incumbent LEC switches, or by installing new high-bandwidth access technologies such as asymmetric digital subscriber line ("ADSL") or wireless solutions.

2.8 First Orders on Access Charge and Universal Service Reform

On May 7, 1997 the FCC approved three Report and Orders that collectively implement the Telecommunications Act of 1996 with respect to universal service and charges imposed on interexchange carriers and consumers for using the LEC facilities to originate and terminate interexchange services.[72] To foster efficiency and to remove market distortions the Commission generally removed from minute-of-use access charges non-traffic-sensitive, "sunk" costs that are not incurred on a per-minute-of-use basis. The Order affirmed the Commission's tentative conclusion that incumbent LECs may not assess interstate access charges on information service providers ("ISPs"). The Commission found that its existing policy promotes the development of the information services industry, advances the goals of the 1996 Act, and creates significant benefits for the economy and the American people. With respect to access charge rate levels, the FCC rejected proposals immediately to reset all interstate access rates to the significantly lower levels that forward-looking Total Service Long Run Incremental Cost ("TSLRIC") would impose. Instead, the Commission adopted the marketplace approach discussed in the NPRM wherein it would rely primarily on competitive forces to drive interstate access prices to cost-based levels. However, the Commission adopted a prescriptive safeguard to bring access rates to competitive levels even in the absence of competition. For all services then still subject to price caps and not deregulated in response to competition, the Commission required incumbent LECs subject to price caps to file TSLRIC studies no later than February 8, 2001.

---

[71] *Id.* at ¶311.

[72] Access Charge Reform, Price Cap Performance Review for Local Exchange Carriers, Transport Rate Structure and Pricing and End User Common Line Charges, CC Docket Nos. 96-262, 94-1, 91-213, and 95-72, First Report and Order, FCC 97-158 (adopted May 7, 1997)[hereinafter cited as Access Charge Report and Order]; Price Cap Performance Review for Local Exchange Carriers, CC Docket No. 94-1, Fourth Report and Order, FCC 97-159 (adopted May 7, 1997); Federal-State Joint Board on Universal Service, Report and Order, CC Docket No. 96-45, FCC 97-157 (rel. May 8, 1997)[hereinafter cited as Universal Service Report and Order].

2.9 Immediate and Future Financial Consequences

The FCC touted its Access Charge Order as generating almost immediate consumer savings by way of lower long-distance bills. However, the Order does impose new or higher fees for residential and business users having more than one phone line. The Commission now considers such additional lines a luxury for which users must bear the full cost of LEC facilities access. For second and additional residential lines, in 1998 the Subscriber Line Charge, which currently is \$3.50 per line and will remain so for single-line residences and businesses, will increase to the lesser of \$5.00 or one-twelfth of annual common line revenues permitted under the Commission's price cap rules divided by total access lines. In 1999, the SLC ceiling for these additional residential lines will increase to the lesser of \$6.00 per line or one-twelfth of annual common line revenues permitted under the price cap rules divided by total access lines. In subsequent years an inflation-adjusted increase can occur up to a cap of \$9.00 per month. The SLC ceiling for some multi-line business customers will increase in July 1997 from \$6.00 to the lesser of \$9.00 or one-twelfth of annual common line costs allocated to the interstate jurisdiction divided by total access lines. However, the Commission estimates that after phased-in increases the SLC payments to most incumbent LECs will stabilize at about \$7.60.

2 10 The Presubscribed Interexchange Carrier Charge

The FCC also replaced the per-minute Carrier Common Line charge imposed on IXCs[73] with a flat-rated, per-line Presubscribed Interexchange Carrier Charge ("PICC") as discussed in the NPRM. The PICC will recover LEC common line revenues not otherwise recovered through SLCs, and in the future also may recover a portion of costs currently collected through the per-minute Transport Interconnection Charge ("TIC"). In 1998, the ceiling on the monthly PICC associated with primary residential and single-line business lines will be the lesser of 53 cents or one-twelfth the annual common line revenues permitted under the FCC's price cap rules divided by total access lines minus \$3.50 and any applicable universal support payments received by the incumbent LEC. IXCs currently pay a per-line charge of \$.53 to contribute to existing universal high cost and low-income support programs and this amount will thus remain unchanged. In 1999 and subsequent years, the Commission will adjust the monthly PICC ceiling for inflation and increase the charge \$.50. The monthly PICC ceiling may never exceed the sum of one-twelfth of the annual permitted common line revenues and residual TIC revenues divided by total access lines, minus \$3.50 and all universal service support received by the incumbent LEC for such lines.

[73] The CCL elimination will occur only after a transition to full PICC recovery. Initially the per-minute CCL charge shall be assessed on originating minutes and shall not exceed annual common line revenues permitted under the Commission's price cap rules less the common line revenues that incumbent LECs are permitted to recover from SLCs and PICCs, divided by forecasted originating interstate access minutes. To the extent that the sum of a LEC's per-minute local switching charge and any residual per-minute CCL, TIC, and marketing expense charges exceeds the sum of its local switching, CCL, and TIC charges on originating access on December 31, 1997, the excess shall be collected through a per-minute charge on terminating access. This circumstance is expected to affect only a few price cap LECs and none beyond 1998. As a result of these policies, the average price cap LEC will have to charge lower rates for terminating than originating access. Originating access charges are more likely to be subject to competition than terminating access.

Effective January 1, 1998, the PICC for second and additional residential lines will be the lesser of $1.50 or one-twelfth of the annual common line revenues permitted to be recovered under the Commission's price cap rules minus the annual common line revenues permitted to be recovered through all SLCs and the PICC assessed on primary residential and single-line business lines, divided by the number of second and additional residential lines and multi-line business lines. In 1999, the monthly ceiling on the PICC for such multiple-line users will be adjusted for inflation and increased by an additional $1.00. In each subsequent year, the monthly ceiling shall be adjusted for inflation and again increased by $1.00. However, the Commission noted that the actual PICC for second and additional residential lines paid by IXCs will decrease over time, as the ceilings on the PICC for primary residential and single-line business lines gradually increase. It estimated that the average second residential line PICC will never exceed $2.00, and that this average will decline from 1999 onwards. Decreases are expected to begin for some incumbent price cap LECs in 1999 and for all LECs no later than 2001. Eventually, the PICC for second and additional residential lines will be one-twelfth of the annual common line and residual TIC revenues permitted under the Commission's price cap rules divided by total access lines, minus the ceiling on the SLC for those lines, plus certain expenses that LECs incur in marketing their retail services. In 1998, the PICC for multi-line business lines will be the lesser of $2.75 or one-twelfth of the annual common line revenues permitted under the Commission's price cap rules minus the annual common line revenues permitted to be recovered through all SLCs and the PICC assessed on all residential and single-line business lines divided by the number of multi-line business lines. In 1999 and 2000, the PICC ceiling shall be adjusted annually for inflation and, if necessary for certain carriers, increased by an additional $1.50 on average. However, even for multi-line business users the FCC estimates that the PICCs will begin to decline and reach a level below $1.00 in 2001.

2.11 Switching

Effective January 1, 1998, the Commission ordered price cap-regulated LECs to recover the non-traffic-sensitive ("NTS") costs associated with local switching's line ports to the common line charges discussed above. Price cap LECs also may assess a monthly flat-rated charge directly on end-users that are subscribing to ISDN, digital subscriber line or other services that have higher line port costs than basic, analog service. This charge will recover the amount by which the cost of the line port exceeds the cost of a line port for basic, analog service.

Also effective January 1, 1998, the FCC directed price cap-regulated LECs to move the NTS costs of local switching attributable to dedicated trunk ports to the trunking basket and to recover them through flat-rated monthly charges collected from the user of the trunk port. The NTS costs attributable to shared trunk ports continue to be recovered through per-minute local switching charges. In addition, price cap LECs may, but are not required to, recover some of their local switching revenue through a call set-up charge that is assessed on a per-call basis.

2.12 Transport

Effective 1998, the FCC eliminated the unitary rate structure option for tandem-switched transport and created a three-part rate structure for recovering the costs of tandem-switched transmission. The NTS costs of tandem switching attributable to dedicated ports will be recovered through a new flat-rated monthly charge, with multiplexing of high-capacity lines linking tandem switches and end offices recovered through a new per-minute rate element. The FCC also ordered the calculation of tandem-switched transport volume rather than the previously used assumption of 9000 minutes.

3. Short-term Beneficiaries of the Legislation

Short-term beneficiaries of the '96 Act enjoy an immediate opportunity to economize or to generate more revenues. While the '96 Act has the potential for generating more substantial marketplace impacts in the future, real and significant opportunities already have begun to appear.

3.1 Media Brokers, Bankers and Transactional Lawyers

By far the greatest short-term beneficiaries of the '96 Act have little if anything to do with local or long-distance telephony.[74] Instead broadcast media brokers, bankers and attorneys handling such transactions have accrued the most immediate positive dividends resulting from their involvement in the proliferation of mergers and acquisitions triggered by the '96 Act.[75] In 1996 the total market value of terrestrial broadcast, cable and satellite television and radio sales amounted to $80.1 billion, a 74 percent increase over 1995.[76] Section 202 of the '96 Act of 1934 directs the FCC to revise Section 73.3555 of its Rules[77] to permit a single entity to own or control an unlimited number of television and radio stations nationwide, with television station ownership subject to a cap of 35 percent of the nationwide market penetration and ownership of radio stations in an individual market subject to a nu-

---

74 Analyses in the general circulation press of the '96 Act's one-year anniversary generally note a negative or neutral impact: "To consumers, who were promised more choices, new services and lower prices as a result of the Telecommunications Act of 1996, the sweeping measure must seem like a case of legislative bait-and-switch. A year after Congress enacted the landmark law to open the nation's heavily regulated telephone and television industries to greater competition, long-distance telephone rates have risen, cable TV rates have remained high and little or no local telephone competition has emerged in most residential markets." Pradnya Joshi and Elizabeth Sanger, Telecom Reform: Big Talk, Little Action, *Newsday* Business Section at A39, Feb. 4, 1997, available in LEXIS Curnws data base; see also Mike Mills and Paul Farhi, This is the Free Market? The Telecommunications Act So Far: Higher Prices, Few Benefits, *The Washington Post*, Financial Section p. H1, Jan 19, 1997, available in LEXIS Curnws data base.

75 FCC regulatory practice lawyers suffer somewhat by the extension of the maximum license term for both radio and television stations to eight years, an increase of one and three years respectively. 47 U.S.C. §307(c)(1997).

76 Tim Jones, New Deal in Media: One AM for an FM; With Rules Loosened, More Stations Traded, *Chicago Tribune*, Business Section, p. 1, March 31, 1997 available in LEXIS Curnws data base. See also Neil Hickey, So Big: The Telecommunications Act at Year One, 35 *Columbia Journalism Review* No. 5 p. 23 (Jan 11, 1997) (providing a report on, and analysis of, the major media mergers and acquisitions of 1996).

77 See 47 C.F.R. §73.3555(1997).

merical cap based on the total number of stations in the particular market. The '96 Act has enabled broadcasters to cluster ownership in individual markets with an eye toward accruing economies of scale, e.g., use of a single advertising sales force for several stations.

3.2 Cable Television Operators

Cable television operators collectively appear to benefit from the '96 Act with greater freedom to raise rates. All forms of rate regulation ended immediately upon enactment for small systems, and all but basic-service tier rate regulation will sunset on March 31, 1999 for larger systems, with earlier deregulation available in locales facing effective competition. However, the market entry of direct broadcast satellite systems may blunt some of the upward pressure on cable television rates despite the freedom to raise rates.

3.3 '96 Act Implementers and Interpreters

The '96 Act does anything but deregulate in the short term. It will have required the Commission to initiate dozens of proceedings and does present a justification for an increased budget. Likewise the '96 Act's support for federalism and the role of states in setting rates, interconnection terms and conditions and new universal service funding rules creates similar upward budgetary pressure for state regulatory authorities. Lawyers and consultants have seemingly endless opportunities to offer diverging interpretations of the law and to dispute the FCC's implementation in court.[78]

3.4 State Public Utility Commissions

The '96 Act set out a blueprint for shared telecommunications policy-making among the states and the FCC. While allowing public utility commissions with the resources and inclination to determine state-specific interconnection and reseller discount rates, the '96 Act also appeared to make it clear that the FCC could provide pricing proxies and other economic and regulatory models for states unable or unwilling to establish their own. To avoid pricing and policy "balkanization" of key procompetitive objectives, the '96 Act appeared to favor FCC leadership, provided the Commission established federal-state Joint Boards to generate formal advice to the FCC and generally worked collaboratively with state public utility commissions.

In the face of sweeping rejection of such shared and collaborative policy-making, at least as to many key local competition rules by the Eighth Circuit,[79] the state commissions, by choice or by judicial decree, have won the right to dominate the process. However, this endorsement of federalism may have the unintended consequence of delaying the onset of

---

[78] See American Civil Liberties Union v. Reno, 929 F.Supp. 824 (E.D.Pa.1996), appeal pending (granting temporary restraining order on implementation of provision in the '96 Act commonly referred to as the Communications Decency Act that restricts transmission of obscene or indecent materials via information services like the Internet).

[79] Iowa Utilities Board et al., v. Federal Communications Commission, File Nos. 96-3321 et al, slip op. (8th Cir. rel. July 18, 1997) available at: http://ls.wustl.edu/8th.cir/fcc.html (finding the FCC exceeded its jurisdiction in establishing pricing rules for local exchange access, including resale discounts and unbundling of network elements).

local exchange competition with some states possibly disinclined to stimulate competition with the kind of "fire sale" reseller discounts and aggressive forward-looking pricing models crafted at the FCC.

3.5 Competitive Local Exchange and Wireless Carriers

Almost overnight certain carriers experienced a double-digit reduction in interconnection costs, a significant cost component.[80] The '96 Act requires reciprocal charges for terminating traffic generated by another carrier.[81] This provision presents the possibility of zero cost, "sender keep all" termination charges, particularly when traffic volumes reach near parity between an incumbent carrier and market entrant.[82] At the very least, wireline market entrants and wireless carriers no longer will face a scenario where they must compensate the incumbent wireline carrier for call terminations, but receive nothing for terminating traffic originating on the incumbent carrier's wireline network.

The FCC's controversial economic analysis of incumbent carriers' infrastructure investment also provides near-term opportunities for cost savings, and perhaps even opportunities to arbitrage the difference between discounted access to unbundled local exchange service elements and the complete, or retail, versions. As directed by the '96 Act[83] the FCC used a "forward-looking" analysis of incumbent local exchange carrier costs when determining the appropriate cost of access.[84] To determine the appropriate discount for resale of local exchange services the Commission applied an "avoided cost"[85] making it possible for effective competition without the time and investment occurring when the competitor must undertake some degree of facilities investment.

---

80 For example, "the long-distance companies currently pay 37 cents of each long-distance dollar to the local telephone companies for access to local networks." Michael A. Noll, The Effectiveness of Long-Distance Competition: Who Really Benefits; Industry Trend or Event, 31 *Telecommunications* No. 3 at p. 28 (March, 1997).

81 All LECs must "establish reciprocal compensation agreements for the transport and termination of telecommunications." 47 U.S.C. §251(b)(5).

82 The '96 Act directs state regulatory commissions to examine the reasonableness of reciprocal compensation arrangements and to ensure the "mutual and reciprocal recovery by each carrier of costs associated with the transport and termination on each carrier's network facilities of calls that originate on the facilities of the other carrier [with a determination of]... such costs on the basis of a reasonable approximation of the additional costs of termination such costs." 47 U.S.C.§252(d)(2). In evaluating the costs of such switching and routing the FCC provided the states with three options: 1) set rates on the basis of total element long-run incremental cost; 2) apply an FCC-calculated range of rates; or 3) permit a "bill and keep", "sender keep all" system.

83 Section 252(d) of the '96 Act requires interconnection and network element charges to be just and reasonable, cost-based plus a reasonable profit and nondiscriminatory.

84 ILECs can recover "forward-looking economic cost . . . based on the most efficient technology deployed in the incumbent LEC's current wire center locations." Implementation of the Local Competition Provisions in the Telecommunications Act of 1996, CC Docket No. 96-98, First Report and Order, 11 FCC Rcd. 15499, ¶685 (1996).

85 Reseller access to wholesale prices will be determined by state regulatory commissions "on the basis of retail rates charged to subscribers . . . excluding the portion thereof attributable to any marketing, billing, collection, and other costs that will be avoided by the local exchange carrier." 47 U.S.C. §252(d)(3).

3.6 Explicitly Identified Universal Service Beneficiaries

The '96 Act codifies a universal service mission, and requires the establishment of an explicit subsidy mechanism to achieve service goals. The '96 Act also identifies categories of telecommunication users entitled to discounted access to both Plain Old Telephone Service ("POTS") and advanced services, commonly referred to as Pretty Advanced New Stuff ("PANS") that carriers might otherwise make available initially in urban or high-income locales. While society as a whole may benefit from this initiative, the direct and near immediate beneficiaries are: libraries, schools, clinics, hospitals, and rural residents seeking access to leading-edge services.

3.7 Internet and Enhanced Service Providers

The '96 Act contains a requirement that all telecommunication carriers bear a fair burden in underwriting a bolstered universal service mission.[86] Internet service providers and other enhanced service providers, which currently are exempt from paying universal service subsidies, potentially can migrate significant traffic volumes that otherwise would be subject to such payments, i.e., the Carrier Common Line charge incurred by IXCs. For example, the use of the Internet to send an e-mail message in lieu of a facsimile transmission, or to use the Internet for telephony in lieu of the dial-up, public switched telephone network, means fewer metered minutes are generated that trigger a subsidy for lifeline assistance to individuals with low incomes and telephone companies serving high-cost regions.

In 1996 the FCC opted to ignore the potential of Internet and enhanced services to substitute for basic interexchange services subject to access charges and universal service subsidy requirements. Until such time as the Commission reinterprets the '96 Act's definition of telecommunications carriers to include Internet-mediated telephony[87] and other services, Internet and enhanced service providers will accrue a substantial and immediate financial windfall.

4. Longer-term Winners

The longer-term beneficiaries collectively are most likely to accrue cost savings, generate operating efficiencies, serve new markets or bundle a larger collection of services. The authors of the '96 Act contemplated increased resale and facilities-based competition resulting from the dismantling of legal and regulatory barriers to market entry, but an equally possible long-term outcome may be marketplace rewards accruing to deep-pocketed players who can afford to enter new markets and package a winning blend of one-stop shopping opportunities and cost savings.

---

[86] "All providers of telecommunications services should make an equitable and nondiscriminatory contribution to the preservation and advancement of universal service." 47 U.S.C. § 254(b)(4).

[87] For an analysis of the regulatory, policy and financial implications of using the Internet for telephone services see Robert M. Frieden, Dialing for Dollars: Will the FCC Regulate Internet Telephony? 23 *Rutgers Comp. & Tech. L. J.* 47 (1997).

4.1 Incumbents Get Bigger

Rather than generating a substantially more robustly competitive marketplace replete with lots more players, the long-term impact of the '96 Act appears to favor incumbents, notwithstanding favorable interconnection terms and conditions for market entrants. Incumbents like the Bell Operating Companies and the three major IXCs (AT&T, MCI and Sprint) are best able to generate internally or externally the funds necessary to upgrade existing infrastructures to exploit new market entry opportunities. Some of these companies, e.g., MCI with the investment of British Telecom and Sprint with the investment of Deutsche Telekom and France Telecom, already have begun to bolster the marketplace prospects by securing additional capital from foreign companies to reach the needed scale.

As to broadcasting incumbents the '96 Act contained language requiring television station operators to pay for the use of new spectrum earmarked for high-definition or advanced television, but also suitable for other commercial services.[88] Current political compromises point to a free loan of UHF spectrum for digital television broadcasts, making it possible for broadcasters to adapt to an age of high-definition television perhaps coupled with other multi-media, Internet-based services.

If bigness matters even more so in the future, the small- and medium-sized ventures perceiving an inability to reach the needed size nevertheless will enjoy a possibly large windfall for selling out to someone that can. If such market consolidation occurs, just about all independent telephone companies, and second-tier cable television ventures and wireless operators, are candidates for a sell-out.

4.2 Exploiters of Regulation-created Market Niches and Competitive Opportunities

The '96 Act presents incumbents and newcomers alike with strategic opportunities to exploit regulation-created opportunities. Quite possibly IXCs will achieve cost-of-access parity with LECs in view of the fact that local and interexchange carrier interconnections are physically alike and typically generate the same costs. Should such parity not result, one can anticipate a plethora of new LECs appearing on the landscape to qualify for unbundled network elements and lower access charges.

Possibly a new market opportunity will exist in the video program distribution marketplace through LEC-provided Open Video Systems, a hybrid of private and common carriage.[89] A programming entrepreneur, at least in concept, could secure access to channel

---

[88] See 47 U.S.C. §336(e)(2)(A)-(C) (requiring the recovery for the public of a portion of the value of spectrum allocated for advanced television services but also used for commercial services at a rate commensurate with what competitive bidding would generate).

[89] See 47 U.S.C. § 651(a)(3)-(4). An open video system operator's certification request must certify that it complies with the Commission's regulations implementing the requirements in Section 653(b), which: (1) prohibit the operator from discriminating among video programmers regarding carriage on its system; (2) require the operator to establish rates, terms and conditions of carriage that are just, reasonable and not unjustly or unreasonably discriminatory; (3) prohibit the operator or its affiliate, if carriage demand exceeds capacity, from selecting the video programming on more than one-third of its activated channels; (4) permit the operator to use channel sharing arrangements that provide subscribers with ready and immediate access to programming; (5) extend the Commission's sports exclusivity, network non-duplication and syndicated exclu

(continued)

capacity in a telephone-company-operated cable television delivery service without having to secure a municipal franchise, and largely free of traditional regulatory responsibilities.

4.3 Content Creators

If we accept the view that incumbents or others ultimately will invest in a broadband infrastructure, then it follows that demand will increase for content to travel through the larger channel capacity. On a per-minute basis, content typically exceeds the switching and transport costs, and an upturn in demand probably will maintain this comparative cost differential.

5. Short-term Losers

The '96 Act has generated an immediate impact on the strategic plans of operators, because the legislation creates opportunities for entry into desirable markets, some perhaps more attractive than ones already penetrated. For example, the '96 Act creates a blueprint for the Regional Bell Operating Companies ("RBOCs") to enter the two remaining lines of business, inter-LATA long distance and equipment manufacturing, prohibited by the Modification of Final Judgment that governed AT&T's divestiture. All the RBOCs will pursue these two market opportunities, and at least some will reduce investments and efforts directed at other markets now perceived as comparatively less attractive.

It appears that in the short term the RBOCs have cooled their zeal to exploit video program production and distribution opportunities. Several RBOCs have downsized their funding and expectation of Americast and Tele-TV, state-of-the-art video production enterprises heralded not so long ago as proof of the RBOCs' commitment to video. Some RBOCs have not acted on options to invest in or acquire Multichannel Multipoint Distribution Systems, omni-directional microwave systems commonly referred to as wireless cable. The allure of long-distance telephone markets made accessible by the '96 Act appears to have prompted some of the RBOCs to shift investment emphasis away from video pursuits.

5.1 ILECs

The LEC interconnection pricing structure required by the '96 Act all but requires ILECs to lose market share immediately. This may constitute a small price to pay for the RBOCs to be certified as having satisfied the 14-point competitive check that serves as a precondition to having authority to provide inter-LATA long distance and to manufacture equipment. On the other hand discounted, avoided-cost resale opportunities, and access to ILEC facilities based on optimized pricing assumptions rather than actual carrier investment, guarantee that ILECs will lose both residential and business subscribers. Several ILECs have challenged the Commission's interpretation and implementation of the '96 Act regarding LEC interconnection. But even with refinement after a court reversal and remand, the FCC's orders on LEC interconnection and parallel access charge reform for rates imposed on IXCs will result in diminished revenues from core business.

(continued)
sivity regulations to open video systems; and (6) prohibit the operator from unreasonably discriminating in favor of its affiliates with regard to information provided to subscribers for the purpose of selecting programming.

5.2 Rights-of-way Holders

At a time when demand increases for access to rights of way, poles and conduits the '96 Act diminishes leverage enjoyed by such owners. The '96 Act requires the FCC to establish procedures for nondiscriminatory access by cable television systems and telecommunications carriers to poles, ducts, conduits, and rights of way owned by utilities or LECs. This means that holders of such valuable assets cannot conduct private auctions or maximize revenues with exclusive or preferential arrangements.

5.3 Consumers

The drafters of the '96 Act touted the legislation as a major enhancement of consumer welfare. Most consumers have not experienced any short-term gain in terms of a proliferation of services or downward rate pressure. If anything, local and "retail" long-distance rates have inched up in the last few years. The '96 Act imposes greater penalties on IXCs who change user selection of preferred long-distance carrier without authorization, but heretofore the Commission with rare exceptions has not severely penalized such behavior.

6. Long-term Losers

In the longer term incumbent telecommunications players should continue to consolidate their market share in core areas while selectively exploiting market entry opportunities. While incumbent LECs have temporarily abandoned or downsized efforts to capture market share in video production and distribution, the longer-term prospects seem likely that they will renew such efforts to the detriment of cable television operators. The major longer-term losers appear to be enterprises that have exploited captive or non-competitive markets without taking adequate steps to establish a superior product or service, despite the likely prospects for future competition. For example, the top incumbent IXCs may face declining margins as a result of BOC entry into inter-LATA long-distance markets, particularly if such entrants refuse to go along with the incumbents' predilection for limited competition and consciously parallel pricing. Likewise, if an Internet access or service provider has done nothing but exploit an exemption from having to pay access charges, a future universal service or access payment will raise its cost of doing business at a time when other ventures may offer a better, value-added service.

7. Summary and Conclusions

In an environment as volatile and uncertain as telecommunications and information processing, one cannot easily predict the overall impact of the '96 Act. Currently, the '96 Act sweepstakes has rewarded carriers who can use the judicial process to delay competition while elsewhere exploiting market entry opportunities. In the longer term no enterprise can rely on the retention of legal and regulatory barriers to solve a failure to respond to changed circumstances. Perhaps in the longer term consumers will join the list of '96 Act winners.